MODERN DAIRY TECHNOLOGY

Volume 2

Advances in Milk Products
Second Edition

Edited by
R. K. Robinson

*Department of Food Science and Technology,
University of Reading, UK*

BLACKIE ACADEMIC & PROFESSIONAL
An Imprint of Chapman & Hall
London · Weinheim · New York · Tokyo · Melbourne · Madras

Published by
Blackie Academic & Professional, an imprint of Chapman & Hall

Chapman & Hall, 2-6 Boundary Row, London SE1 8HN, UK

Chapman & Hall GmbH, Pappelallee 3, 69469 Weinheim, Germany

Chapman & Hall USA, 115 Fifth Avenue, New York, NY 10003, USA

Chapman & Hall Japan, ITP-Japan, Kyowa Building, 3F, 2-2-1 Hirakawacho, Chiyoda-ku, Tokyo 102, Japan

DA Book (Aust.) Pty Ltd, 648 Whitehorse Road, Mitcham 3132, Victoria, Australia

Chapman & Hall India, R. Seshadri, 32 Second Main Road, CIT East, Madras 600 035, India

Second edition 1993
Reprinted 1996

Softcover reprint of the hardcover 2nd edition 1996

Typest by Interprint Ltd., Malta

ISBN 978-1-4684-8174-7 ISBN 978-1-4684-8172-3 (eBook)
DOI 10.1007/978-1-4684-8172-3

A Catalogue record for this book is available from the British Library

Library of Congress Cataloging-in-Publication Data available

Printed on acid-free text paper, manufactured in accordance with ANSI/NISO Z39.48-1992 (Permanence of Paper).

Preface to the Second Edition

As with the products and processes described in Volume I of this book, many of the technical changes associated with, for example, the manufacture of cheeses or fermented milks have been subtle rather than dramatic. Nonetheless, the importance for the dairy industry has often been profound. The market demand for dairy products containing 'health-promoting' cultures is a development that was barely discernible 10 years ago, and yet many manufacturers are now generating a whole range of bio-yoghurts and similar retail items.

Similarly, the legislation covering food hygiene has been modified to place additional demands upon manufacturers, a move that has in turn encouraged the further development of analytical methods for quality control.

These modifications to manufacturing practices are, along with many others, reflected in this second edition, and I acknowledge with gratitude the enthusiastic co-operation of all the authors associated with this project in bringing their disparate contributions up-to-date.

R. K. Robinson

Preface to the First Edition

Retail sales of most dairy products are still on the increase world-wide, and this expansion is, at least in part, a reflection of the fact that prices have tended to remain at a competitive level. This relative stability has been achieved either through the introduction of major changes in technology, as in the case of the territorial cheeses, or by a massive scale-up of a traditional process like yoghurt making, but whatever the chosen route, the enhanced productivity has been to the benefit of the consumer. Improved methods of product control have also been instrumental in raising the efficiency of the various manufacturing procedures, and the intention of this second volume is to record the current 'state of the art' in respect of the major dairy products. Obviously, processes will continue to become sophisticated, but if this text can provide a background to future developments, then the endeavours of the contributors will have been worthwhile.

R. K. Robinson

Contents

List of Contributors

Mr T. Anderson
A/S N. Foss Electric, 69 Slangerupgade, DK 3400, Hillerod, Denmark.
Mr M. M. Borglum
A/S N. Foss Electric, 69 Slangerupgade, DK 3400, Hillerod, Denmark.
Mr N. Brems
A/S N. Foss Electric, 69 Slangerupgade, DK 3400, Hillerod, Denmark.
Mr E. Hansen
A/S N. Foss Electric, 69 Slangerupgade, DK 3400, Hillerod, Denmark.
Dr H. J. Hess
Nestec Ltd, Avenue Nestlé 55, 1800 Vevey, Switzerland.
Mr J. T. Homewood
Nestec Ltd, Avenue Nestlé 55, 1800 Vevey, Switzerland.
Mr J. H. Jorgensen
A/S N. Foss Electric, 69 Slangerupgade, DK 3400, Hillerod, Denmark.
Mr S. Kold–Christensen
A/S N. Foss Electric, 69 Slangerupgade, DK 3400 Hillerod, Denmark.
Dr M. J. Lewis
Department of Food Science & Technology, University of Reading, Whiteknights, Reading, Berkshire, UK, RG6 2AP.
Mr H. L. Mitten
Apartment 726, 400 West Butterfield Road, Suite 726, Elmhurst, IL 60126, USA.
Mr J. Neirinckx
APV Pasilac AS, Pasteurvej, DK 8600 Silkeborg, Denmark.
Mr L. Nygaard
A/S N. Foss Electric, 69 Slangerupgade, DK 3400, Hillerad, Denmark.

Dr G. L. Pettipher
Reading Scientific Services Ltd, The Lord Zuckerman Research Centre, University of Reading, Whiteknights, Reading, Berkshire, UK, RG6 2LA.

Dr R. K. Robinson
Department of Food Science & Technology, University of Reading, Whiteknights, Reading, Berkshire, UK, RG6 2AP.

Mr M. B. Shaw
Dairy Crest Ltd, Portsmouth Road, Surbiton, Surrey, UK, KT6 5QL.

Dr A. Y. Tamime
SAC — Auchincruive, The National College of Food, Land and Environmental Studies, Food Science & Technology Department, Auchincruive, Ayr, UK.

Chapter 1

Manufacture of Yoghurt and Other Fermented Milks

R. K. Robinson

Department of Food Science & Technology, Univeristy of Reading, UK

and

A. Y. Tamime

SAC—Auchincruive, The National College for Food, Land and Environmental Studies, Food Science & Technology Department, Ayr, UK

INTRODUCTION

Although yoghurt and other fermented milks can be manufactured on a small scale, and with modest levels of technology, the growing demand for products within the industrialised countries has lead to the installation of plants capable of handling thousands of litres of milk per day. As a result, sophisticated, automated systems of manufacture are now commonplace, and yet the basic approach to production is still reminiscent of the traditional procedures that have been associated with the peoples of the Middle East for hundreds of years. The reason for this linkage lies in the fact that success with any fermented product lies with the microflora and, in many cases, the microorganisms employed in a modern factory are basically derived from traditional cultures. Obviously, the manufacturers of starter cultures have become heavily involved with the selection of strains with special attributes, but at generic or species level, many of the bacteria enjoy a long association with the dairy industry (Tamime, 1990). Although historical in origin, the association of certain cultures with specific products is now well established, and the link is maintained either by

(a) custom, e.g. the characteristic flavour of yoghurt depends, in the main, on the presence of detectable levels of acetaldehyde, a compound released in quantity during the synergistic growth of

Lactobacillus delbrueckii subsp. *bulgaricus* (*Lac. bulgaricus*) and *Streptococcus salivarius* subsp. *thermophilus* (*Str. thermophilus*); or by

(b) legal or non-legal requirements. In most countries, it is usual to insist that yoghurt contains at least a small percentage of *Lac. bulgaricus*, and similarly that the retail product must contain an 'abundant and viable microflora of starter origin'; products that do not meet these requirements should not be called 'yoghurt'.

It is against this background that modern systems for fermented milks have been derived, and as yoghurt remains the most popular item in the group, it is appropriate that its method of production should serve to exemplify the associated industrial processes.

PREPARATION OF THE BASIC MIX

The overall procedure for the manufacture of yoghurt is shown in Fig. 1, and the first step is the production of the basic mix. Most factories employ liquid milk as the main ingredient, and this material will be delivered, in bulk, employing road tankers. The reception tests applied to such milk, and the standards of hygiene expected in the unloading bays, are similar to those associated with any milk-processing plant (Luck and Gavron, 1990), and assuming that the normal criteria are met, the milk will be transferred to storage silos at <5°C. Beyond this stage, however, the processing of the milk for yoghurt follows a distinctive pattern of its own. Although there is a definite market for full-fat yoghurts — both natural and fruit yoghurts — the major share of the market is taken with low-fat yoghurts. This trend, while driven in the main by the diet-conscious consumer, does allow the manufacturer to market the cream as a separate entity. The next step in manufacture usually involves, therefore, passage of the milk through a centrifugal separator to give a stream of skim-milk (0·5–0·7% fat) and cream. If the manufacturer wishes to produce a medium-fat yoghurt, then some of the separated cream will be fed back into the skim-milk, and automatic monitoring will ensure the constant composition of the final process stream (see Fig. 2). Once the fat content has been standardised, the next stage is to raise the level of milk-solids-non-fat (MSNF), for without this additional protein, the 'gel' produced during the fermentation stage will be thin and 'watery', and prone to syneresis, i.e. separation of free whey from the coagulum

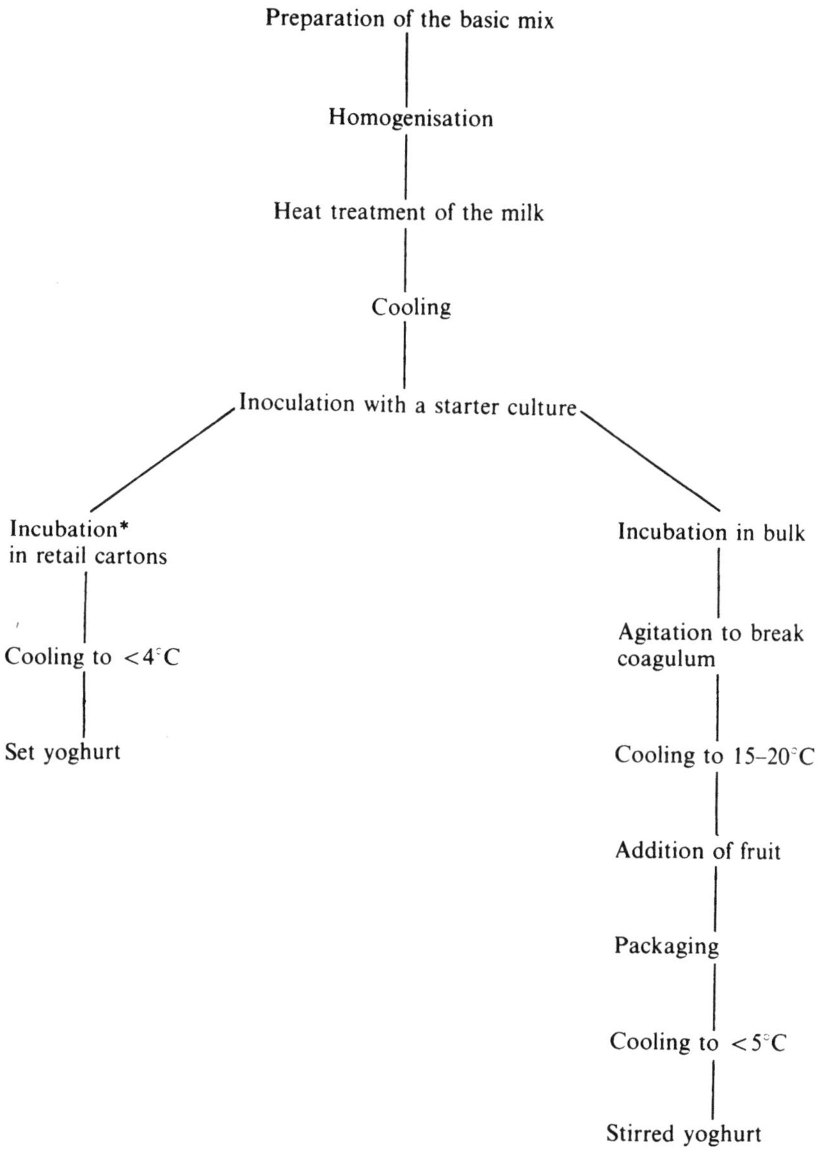

*Sucrose and/or fruit (fruit flavours) can be added at this point.

Fig. 1. An outline of the principal steps necessary for the production of yoghurt.

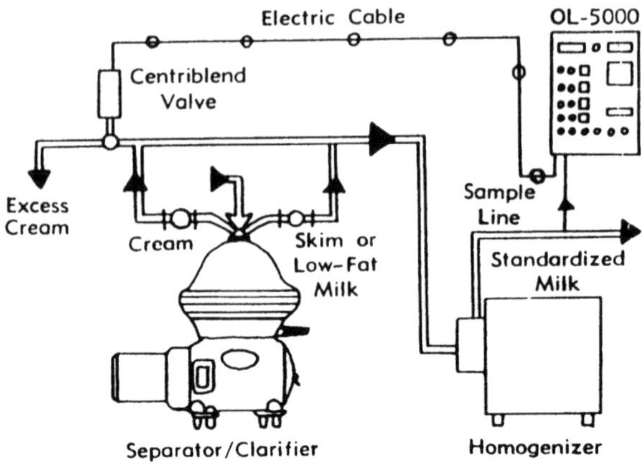

Fig. 2. A schematic diagram of a system for the automatic in-line standardisation of process milk. Samples are taken after homogenisation, the fat content measured and compared with the desired value. Any changes are brought about by the electronically controlled 'Centriblend' valve. (Reproduced by courtsey of On-Line Instrumentation Inc., New York, USA.)

(Robinson and Tamime, 1986). Some yoghurts with low levels of MSNF (11–12%) are produced for cooking purposes, but for consumption as a snack-food, a stronger gel is essential. In fruit yoghurts, stabilisers, such as selected blends of starches, pectins or gelatine, can be employed to increase the viscosity of the product rather than relying on the milk protein alone, but in natural set yoghurts, this practice, if not banned by legislation, is certainly discouraged by a variety of non-legal specifications. It is also relevant that the nutritional value of yoghurt is enhanced by the raising of the non-fat solids (see Table I), and most manufacturers achieve the necessary adjustment by one of two alternative approaches: (i) the addition of milk powder, or (ii) the concentration of the milk.

TABLE I
Some typical values ($g100\,g^{-1}$) for the major components of commercial yoghurt

Component	*Natural yoghurt*	
	Full fat	*Low fat*
Protein	3·9	5·0
Fat	3·4	1·0
Carbohydrate	4·9	6·5

After: Deeth & Tamime (1981)

Addition of Milk Powder

The powder most widely used in this context is antibiotic-free, skim-milk powder, because apart from being an adequate source of the types of protein required, it is readily available in most countries. The description 'antibiotic-free' is usually rephrased into the form 'CYP', i.e. Cheese and Yoghurt Powder, because antibiotic-free means, in practice, below the level of detection by comparatively simple techniques, such as the 'Intertest System' or the 'Disc Assay'. Thus, the description 'CYP' means that the powder will contain less than 0·02 IU g^{-1} of inhibitory substances, and as levels below this point are unlikely to affect any starter cultures employed, adherence to this specification is reasonable enough.

It is important also that the powder should be:

(a) easily dispersed into the aqueous phase;
(b) soluble to the extent that the solubility index should be less than 0·1 ml (ADMI, 1971);
(c) free from scorched particles, i.e. Grade A on the ADMI scale, and certainly any black specks would cause considerable problems with natural set yoghurt; and
(d) a low- to medium-heat powder of the type which is also suitable for the production of fermented milks from recombined milk (Wilcek, 1990). The specifications for other materials used during recombination, e.g. anhydrous milk fat, have been reported by Sjollema (1988).

In order to meet these disparate criteria, the selection of most manufacturers would be a 'semi-instantised' powder, particularly as such powders can be manufactured from certain types of spray-drier without additional processing or cost.

The method of incorporation of the powder varies with the scale of the operation but, in principle, the aim is always to achieve rapid mixing of the powder/aqueous phases, and with the minimum of opportunity for the formation of lumps or excessive production of foam. The most efficient system for the smaller plant involves the rapid circulation of milk, at 20–25°C, beneath a funnel into which the powder is poured from a bag or other container (see Fig. 3). The milk is then circulated back into a multi-purpose tank that will serve later for heat treatment of the milk and for the fermentation stage; the pipe returning the milk to the tank must be below the level of the liquid to avoid foaming. The pump can be located on either side of the funnel, and if on the suction side of the

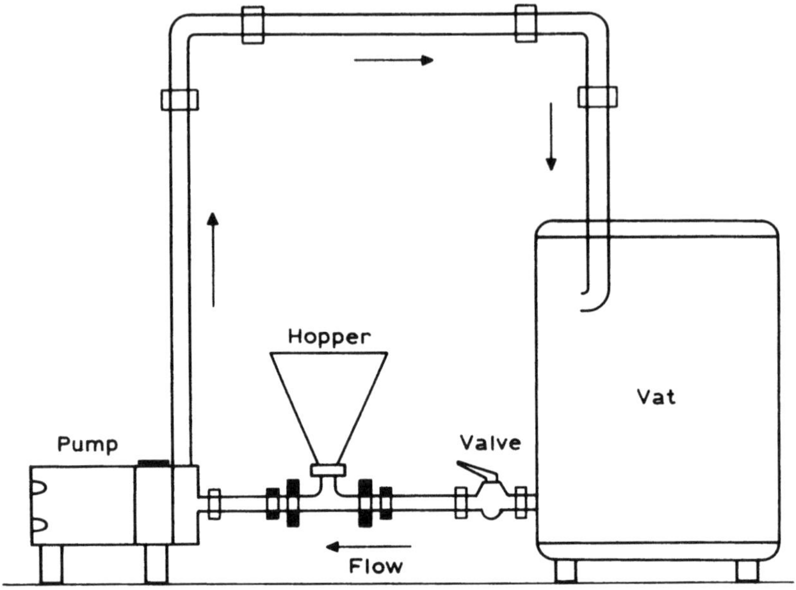

Fig. 3. A typical layout including a high-speed mixer/pump and 'funnel unit' for incorporating milk powders into liquid milk.

centrifugal pump, the dispersal and solution of the milk powder is rapid due to the action of the blades rotating at high speed. However, if the powder is not totally free-flowing, then it may have to be mechanically assisted into the flowing liquid. It is to avoid this problem that some manufacturers recommend that the funnel should be placed on the discharge side of the pump, for if, in addition, a restriction is placed upon the flow of liquid below the funnel, a vacuum can be built-up which, in effect, sucks the powder into the circulating liquid. Dispersal of the powder tends to be slower with the latter system, as any lumps are only broken down during subsequent passes through the pump; nevertheless, the approach can be useful if the funnel unit is large.

Alternatively, the powder can be added directly to the multi-purpose tank and incorporated with a high-speed agitator. Such a blender can be designed for immersion into the tank or as an in-line mixer but, in either case, the intense shearing action of the rotor against the walls of the chamber ensures total dispersion of the powder. A modification of this latter approach is suitable for larger factories, and the Crepaco 'blender' shown in Fig. 4 is a popular example. The expanded casing around the

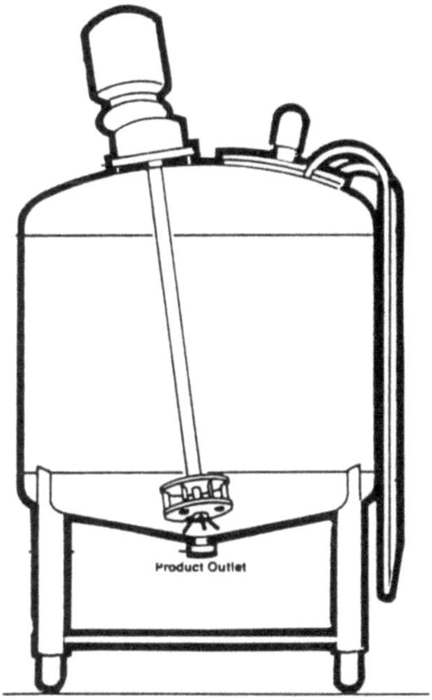

Fig. 4. This Crepaco 'Blender' is designed to mix milk powder or other ingredients into a liquid base. (Reproduced by courtesy of APV International Ltd, Crawley, Sussex, UK.)

blades has the advantage of creating a massive vortex which rapidly draws the powder into the milk, and with very little formation of foam. The cone-shaped base of the tank allows for easy discharge of the process mix, and hence a large volume of several hundred gallons can be produced within the first hour of a working day (1 gallon $\approx$ 4·5 litres). A number of variants of this system are available (Tamime and Robinson, 1985), but whatever the precise engineering of the process, total solution of the powder cannot always be guaranteed.

The presence of particulate material, especially in a set yoghurt, would be quite unacceptable, and hence some form of filtration system must be included. Centrifugal clarifiers can be employed, but the less expensive option is to include stainless-steel or nylon filters in the lines carrying the process milk onto the next stage. If large volumes of milk are involved, the use of more than one filter is essential, so that if one becomes clogged, the alternate route can come into operation whilst the first filter unit is being cleaned (see Fig. 5).

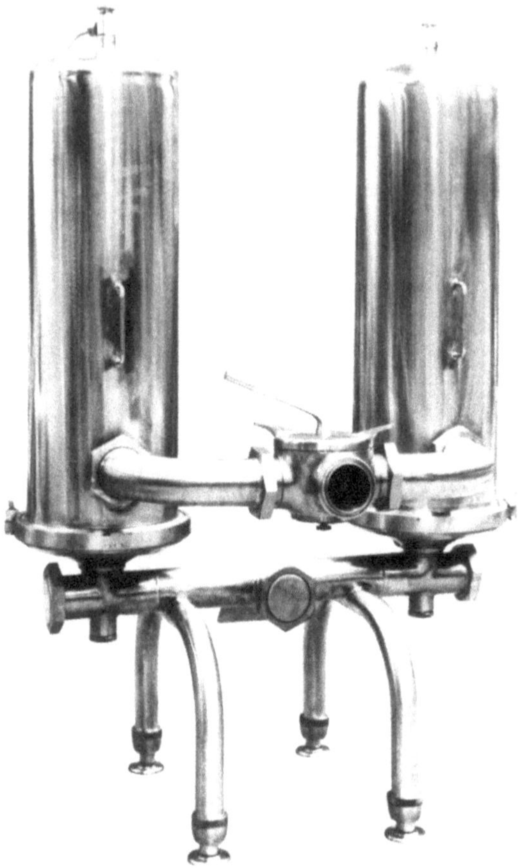

Fig. 5. A filter unit capable of removing particulate material from large volumes of fresh or reconstituted milk. (Reproduced by courtesy of APV International Ltd, Crawley, Sussex, UK.)

Although skim-milk powder is the obvious choice for most manufacturers, there is an extensive range of milk-based powders that could be considered as alternatives. As it is the casein level that is critical in relation to coagulum strength, caseinates or high-protein skim-milk powders (see Table II) are the preferred option. The inclusion rate for these special powders tends to be kept low, perhaps 1–1·5%, together with an equal or slightly larger volume of skim-milk powder, and this usage is a reflection of the following:

TABLE II
Some milk-based powders that could be employed to raise the total solids of yoghurt milk, or selectively increase the level of casein[a]

Product	*Protein* (%)	*Fat* (%)	*Ash* (%)	*Lactose* (%)	*Moisture* (%)
Milk powder (full-cream)	27·0	25·6	7·8	37·6	2·0
Milk powder (skim-milk)	35·6	0·6	10·3	49·5	4·0
Whey powder	13·0	0·8	11·0	71·0	4·2
Whey powder (demineralised)	12·5	1·0	4·5	79·0	3·2
Whey protein	30·0	—	7·0	58·0	5·0
Concentrates	81·5	4·0	3·2	7·3	4·0
Retentate powders					
(full-cream milk)	41·7	41·7	7·8	37·6	2·0
(skim-milk)	74·3	1·4	8·5	11·8	4·0
Casein powders					
Casein (rennet)	85·8	1·5	2·5	0·2	10·0
Caseinate (sodium)	89·3	0·9	4·5	0·2	5·1
Co-precipitate	86·5	1·5	3·0	1·0	8·0

[a]Data taken from Robinson (1981), Bjerre (1990) and Tamime and Kirkegaard (1991).

(a) the functional properties of these high-casein powders endows the product with marked viscosity *vis-à-vis* an equivalent addition of skim-milk powder; and
(b) the cost of the powders is much higher than skim-milk powder, and hence a lower inclusion rate is essential to make the process cost-effective.

Concentration of the Milk

Elevation of the total solids level by 2–4% can also be achieved by evaporation under vacuum, and single-effect evaporators are widely popular. In this system, the milk is usually pre-heated by passage through a plate heat exchanger and, at a pre-set temperature, the milk enters the evaporation unit where water vapour is removed. By circulating the milk through the unit on a number of occasions, the desired degree of concentration can be reached, and with a high level of thermal efficiency. Although the move to evaporation does raise the cost of plant installation, a typical unit is quite capable of handling up to 8000 litres h^{-1}, so that the cost per unit volume of process milk may well be lower than if fortified with milk powder. It has been argued also that the quality of the end-product is enhanced when heat-concentrated milk provides the base (Robinson, 1981), but objective evidence has not been established.

An alternative approach is to concentrate the milk by ultrafiltration or reverse osmosis, with ultrafiltration providing the more popular route. Thus, whilst reverse osmosis brings about a concentration of all the constituents in the milk, ultrafiltration selectively raises the levels of protein and fat at the expense of lactose and minerals. The influence of these changes on the quality of yoghurt has been assessed by Glover (1971, 1985) and Glover *et al.*, (1978), and the conclusions tended to support the view that membrane processing does offer a possible alternative. Certainly, ultrafiltration has been used in Denmark for the production of both yoghurt and ymer, and Tamime *et al.* (1989*b*) have successfully employed a similar process for the production of concentrated yoghurt (see later). Nevertheless, in many countries, the balance does seem to be in favour of evaporation or the addition of milk powder(s).

Homogenisation

In some countries, notably Greece, yoghurts are often made with full-cream milk that is not homogenised. As a result, the cream rises to the surface during incubation and, on cooling, sets as a well-defined 'crust'. Salt is then sprinkled over the surface, and the product appears in the market-place as 'crusty yoghurt'. However, if normal full-fat, set yoghurt is being made, i.e. above 3·0% fat, then homogenisation is essential to prevent separation of the cream during incubation, and even for low-fat varieties, homogenisation offers a number of advantages.

The basic process involves passing the milk under high pressure through a small orifice, and the shearing effect reduces the average diameter of the fat globules to $<2{\cdot}0\ \mu m$. As a result, the globules are less inclined to coalesce into large units and rise to the surface. At the same time, the increased number of small globules enhances the ability of the milk to reflect light, and hence the yoghurt appears 'whiter' and more attractive to the consumer. Further benefits are observed in that

- many of the small fat globules adsorb onto the casein micelles, an effect that may increase the viscosity of the end-product;
- the risk of syneresis, i.e. the separation of free whey onto the surface of set yoghurt, tends to be reduced; and

— any residual particles of added ingredients, such as milk powder or stabilisers, will be totally disintegrated.

In practice, the yoghurt milk is normally heated to 60°C, and passed through a suitable homogeniser, such as the APV-Gaulin unit shown in Fig. 6, at a pressure of 15–18 MPa. The end result is a stable homogeneous milk that can then be further heated to 85–95°C.

Fig. 6. An APV-Gaulin homogeniser (MC45) capable of handling in excess of 20 000 litres of milk per hour. (Reproduced by courtesy of APV International Ltd, Crawley, Sussex, UK.)

HEAT TREATMENT OF THE YOGHURT MILK

The aims of this stage of the process can be summarised as follows:

(a) destruction of any microorganisms present in the vegetative state, so avoiding the risk of competition between the starter and any adventitious bacteria; potential pathogens are destroyed also, as are yeasts or mould spores that could cause spoilage of the end-product. The surviving spore-forming bacteria are not likely to cause problems, because the endospores of *Bacillus* spp. and *Clostridium* spp. do not germinate at the pH of most fermented milks ($<$pH 4·5).

(b) expulsion of oxygen from the milk, so providing the micro-aerophilic conditions necessary for the growth of the starter organisms; and

(c) redistribution of minerals, especially calcium, between the soluble and the colloidal states, so leading to a decrease in the time taken for the milk to coagulate.

Related to this latter change is an alteration in the structure of the protein, and it is important that the severe heat treatment received by the yoghurt milk

(a) denatures the whey proteins, so that both the β-lactoglobulin and the α-lactalbumin become attached to the surfaces of the casein micelles (Davies *et al.*, 1978; Haque and Kinsella, 1988; Mottar *et al.*, 1989); a change that is believed to both increase the viscosity of the end-product, and enhance the water-binding capacity of the coagulum; and

(b) encourages the physical expansion/uncoiling of the tertiary structure of the casein micelles, so giving rise to a 'softer' coagulum than would result from the aggregation of casein in unheated milk.

These latter changes combine to give a firm coagulum that effectively immobilises the aqueous phase within the protein matrix, so all but eliminating any separation of whey. Maximum benefit in this respect is best achieved when the milk is heated to 85°C and held at this temperature for 30 min (Tamime and Robinson, 1985; Danneberg and Kessler, 1988*a*, *b*). This approach is feasible for small-scale production—batch process—but for large volumes of milk, a high temperature – short time system has to be employed, even though it represents something of a compromise.

The Batch Process

The tanks employed at this stage are usually water-jacketed, stainless-steel vessels, ranging in capacity from a few hundred up to several thousand litres (see Fig. 7). In practice, a series of these tanks could be

Fig. 7. These multi-purpose tanks can be employed for heat-treating the process mix, the incubation stage and, finally, cooling the finished product. (Reproduced by courtesy of Goavec SA, Alencon Cedex, France.)

used, with a planned schedule of filling/emptying, for the production of yoghurt on a semi-continuous basis, and a typical usage cycle might be

(a) fill the tank with milk that has already been standardised, fortified to the desired total solids, perhaps homogenised, and finally heated via a plate heat exchanger to 80–85°C;
(b) circulate hot water through the jacket to maintain the temperature at 80–85°C for 25–30 min; alternatively, if the milk is not homogenised, the entire heating process can be carried out by circulating hot water through the jacket of the tank;
(c) cool the milk to incubation temperature either by circulating chilled water (2°C) through the jacket of the tank and/or any internal cooling coils, or by running the process milk through a plate heat exchanger against a counter-flow of chilled water;
(d) once the incubation temperature has been achieved (22–25°C) for mesophilic milks or 42°C for yoghurt (30°C for overnight incubation of yoghurt), the milk can be inoculated with the appropriate starter culture;
(e) incubate in the same tank or a separate insulated vessel for the production of stirred yoghurt, or transfer to retail cartons for the

set variety; in both cases, allow the fermentation to proceed to the desired acidity. The advantage of using a water-jacketed vessel for the incubation of stirred yoghurt is that the temperature can be controlled accurately. However, the benefit of investing in a series of inexpensive, insulated vessels derives from the fact that the water-jacketed vessel can be in continuous use for the production of the yoghurt milk;

(f) cool for further processing, e.g. to <20°C for a yoghurt coagulum that is to be stirred prior to the addition of fruit. It is important that tanks of this type are fitted with effective, slow-speed agitators, and total-sweep blades that 'scrape' the walls of the tank offer the most appropriate design. Propeller-style agitators can be employed for low-viscosity products in volumes of 450–1000 litres, but with yoghurt, the high rotational speeds required to move large volumes can result in some damage to the product. A cone-shaped base to the tank provides a covenient aid for discharging the coagulum, and may also improve the heat-transfer characteristics of the tank. This latter aspect can be further refined by fitting a standard, jacketed tank with internal cooling coils and/or a hollow agitator through which chilled water can circulate. Such additions to the basic design can bring the cooling time for a 1000-litre batch of yoghurt at 42°C down to some 30–40 min, which means that, even if the tank is operated as a multi-purpose unit, i.e. every one of the above stages is completed in the one tank, the entire operation can still be completed within a normal working shift.

The Continuous Process

The unit that is most widely employed for heating milk on a continuous basis is the plate heat exchanger, in which the heating medium (hot water) flows along a channel separated from the yoghurt milk by a thin partition of stinless-steel. Details of such a unit are described elsewhere (Lewis, 1993), but the advantages to the producer are:

(a) lower energy costs compared with heating the milk in a jacketed vessel;
(b) considerable saving in time compared with heating in bulk.

An alternative design is the tubular heat exchanger, in which a tube(s) containing the yoghurt milk is suspended in an outer tube filled with the heating medium. Although this design may be more suitable for heating particulate foods, it offers little advantage with respect to milk. However, if the same plant is to be used to cool the finished yoghurt as well, then the tubular configuration is generally believed to be less damaging to the coagulum. To allow for the essential holding time, a plate exchanger for yoghurt milk must have an extended holding section, so that the milk, usually at 90–95°C, can be held at elevated temperature for at least 10 min. This approach allows the milk to be heat treated on a continuous basis (see Fig. 8), and although not the ideal time/temperature combination, it is much better suited to the handling of large volumes of milk.

Fig. 8. A heating unit for yoghurt milk, including a balance tank (centre), a plate heat exchanger and a 'zig-zag' holding tube (foreground) that provides a residence time of 8 min. (Reproduced by courtesy of the Northern Ireland Milk Marketing Board.)

If the intention is to heat treat the finished yoghurt as well, so as to produce a 'near-sterile' dessert (pasteurised yoghurt) for distribution at ambient temperature, the particulate nature of the material—only fruit yoghurts are usually treated in this fashion—may necessitate a switch to a scraped/swept surface heat exchanger. In essence, this plant consists of a jacketed cylinder fitted with a rotating scraper blade, and the action of the blade is to remove the product from the heated surface in a continuous motion, so preventing heat damage to the ingredients. This movement means that the barrel of the tube can have a larger diameter than is possible with other tubular heaters, a factor that makes it eminently suitable for particulate products. If the heating tube is run in series with similar cooling tubes, then large volumes of yoghurt or other fermented milk can be heated and cooled within a short space of time.

Although not suitable for yoghurt, the process milk for some fluid products, such as acidophilus milk, could be heat treated by the ultra-high temperature (UHT) process. In this case, the milk is exposed to a temperature of 115°C for 3 s, and whilst the conditions will ensure virtual sterility, the holding time is too short to bring about the changes in the milk protein that are essential for the formation of a firm coagulum during fermentation. However, for fermented milk drinks, this lack of natural viscosity is irrelevant and, certainly with acidophilus milk, the essential is to provide a process milk devoid of bacteria that might compete against *Lactobacillus acidophilus*.

The final choice of a system for this essential heat treatment of the milk will be based on considerations of installation costs/anticipated throughput, but however it is carried out, avoidance of post-heating contamination is essential. A detailed consideration of the cleaning process has recently been compiled by Romney (1990), and Luck and Gavron (1990) suggests that the total bacterial count on plant surfaces carrying pasteurised products should be <50 cfu 100 cm^{-2}. This figure could be applied also to the milk for fermentation, but a lower figure of <10 cfu 100 cm^{-2} should be employed for lactose-fermenting yeasts, e.g. *Kluyveromyces marxianus* var. *marxianus* (*Kluy. fragilis*).

INOCULATION AND INCUBATION

After the heat treatment stage, the milk is cooled to incubation temperature and, prior to inoculation, either remains in the multi-purpose tank or is pumped to insulated tanks to be held throughout the entire period of fermentation. The temperature selected depends upon both the system of production and type of fermented milk, and some typical examples are shown in Table III.

As mentioned earlier, yoghurt is manufactured almost universally with a culture composed of *Str. thermophilus* and *Lac. bulgaricus* for the unique synergism between these two organisms ensures not only that the required level of lactic acid is reached within the allotted time, but also that the

TABLE III
Some typical examples of starter cultures employed in the manufacture of fermented milks[a]

Type of culture	*Product*	*Microorganisms involved*
Mesophilic	Taetmjolk	*Lactococcus lactis* subsp. *lactis*
	Filmjolk	*Lactococcus lactis* biovar. *diacetylactis*
		Leuconostoc mesenteroides subsp. *cremoris*
	Ymer	*Lact. lactis* subsp. *cremoris*
		Lact. lactis biovar. *diacetylactis*
	Kefir	Kefir grains—thermophilic lactobacilli and *Kluyveromyces marxianus* var. *marxianus*
Typical fermentation temperature 20–22°C		
Thermophilic	Yoghurt	*Streptococcus salivarius* subsp. *thermophilius*
		Lactobacillus delbrueckii subsp. *bulgaricus*
	Yakult	*Lactobacillus casei* subsp. *casei*
	Acidophilus milk	*Lactobacillus acidophilus*
	A/B milk[b]	*Lac. acidophilus*
		Bifidobacterium bifidum[c]
	A/B yoghurt	As above plus yoghurt culture
Typical fermentation temperature 37–42°C		

[a]Adapted from Tamime and Robinson (1988*a*).
[b]'Mild yoghurt' (Germany) or 'bio-yoghurt' (UK) have the same microflora.
[c]Other species of bifidiobacteria of human origin, e.g. *Bif. longum* or *Bif. adolescentis* may be used instead.

end-product has the essential flavour and consistency. Obviously, these attributes vary with the precise combination of strains that is being employed, but once the behaviour pattern of a particular combination has been established, its performance within a factory should be entirely predictable. The same expectation applies to other cultures as well; for example, the mesophilic strains employed for filmjolk or ymer, but these demands can only be met if the system for culture preparation is totally reliable.

Although it is possible to maintain the necessary cultures in a laboratory and then grow them on-site to the volume needed to inoculate the process milk (Tamime, 1990), the system can prove:

(a) expensive in terms of laboratory facilities, culture vessels and time of the creamery personnel;

(b) prone to problems of infection; and
(c) liable to lead to changes in culture, in respect to

   (i) the balance between constituent species—as might occur with a yoghurt culture;
   (ii) the balance between strains of given species; and/or
   (iii) the inadvertent selection of a mutant of a single strain, or a variant that has lost a plasmid-controlled characteristic during replication.

For these reasons, most manufacturers rely on culture suppliers for their routine requirements, and purchase either deep-frozen or freeze-dried cultures with specified properties. Cultures of this type can be stored on-site for several months, and then be employed either for the production of a bulk starter or for direct-to-vat inoculation (DVI) of the yoghurt milk. The production of bulk starters is the popular option for factories with the tanks already available, for although costly to maintain and run—for example, the milk for bulk starter production has to be heat treated at 95°C for 30 min to ensure the absence of bacteriophages—the system does have some advantages:

(a) the manufacturer can check the culture, either microscopically or by means of a simple activity test, prior to inoculating the process milk;
(b) by halting the growth of the bulk starter during its log phase, growth in the process milk will be more rapid than if the starter is frozen or freeze-dried. This difference in performance is even more noticeable if the bulk starter is warmed to fermentation temperature prior to addition to the process vat, and the time saved over the period of a typical yoghurt fermentation could well be between 60 and 120 min; and
(c) it will enable a manufacturer to employ specially selected substrates to grow starter bacteria with fastidious growth requirements (Merilainen and Dellaglio, 1990).

Nevertheless, the sheer convenience of adding a concentrated culture directly to a vat of process milk has much in its favour (Puhan and Zambrini, 1990), as does the facility to change cultures from vat to vat. Thus, as consumers look for greater variety on the retail shelves, so manufacturers have to become more adaptable, and with direct-to-vat cultures, the production of small volumes of a specially fermented milk becomes a feasible option. Yoghurt provides a good example of this trend, for while the normal product is fermented with a balanced culture of one chain of *Str. thermophilus* to one chain of *Lac. bulgaricus*, direct-to-vat cultures are now available in which the streptococci outnumber the lactobacilli by a considerable margin (Anon., 1991). The use of such cultures gives rise to a yoghurt with an extremely mild flavour *vis-à-vis* the standard type, and yet with the same physical characteristics in respect of viscosity and 'mouthfeel'. Consequently, a manufacturer has the facility to adjust, in response to consumer demand, the volumes/types of yoghurt leaving the factory, and hence it is not surprising that, as bulk starter tanks come up for replacement, many medium-sized plants are turning to the direct-to-vat approach.

Ultimately, of course, the choice will depend on a number of factors, but whatever the route, the process milk must be furnished with an abundant and viable microflora. Once this goal has been achieved, the fermentation stage can begin. The plant required at this point in the process is designed to provide and maintain the temperature conditions essential for growth and metabolism of the bacteria, but its design will reflect also the nature of the end-product, e.g. set yoghurt as against stirred yoghurt.

Production of Set Yoghurt

Set yoghurt, which may be natural or flavoured, should have a distinctive, gel-like structure when cut with a spoon, and hence it follows that the formation of the coagulum must take place in the retail container. Sizes of carton tend to vary from 500 ml, down to 150 ml, but in every case, the process involves the following:

(a) cooling the yoghurt milk to incubation temperature;
(b) adding the starter culture along, perhaps, with selected flavours and colours;
(c) filling the retail cartons; and
(d) incubation to achieve the acidity dictated by the consumer market in question.

If fruit pieces are to be included, then the fruit can be added to the carton prior to filling with the milk, and then it is for the consumer to stir in the fruit during consumption. If sold in a clear carton, then such products can look most attractive, particularly if the fruit purée contains a stabilising system that prevents migration of the colour into the yoghurt gel (Anon., 1990).

Assuming that the inoculated milk will be filled into the cartons at incubation temperature, then the temperature for fermentation can be maintained most easily by circulating air.

Water has been employed in the past (Crawford, 1962), and obviously it is an excellent medium for heat transfer, both into the product during incubation, and for subsequent cooling by changing to cold/chilled water. However, the problem is that glass bottles/jars with sealed caps are the preferred containers for use with this system, and this limitation has curtailed widespread adoption in Western Europe and North America. In Eastern Europe where fermented drinks in glass bottles are popular, this restriction does not apply, but for set yoghurt, air has become the medium for temperature regulation.

Incubation Cabinets and Tunnels

For small volumes of yoghurt, e.g. 450–750 litres, cabinets of the type shown in Fig. 9 can prove extremely useful. The cartons of yoghurt are stacked into cardboard trays holding nine or 12 units—the number is varied depending upon the volume of the cartons—and then placed on a shelf in the incubator. Warm air is circulated around the cabinet during the fermentation stage, and once the product has reached the desired acidity, cold air is employed for rapid cooling to below 20°C. Below this temperature, the metabolism of the bacteria declines to the extent that acid development almost ceases, and the product can be transferred to a refrigerated store for final chilling to 2–4°C. Further acidification does take place, of course, even at this final temperature, but it is sufficiently slow to allow the product to be given a 2–3 week shelf-life without running the risk of consumer complaint. The other reason for initially chilling the product within the incubation cabinet relates to the structure of the actual gel. Thus, at 42°C, or even 30°C, the gel consists of a delicate network of interlinking protein chains with whey filling the interstices, and if handled roughly in this condition, the whey exudes onto the surface of the product—syneresis. However, if the gel is cooled to <20°C, then the slight shrinkage and hardening of the chains traps the whey in the coagulum, and the consumer, upon opening the carton, is presented with a smooth, clean surface with only a hint of moisture.

Success with cabinets of this type depends upon the manufacturer ensuring that

(a) the air circulates throughout the chamber, so avoiding the risk of stagnant pockets of air;

Fig. 9. An electrically operated cabinet that can be used for the incubation of fermented milks on a small scale. These units can be provided also with a refrigeration unit to allow subsequent cooling of the product.

(b) the thermostat is so placed that it accurately records a true air temperature; and
(c) the trays of yoghurt can be stacked quickly at the start of the incubation period, so avoiding too long a time lag between the first and last cartons entering the chamber.

Although the cabinet shown in Fig. 9 is designed for small volumes, the principle of operation can be extended to cover incubation rooms capable of holding several thousand litres of product. In this latter situation, a conveyor system can be employed to assist with rapid loading at the start of incubation but, even so, it may well become necessary to load the product on a 'batch system'. While this approach does ensure that one entire group of trays is fermenting at roughly the same pace, and hence can be removed from the incubator at the same time, it does necessitate moving the yoghurt at a higher than desirable temperature. However, so long as the conveyor system is smooth-running, then such movements to the cold-store should be possible. Indeed, the use of incubation rooms is sufficiently widespread to suggest that if the yoghurt has been correctly made, i.e. adequate total solids and homogenisation, syneresis may not pose too many problems.

As an alternative, large-scale manufacturers can resort to the installation of a tunnel system. The tunnel consists of two sections, one with circulating warm air and the other with chilled air, and stacks of retail cartons are placed on a conveyor that runs throughout the length of the unit. The speed of the conveyor is regulated to reflect the rate of acid production in the yoghurt milk, so that entry to the cooling section only occurs at a pH of 4·5–4·6. After passage through the chilling section, the cartons are transferred to a cold store for the final temperature reduction to 2–4°C. The fact that the retail cartons can be stacked into wire cages or onto pallets offers great advantages for mechanical handling, but one cautionary note centres on the need for the coagulating milk to be in constant motion. A smooth-running conveyor is essential to avoid jarring the delicate structure of the gel, but despite this reservation, Cottenie (1978) and Norling (1979) claim that the system is finding increased popularity.

Production of Non-gelled Products

The base milk for stirred yoghurt and many other types of fermented milk is fermented in bulk, and the gel structure is then broken to provide a smooth, homogeneous product. A standard, water-jacketed tank of the type shown in Fig. 7 is often used for this purpose, so that after heat treatment of the milk, the same tank can be employed for both incubating the milk until the desired acidity and final cooling to <20°C. This same approach can be employed irrespective of whether the product is formed by a thermophilic culture or a mesophilic one, but the limitation comes in relation to volume. Thus, for outputs of 500–1000 litres day^{-1}, the system is quite adequate, but beyond that level, separate tanks must be employed for incubation. These latter tanks—each with their own agitation system—rely upon polystyrene or some similar material for insulation rather than a water-jacket, and hence are less expensive to manufacture compared with a multi-purpose tank. It is quite feasible, therefore, to increase the output of a factory by servicing a number of fermentation vessels from just one heat-treatment unit, and at a cost that is not prohibitive.

The disadvantage of this approach is that the yoghurt or other product has to be removed from the tank for cooling, and while this may not be a problem for products that will ultimately be fluids anyway, for viscous milks: damage to the coagulum needs to be minimised. The overnight incubation of yoghurt at 27–30°C may offer one solution—the coagulum is less susceptible to damage than at 42°C—as may the careful selection of an agitator system (see below), but either way, the warm product is subject to considerable stress. These potentially destructive forces increase as the product is then pumped through a plate or tubular heat exchanger to achieve rapid cooling. According to Piersma and Steenbergen (1973), minimal structural damage to the coagulum will occur in a tubular cooler, but conventional plate heat exchangers can be modified to achieve the same end. The essential features are the provision of large gaps, *vis-à-vis* the normal situation, between the plates and the avoidance of high back-pressures. This reduction in pressure can be sought by restricting the flow-rate of the product, or by progressively increasing the gaps between the plates across the unit, and this approach, together with the use of a number of small exchangers rather than one large one, may lower the risk of product damage. It is usually suggested also that product flow should be restricted to one upward and one downward pass across the length of the exchanger. However, success in the production of viscous products will depend also upon

(a) the formulation of the initial mix, in that stabilisers will almost certainly be needed to both protect the milk protein, and mask any deleterious decline in viscosity; and
(b) the selection of a starter culture capable of secreting extracellular polysaccharides capable of inducing appreciable 'set-back', i.e. once the product has been packaged, a distinct reformation of the coagulum is observed (Tamime, 1990).

Recently, Driessen and Loones (1990) have suggested that this 'set-back' phenomenon may, at least in full-fat yoghurts, be associated with a rearrangement of the fat/casein structure in the yoghurt after cooling. The application of pressure attained by passing the yoghurt through a reducing valve or sieve induced this increase in viscosity in a range of samples, and it was concluded that bacterial polysacchides were not important. However, it must be assumed that, for low-fat products, this lipid–protein interaction has only a minor role, and polysaccharide secretion by the culture remains the important factor.

In-Tank Cooling

If multi-purpose tanks are employed, then the cooling operation can be achieved without moving the product elsewhere, but design of the system needs careful thought. Thus, the agitation system in such a tank must be capable of ensuring efficient mixing of the ingredients into the mix, e.g. incorporation of milk powder and/or stabilisers, but at the same time must be suitable for mixing the finished product with minimal damage. According to Tamine and Greig (1979), efficient mixing of the ingredients is related to

(a) the shape, size and position of the agitator;
(b) the speed of rotation of the agitator;
(c) the shape of the processing tank; and
(d) the effect of the agitator on the product, e.g. whether or not a vortex is created, and the degree of shearing.

However, many of the requirements for optimum mixing are, of course, incompatible with those for cooling the end-product. The creation of vortex, for example, is neither necessary nor desirable during the cooling stage, and the elimination of excessive shear is vital.

Obviously, stirring the warm coagulum does involve some degree of shearing, but the extent can be influenced by the design of the agitator. The speed of rotation of the agitator and, in particular, the difference in velocity between the bulk liquid and the tips of the agitator blades, is crucial. One approach to minimising potential damage to the coagulum is offered by the APV Pasilac AS system illustrated in Fig. 10, in which the speed of the scraped-surface agitator is 7 rpm, and that of the centrally mounted helical paddle is 15 rpm (Tamime and Greig, 1979); this configuration provides both efficient mixing of the yoghurt and minimum shear. This use of scraped-surface agitators is widespread, and efficient ingredient mixing is usually achieved by an additional mixing system, either within the tank or located in the adjacent pipework, e.g. the Silverson High-Speed Mixer. A conical base to the vessel or the inclusion of fixed baffles also helps to improve the mixing of the yoghurt during the cooling stage (Tamime and Robinson, 1985).

The reduction in temperature to <20°C relies upon the passage of chilled water through the jacket of the tank, and the efficiency of the process is governed by

(a) the speed of agitation of the bulk yoghurt, and hence the contact time between the yoghurt and the cooled surface;

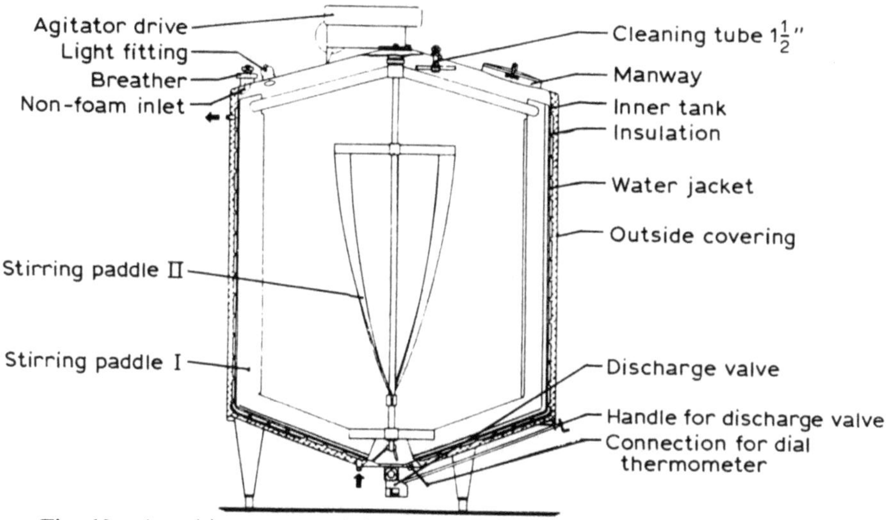

Fig. 10. A multi-purpose tank from APV Pasilac AS, with the design emphasis placed upon ensuring efficient heat transfer, especially during the cooling stage. (Reproduced by courtesy of Dairy Industries International.)

(b) the contact area between the yoghurt and walls of the tank; and
(c) the temperature differential between the product and the cooling medium, together with rate of flow of the latter.

Obviously, there is a degree of inter-relationship between each of these factors and, at the same time, the properties desired in the end-product, e.g. viscosity, have to be considered. However, the impact of both contact area and temperature difference can be assessed in isolation, and a number of options for improving the rate of heat removal are available. Forced circulation of the cooling agent is one obvious improvement, but increasing the contact area offers advantages as well. Thus, whilst some tanks are only jacketed on the side-walls, others also allow for the flow of coolant into a basal jacket. The flow of coolant through a central cylinder which supports the agitator has been proposed, as has the insertion of portable cooling coils, but the latter approach tends to increase the risk of infection. It is usual, therefore, for in-tank cooling to rely on the following:

(a) a wide temperature differential, e.g. coolant at <2°C against yoghurt at 45 or 30°C;
(b) rapid circulation of the coolant through the jacket of the tank; and

(c) the use of a scraped-surface agitator, so that 'cooled' product is constantly being removed from the walls and replaced by 'warm' yoghurt.

This approach should ensure rapid cooling with minimal damage to the coagulum, and it is estimated that such a system could cool 5000 litres of yoghurt from 42°C to 5°C in 4 h; the drop to 20°C could be achieved much more rapidly (J. L. Jay, pers. comm.). As the yoghurt is such tanks may remain fairly viscous, sterile air under pressure may be employed to assist with the transfer of the yoghurt after cooling. However, while this system will avoid wastage due to product remaining in the tank, further movement will involve the use of pumps.

Pumps for Handling Fermented Milks

For moving liquid products/milk, one of the most efficient types of pump is the centrifugal pump, in which a vaned impeller propels the fluid around a circular chamber from the inlet pipe towards the outlet. By spinning at high speed, the impeller is able to generate considerable force, and the fluid is released through the outlet at high velocity. One popular example of this type of pump is the 'W-Series' manufactured by APV Baker Ltd (see Fig. 11), and flow-rates of around 200 $m^3 h^{-1}$ at pressures of up to 0·8 MPa are attainable. Such forces are unsuitable, of course, for a coagulated product, but the value of including a centrifugal pump into the plant lay-out is that it can be employed also for the circulation of cleaning-in-place (CIP) solutions; the 'scouring' action of the high-velocity fluid improves the action of detergent agents to a marked degree. Consequently, a different type of pump has to be employed for moving viscous products, and examples from this group are placed under the heading of 'positive displacement' pumps. The simplest form is the 'piston pump', in which a set volume of yoghurt or other viscous product, such as quarg or cultured cream, is drawn from a reservoir as the piston moves up the barrel, and is then discharged through an exit port as the piston returns. The pressure needed to force the product through the exit pipe does induce a pronounced degree of shear, but as the effect is short-lived, i.e. only for as long as the piston is travelling down the barrel, serious damage is not a problem. However, if considerable pressure is required to be exerted by the piston, then damage could be unacceptable and, for this reason, piston pumps are used widely in situations that

Fig. 11. The complete range of W-pumps consists of a large number of different models and variants based on a modular system of standard components. (Reproduced by courtesy of APV Baker Ltd, Crawley, Sussex, UK.)

— require small, accurately measured, volumes of product to be moved; and
— where the back-pressure against the product leaving the exit port is very low.

These two considerations are met most clearly during packaging operations, and most filling machines designed to handle yoghurt and similar products employ piston pumps to dispense the material into cartons or other containers.

In situations where the product has to travel a long distance, then movement has to be achieved by a rotational displacement pump. In this type of pump, a cavity is formed by the slow rotation of a lobed-cam within a circular chamber. Gravity forces the product into the cavity, and as the cam rotates, so the material is carried around the chamber—trapped between the lobes and the wall—towards the exit port. The product is then forced out into the exit pipe. In general, loss of viscosity should not be too great so long as the speed of rotation is kept as low as conveniently possible, along with a low back-pressure, i.e. short pipe runs, and variants of this approach are widely employed to transport yoghurt around a plant. Success can, however, be impaired by inattention to details of the associated pipework, for unnecessary obstructions like elbows or constrictions, or even poor finishes to the internal surfaces of pipework, can cause friction against the product flow, and hence loss of viscosity. Steenbergen (1971) considered that the average velocity of the product, along with the diameter and length of the pipework, were the critical factors, and his data showed that

— if diameter and velocity are constant, then the viscosity of a product declines with increasing length of the pipe; and
— if velocity and length are fixed, minimum damage occurs in pipework with a large diameter.

and although this relationship is not unexpected, it needs to be borne in mind during the design of a plant for yoghurt or similar product. In particular, these constraints become especially important if

(a) the product has to be pumped through any device to improve texture and/or eliminate lumps. Thus, it is sometimes recommended that stirred yoghurt should be passed through a screen of stainless-steel mesh prior to the addition of fruit, for in this way, the base is rendered smooth and homogeneous (Nielsen, 1972, 1976). However, although this device does impart a considerable back-pressure, correct formulation of the base and the choice of an appropriate starter culture should avoid deleterious damage to the coagulum; and
(b) fruit is to be incorporated into the product, and in a manner that necessitates the use of mechanical agitation/an additional reduction in viscosity; this aspect of production is considered below.

EQUIPMENT FOR THE ADDITION OF FRUIT

Although there is a market in some countries for flavoured, set yoghurts, the most popular type remains, at least outside the Mediterranean countries and the Middle East, stirred fruit yoghurt. In this product, a fruit purée or similar material is added to a base of natural yoghurt that has been mixed to a smooth, even consistency. Fresh fruit can be used for the highly specialised, delicatessen trade or for local markets, but the risks of spoilage are so high that few manufacturers find this option attractive. Similarly, frozen fruits are available to manufacturers who wish to add only 'natural' fruit to their products, but again the market is a rather specialist one, and most yoghurt-makers prefer to employ pasteurised fruit mixes. These products offer the advantages that

— they should be entirely free from all yeasts and moulds,
— the level of colour can be adjusted to ensure that the finished yoghurt has a uniform appearance, and
— sucrose can be added to the fruit by the supplier, so that the degree of sweetness desired in the end-product can be achieved through one combined addition.

Processed fruits of this type are usually received by the dairy in metal cans (5 kg), polypropylene drums or buckets (25 kg) or in stainless-steel tanks holding around 500 kg. The former two containers are well suited for small- or medium-sized dairies, but large dairies will need to obtain supplies in bulk, or if the scale of the operation so demands, even process their own fruit on-site. This latter option does offer economic advantages, but the potential problems of handling fresh fruit, and its attendant rich flora of yeasts and moulds, alongside the facility for making the fermented milk may off-set any gains. If the fruit is contained in metal cans, then hand-operated or automatic openers are needed (see Fig. 12), and the fruit is then tipped into a hopper linked to the yoghurt line (in-line mixing), or directly into a tank of stirred, cooled product and blended with slow agitation. Bulk tanks of fruit are normally metered directly into the yoghurt line but, whatever the approach, the essential is to ensure that the fruit is fully blended into the base. Highly coloured fruits, such as black cherry, pose a particular problem in this respect, for uneven appearance can provide a source of consumer complaints.

The method of fruit mixing can be manually or automatically controlled, and the two approaches are summarised in Fig. 13. When a batch/manual system is in operation, the sequence of events is as follows:

— empty into each tank of yoghurt, the amount of fruit required for a given volume of end-product, usually between 8 and 10%, with the precise figure depending upon the type of fruit purée being used, e.g. level of actual fruit and/or sucrose content;

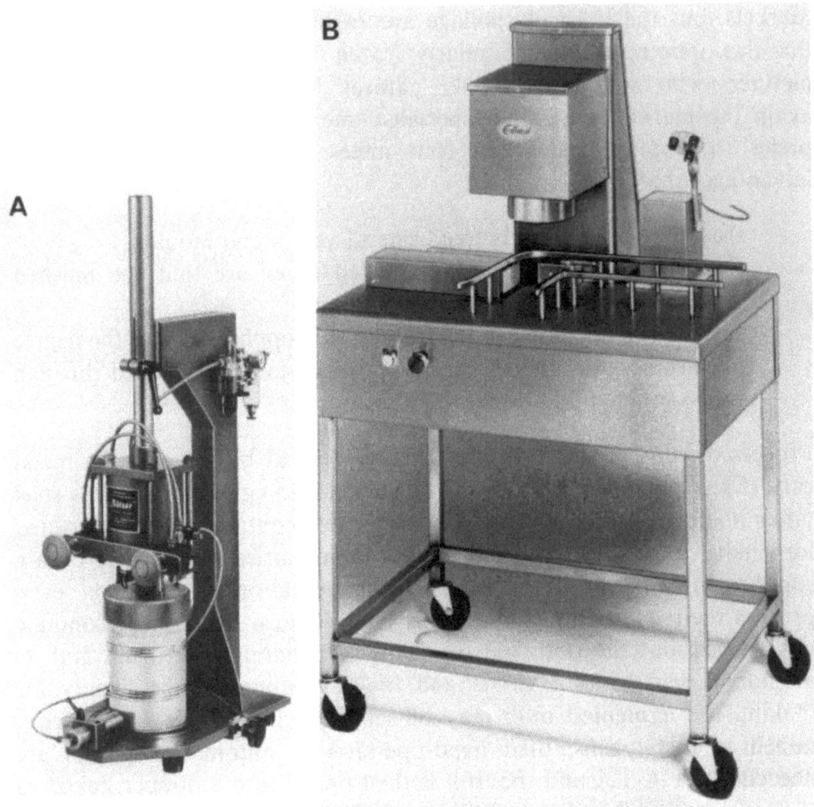

Fig. 12. Different types of opener for metal cans. (A) Pneumatic opener—Blitzer PD-10—a semi-automatic unit operated by compressed air. (Reproduced by courtesy of Karl Engelhardt, Bremen, Germany.) (B) An automatic device that can handle 1500 cans per hour. (Reproduced by courtesy of Peter Holland Food Machinery Ltd, Lincolnshire, UK.)

— mix gently with slow speed agitation until the product is homogeneous, at least to the naked eye; and then
— pump to filling machine,

and if a second tank can be prepared whilst the first one is emptying, the filling machine can be in continuous operation. Alternatively, if the base yoghurt has been produced in a batch of several hundred litres, then a fruit-mixing tank can be interspersed between the yoghurt tank and the filler. If specific volumes of yoghurt are then transferred to the mixing tank, a sequence of filling, fruiting and packaging can be

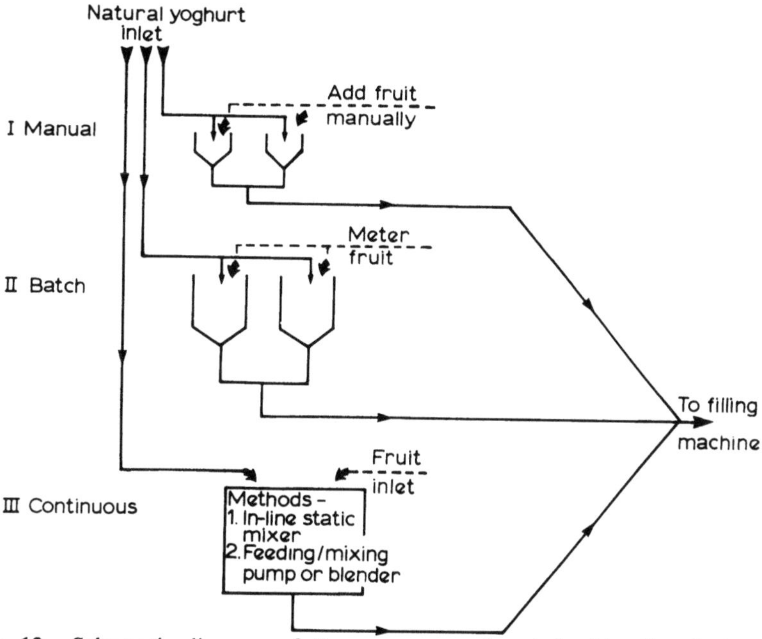

Fig. 13. Schematic diagram of the systems employed for blending fruit and yoghurt.

employed to produce a range of flavoured items with little loss during the change-over.

Automatic Addition of the Fruit

A number of alternatives are available, and the INDAG system (INDAG GmbH, Heidelberg, Germany) employs two positive displacement pumps to introduce the required volumes of product and fruit. Blending is achieved in a mixing chamber, and the final product/fruit mix is fed directly into the hopper of the filling machine. The speed of the fruit pump is adjustable to compensate for the different characteristics of the fruit purées, and by connecting this pump to a series of fruit tanks, the change-over from one type to another can be achieved at the press of a button.

The Burdosa fruit blender from Herwig Burget in Germany employs a duplex metering pump which simulatenously moves the fruit and base product into a mixing chamber (see Fig. 14). The ratio of fruit to product

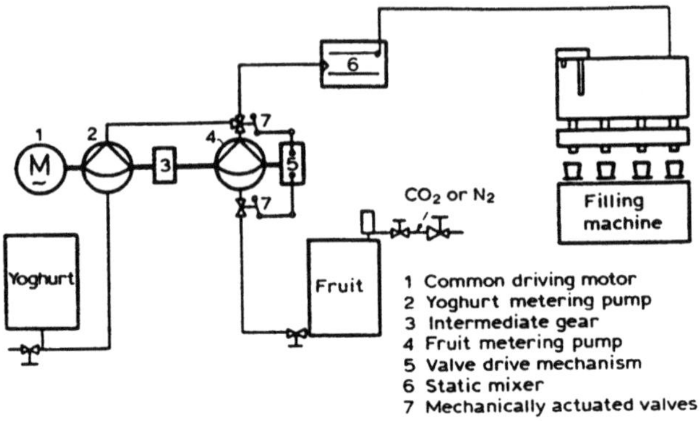

Fig. 14. A typical configuration for the continuous blending of yoghurt and fruit prior to transfer to the filling machine. (Reproduced by courtesy of Burdosa UK Ltd, Wembley, Middlesex, UK.)

can be pre-set, but the maximum level of fruit is set at 250 cm^3 $litre^{-1}$ of dairy base; this inclusion rate is well above any likely requirement for a fermented milk product. The total output from the unit, depending upon the model, is up to 4500 litres h^{-1}. Continuous fruit/yoghurt mixing is achieved using an in-line static mixer. In this system, the two components are mixed by forced circulation within the special chamber, but other designs could be equally suitable. For example, in an alternative approach, a set of metal blades, twisted into the form of a spiral, is welded inside a length of stainless-steel pipe, and this section is then welded into the side-wall of another piece of pipe of similar dimensions to form a T-junction. In operation, the yoghurt base and the fruit are metered through the side-arms of the 'T', and are forced through the blades contained in the 'down-section'. The rotation of the two streams is sufficient to bring about a mixing of the fruit and the base, and to a degree that is satisfactory for most practical purposes. The great advantage is, of course, that there are no moving parts in the system, but because there is little control over the operation, some manufacturers prefer a more mechanised system for the fruit and yoghurt.

Gasti-Verpakungsmaschinen GmbH of Germany manufactures mixing and feeding pumps that can handle fruit and dairy bases, and the DOGAmix 60 unit, for example, consists of two feed pumps that draw the base and the fruit from the storage tanks in pre-set proportions. The mixing chamber is fitted with an agitator of variable speed, and the final product is fed straight from the chamber to the hopper of the filling machine. The maximum feed rates are 75 litres min^{-1}, and ratios of base:fruit of 1:5 to 1:20 can be achieved with an accuracy ±0·5%. The entire product contact area is suitable for CIP cleaning, and can be steam sterilised at 140°C; this latter facility can be advantageous to ensure that yeasts do not build-up at any point. Similar systems are available from other manufacturers, but the overall aim is always to send to the filling machine a blend that is both uniform in appearance and consistency, and free from ingredient damage.

PACKAGING OF FERMENTED DAIRY PRODUCTS

Most filling operations involve the use of piston pumps that dispense a pre-set volume of product, and the main differences arise in respect to

— filling capacity which may range from 5000 up to 50 000 cartons per hour;
— whether the cartons are pre-formed prior to loading into the machine, as against systems that form the cartons and fill them in one continuous operation
— the degree of product protection that is built into the system, i.e. whether or not the filling head is surrounded by a controlled atmosphere, or whether the pre-formed cartons and lids are sterilised before actual filling.

Filling Machines—No Atmosphere Control

Most manufacturers of filling machines offer a system of packaging in which the product is exposed to the atmosphere during the filling stage, and for small-scale operations, this approach is the 'norm'. A typical example of a filling machine with a capacity suitable for small to medium dairies is the Colunio (UK) range, and a throughput of 2000 cartons per hour would be a reasonable aim. Most variants consist of a rotating turn-table onto which, in sequential fractions of a rotation, a carton is dropped into a holder, filled with the correct volume of product, sealed and ejected onto a stacking table. Carton sizes can range from 150 to 500 ml, and the lids (aluminium foil) can be heat-sealed into place or they be of the snap-on type (polypropylene). Changes of carton size do, of course, necessitate a change of table and lidding attachment, and to avoid the inevitable delays, most companies would have a small filling machine to handle limited batches of special lines, and one major machine to fill, for example, single portion cartons of a more popular product, e.g. stirred fruit yoghurt. In selecting a filling machine of this type, a number of points need to be considered:

- versatility, i.e. ability to handle a range of products and/or retail cartons;
- capability for operation to a high standard of hygiene, i.e. all contact surfaces must be stainless-steel, and accessible for cleaning/sterilising;
- accuracy of filling, and elimination of 'drips' between the filling of each carton;
- facility to add a device for date marking or bar coding each carton of product; and
- safety measures, such as smooth ejection of the cartons to avoid damage to lids, or a no carton – no fill system to ensure that the table is not accidently flooded with product (Tamime and Robinson, 1985).

If a larger throughput is needed, then the lay-out is altered so that each filling head serves one line on a conveyor, and lines of cartons—up to eight across a single conveyor—are fed beneath the heads and onto the sealing unit. In this configuration, thousands of cartons per hour can be filled and with a range of flavours, and after sealing, the cartons can be automatically stacked into trays ready for transfer to a cold-store.

Filling Machines—Controlled Atmosphere

In this situation, the basic multi-track system is modified so that the cup filling and closure area is enclosed within a cabinet served with a laminar flow of sterile air (Bruderer and Schicht, 1987). The advantage is that no organisms in the atmosphere, particularly yeasts or moulds, can come into contact with the product, a point that has some importance with the aspetic versions. In this latter case, the cartons are sterilised with hydrogen peroxide (H_2O_2), and then dried with hot, sterile air prior to filling with product. The contact surface of the lids may also be sterilised with H_2O_2 (see Fig. 15), but the use of UV-C lamps offers an alternative. The extent to which the shelf-life of a product is enhanced by these techniques depends upon a number of associated factors, but with potential outputs of 36000 cartons per hour from such machines, any precautions that lower the risk of infection must be worthwhile.

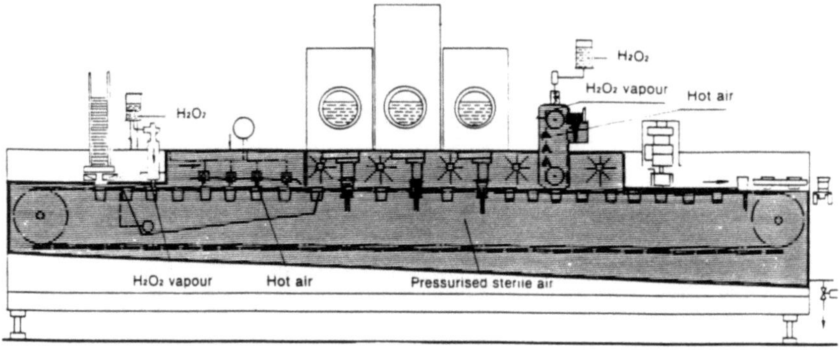

Fig. 15. A diagrammatic representation of the DOGAseptic 42 GASTI—Cup Filling and Sealing Machine. The cups and lids are sterilised with hydrogen peroxide, and the three filling stations above the conveyor can be employed in a variety of ways, e.g. pre-filling a carton with fruit from one station, and completing the fill with natural yoghurt, or direct filling with already fruited product. (Reproduced by courtesy of Jagenberg (London) Ltd, Purley, London, UK.)

Miscellaneous Filling Machines

The distinctive feature of all the filling machines described so far is that they employed pre-formed cartons. These cartons are manufactured by softening a portion of 'plastic' material, and then injecting it under high pressure into a mould for hardening and shaping. After ejection from the moulds, the carton has a rigid structure that is sufficiently robust for an individual portion carton to be handled by a consumer with little fear of breakage. However, this security does come at a price, and less expensive forms of packaging have been sought for products that will, in the main, be consumed in the home.

For such products, the form–fill–seal process, in which the 'plastic' is delivered in the form of a continuous roll, has proved popular. During production runs, one end of the roll is fed into the packaging machine (see Fig. 16), and as it passes through the first region, the sheet of 'plastic' is heat-softened and formed into a cup—either in or around a mould. The cup is then immediately filled with product and heat sealed. The cartons are less rigid that the pre-formed type, but the nature of 'form–fill–seal' operation has opened the way for the introduction of family packs of four or six cups all joined together as one retail unit. As each cup is filled by a

Fig. 16. A form-fill-seal (Formseal) machine of the type that can be used for handling yoghurt under aseptic conditions. The enclosed filling chamber is held under a positive pressure of sterile air to prevent contamination prior to sealing, and both the base material—feeding from the reel on the left—and the aluminium foil (top cover for the pots) are sterilised with hydrogen peroxide. (Reproduced by courtesy of Erca UK Ltd, London, UK.)

separate filling head, a group of cups may be all the same flavour, or each cup can be individual. The lack of rigidity often means that volumes are smaller (125 g) in form–filled cups than in the rigid cartons (150 g), and that the groups of cups must be further wrapped in a cardboard sleeve for retail distribution. Nevertheless, for family purchases, the four or six unit has some attractions, but most retail outlets, mindful of the needs of their customers, insist that individual portions are readily available as well.

As an alternative to the usual 'plastic' containers for fermented milks, some manufacturers are turning to the cartons normally associated with liquid milk, which consist of a paper-board base coated on each side with a suitable 'plastic' material, such as polyethylene. The actual cartons may arrive part-formed, i.e. sealed along the sides but open at both ends, or as a reel of the base material and, if part-formed, the sequence of filling is as follows:

— a carton is forced onto a spoke of a rotating wheel, and this action forces the walls of the carton apart;
— the carton is removed automatically from the wheel and the base is sealed; and
— after filling with product, the top is sealed, and the carton is ready for transfer to the cold-store.

In mechanical terms, this apparently simple sequence of events is extremely complex, but the long experience gained with handling liquid milk and fruit juices has made the transfer of the technology to fermented milks a reasonable proposition. The type of carton selected, i.e. either gable-end or flat-pack, is for the manufacturer to decide, but many large suppliers are opting for the flat-pack design as a means of economising on storage space. The final stage for all these operations is the crating and transfer of the product to the retail outlet, and this process always involves the following:

— placing individual cartons into cardboard trays with spaces for 12 or more cartons; for set yoghurt, for example, these same trays may well have served to move the cartons to the incubation room; and
— grouping the trays in some manner—cardboard outer boxes or shrink-wrapping—so that they can be easily handled mechanically on wooden pallets or in wire cages.

Considerations of Shelf-Life

Once the final packaging has been achieved, storage of fermented milks should be at 2–4°C throughout the distribution chain. For yoghurt, this requirement can be critical, not only to avoid the risk of spoilage from stray yeasts (Robinson and Tamime, 1990), but also to prevent further activity by the started culture. The potential role of *Lac. bulgaricus* is especially important in this context, because while *Str. thermophilus* becomes largely inactive at acidities about 1·0% lactic acid, *Lac. bulgaricus* can generate acid levels in excess of 2·0%.

Although chill temperatures are important for all products which are expected to retain high levels of bacteria of starter origin, the problem of post-production acidification has been eased by the recent introduction of the following:

(a) starter cultures for yoghurt that contain low counts of *Lac. bulgaricus*; perhaps one chain of *Lac. bulgaricus* to nine chains of *Str. thermophilus*, compared with the normal one to one ratio (Anon., 1991). This approach, while meeting the legal/non-legal stipulation that 'yoghurt' should contain *Lac. bulgaricus*, and permiting sufficient synergism between the two species to ensure that normal production times can be achieved, gives a product with little tendency to become over-acidic on storage; and
(b) starter cultures for yoghurt-like products that consist of *Str. thermophilus* and *Lac. acidophilus*. Such products tend to lack the characteristic, aromatic flavour notes of normal yoghurt but, owing to acid intolerance of the cultures, shelf-stability can prove attractive, even under less than desirable refrigeration temperatures; in order to use the name 'yoghurt', some manufacturers may include very low levels of *Lac. bulgaricus* as well.

However, even with selected cultures and chill storage, the shelf-life of a fermented milk is unlikely to exceed 3 weeks, and hence the dairy industry has examined various techniques for extending this time. These procedures do, of course, severely reduce the bacterial count in the product or, at least, have an adverse effect on viability, and hence they should never be considered for products seeking to retain 'an abundant and viable population of bacteria of starter origin'. However, if the aim is to produce a dairy dessert with a long shelf-life, then post-fermentation treatments are available. Freeze-drying, spray-drying or even sun-drying are available options, but as the reconstituted product is either totally unlike a fermented milk, or is employed simply as a flavouring ingredient, this form of processing need not be considered in detail. Similarly, frozen yoghurt has more in common with ice-cream than a fermented milk, so that attention can be focused on two specific treatments—the application of heat and concentration.

Heat treatment of fermented milks

If the product is a gelled-type, then one proposed method for extending the shelf-life is the 'heat shock' process (Tamime and Robinson, 1988a). This procedure was described originally by Driessen (1984), and involves heating cartons of the set, fermented product at 58°C for 5–10 min. A detailed evaluation of the procedure, as applied to yoghurt, was carried out by Waes (1987), who noted that the treatment was sufficient to ensure the elimination of yeasts. The population of starter bacteria was reduced also, but he noted that *Str. thermophilus* was less sensitive to heat than the lactobacilli, and that the reaction of *Lac. bulgaricus* depended upon the strain employed. The results give the impression that the time–temperature relationship is a compromise one, i.e. sufficient to reduce the microflora, but without adverse effect on the gel structure and excess syneresis, and hence it is likely that refrigerated storage will still be essential to assure product quality. Storage at ambient temperature can only be secured if all microorganisms have been eliminated, and the more severe heat treatments involved tend to be confined to stirred or fluid products. The loss of flavour has to be masked by fruit or other ingredients, and stabilisers are needed to prevent serum separation (Foley and Mulcahy, 1989) and/or too great a decline in viscosity, but given these precautions, a shelf-stable, fruit yoghurt can be made as follows:

— cool the yoghurt to 20°C,
— pasteurise at 70°C for 30–40 s, and
— package at 55–60°C and allow to cool to ambient for distribution (Anon., 1979),

and such a product should have a shelf-life of 2–3 months. In countries where this process is legal, the product must be clearly identified as 'pasteurised' or 'heat treated' yoghurt or, if applicable, 'UHT' or 'long-life' yoghurt. In this latter case, the product will have been exposed to a temperature in excess of 100°C for around 2–3 s, and the 'commercial sterility' so imparted will ensure a shelf-life of several months at ambient temperature. However, many of these pasteurised milks lack the essential character of the 'live' product, and it is noticeable that their share of the fermented milk market is, in percentage terms, usually in single figures. This poor performance is a reflection not only of indifferent product quality, but also of the fact that the consumer market in many countries has swung in favour of so-called 'health-promoting' milks (Robinson, 1989)—see later.

Concentrated products

If some of the whey is drained from a fermented milk, the resultant product has a high total solids and an elevated acidity—up to 2·0%, and hence the nature and stability of the material becomes totally different.

The Danish product, Ymer, provides a North European example from this group, and although the total solids is not high (3·5% fat, 5–6% protein, 5–6% lactose and minerals), the consistency does reflect the high protein content. Traditionally, whey drainage took place after fermentation, but the present procedure, according to Samuelsson and Ulrich (1982), is as follows:

— heat the milk to 55°C and separate the cream;
— pasteurise the skimmed fraction at 92°C for 15 s, and cool to 55°C;
— concentrate the milk by ultrafiltration, standardise the retentate to 3·5% fat and homogenise (19·6 MPa) at 65°C;
— heat the mix to 85°C for 5 min, cool to 20–22°C and inoculate with a culture containing *Lactococcus lactis* biovar. *diacetylactis* and *Leuconostoc mesenteroides* subsp. *cremoris*; and
— incubate for 20 h, stir and cool to 5°C, store for 24 h prior to final mixing and packaging.

At present, Ymer is manufactured under factory conditions in Denmark using a different method (J. Kirkegaard, pers. comm.), and the procedure can be summarised as follows. Skimmed milk is concentrated to a protein content of around 6% by ultrafiltration at 50°C. The fat content of the retentate is standardised to 3·5% (Danish standard), and this process milk is then homogenised at 13·7 MPa (74°C), de-aerated and heated to 95–100°C for 1 min. The milk is then cooled to 22°C, inoculated (3% addition rate) with a mesophilic starter culture—including flavour- and aroma-producing strains, and incubated for 20–22 h. Next day, the cultured product is mixed gently for 1 h, homogenised at 4·9 MPa to impart a smooth texture, and then cooled to 12°C for packaging (Tamime *et al.*, 1990). Lactofil (5% fat) is a rather similar product, but neither enjoy the popularity of the products made with thermophilic (yoghurt) cultures.

Concentrated yoghurt has its origins in the Middle East, and was produced originally by leaving normal yoghurt to hang in a cloth or animal-skin bag. As the whey drained out, so the total solids of the remaining milk rose to 20–24%, and a rich, creamy product—acidity around 2·0%—was left to be scraped out of the bag. So popular has this product, Labneh, become that it is now produced on an industrial-scale in many countries, including the Middle East.

At present, there are many different mechanised systems that can be employed for the production of Labneh but, in general, they can be classified as (i) mechanical separation; (ii) ultrafiltration; and (iii) direct product formulation. A typical plant employing mechanical separation is shown in Fig. 17. In this process, warm, low fat, natural yoghurt is concentrated to around 18% total solids using a nozzle separator, and then standardised with cream to >10% fat. After thorough blending, the

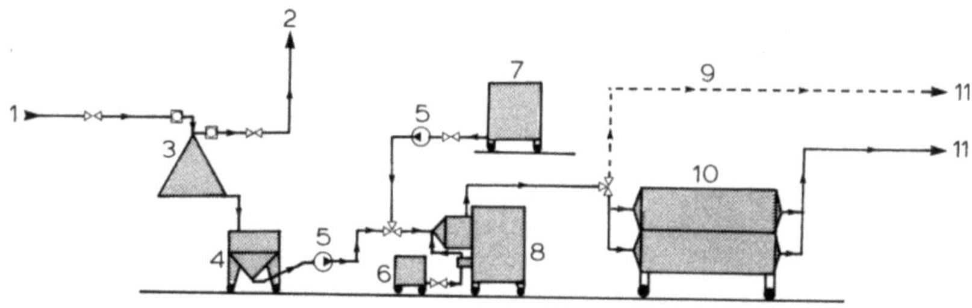

Fig. 17. A mechanised system for the manufacture of 'Labneh'. The yoghurt (40–45°C) (1) enters the Westfalia separator (3), and while the labneh passes to the holding tank (4), the whey is discharged (2). A positive pump (5) sends the labneh to the Westfalia Quarg Mixer (8) for the addition of cream (6), and other ingredients if required (7). The final product is chilled to 6°C in the tubular cooler (10) and filled (11); a pasteurising line (9) is optional. (Reproduced by courtesy of Westfalia Separator Ltd, Milton Keynes, UK.)

smooth, homogeneous mix is packaged ready for chilling and distribution (Tamime and Robinson, 1988*b*). However, recent developments in the design of separators have made it feasible to use fermented, full-cream milk for the production of Labneh (Lehmann *et al.*, 1991). Some analyses of typical commercial products are shown in Table IV.

TABLE IV
Some typical values for the major components of some commercial samples of concentrated fermented milks available in the countries indicated, together with the relevant national standards for composition[a]

Product/Country	*Total solids*	*Fat*	*Protein*	*Ash*	*Lactose*	*Acidity*[b]
Labneh						
Lebanon	22·9	10·5	—	—	—	1·7
	21·8	8·4	—	—	—	2·5
Standard	26·0	10·0				
Saudi Arabia	24·6	8·1	10·4	1·1	4·9	1·1
	23·4	8·3	9·3	1·3	3·5	1·9
Standard	22·0	7·0				
Greece	22·4	10·7	8·2	—	—	1·7
Skyr						
Iceland	20·8[c]	0·4	15·8	1·0	3·6	2·7
	17·5	0·2	12·7	0·7	3·9	1·9

[a]Adapted from Tamime and Robinson (1988*a*).
[b]Acidity as percentage lactic acid.
[c]Traditional product.

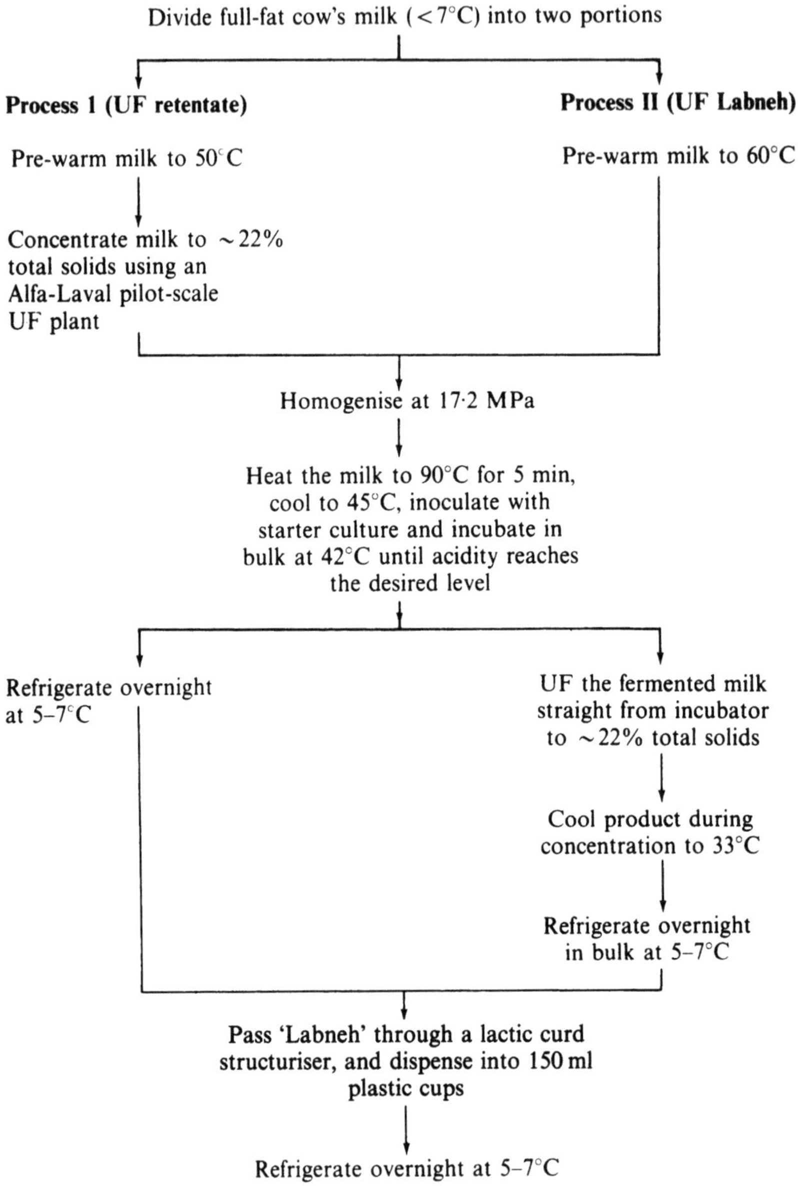

Fig. 18. The alternative methods for producing Labneh by ultrafiltration. Adapted from Tamine *et al.* (1989*a*).

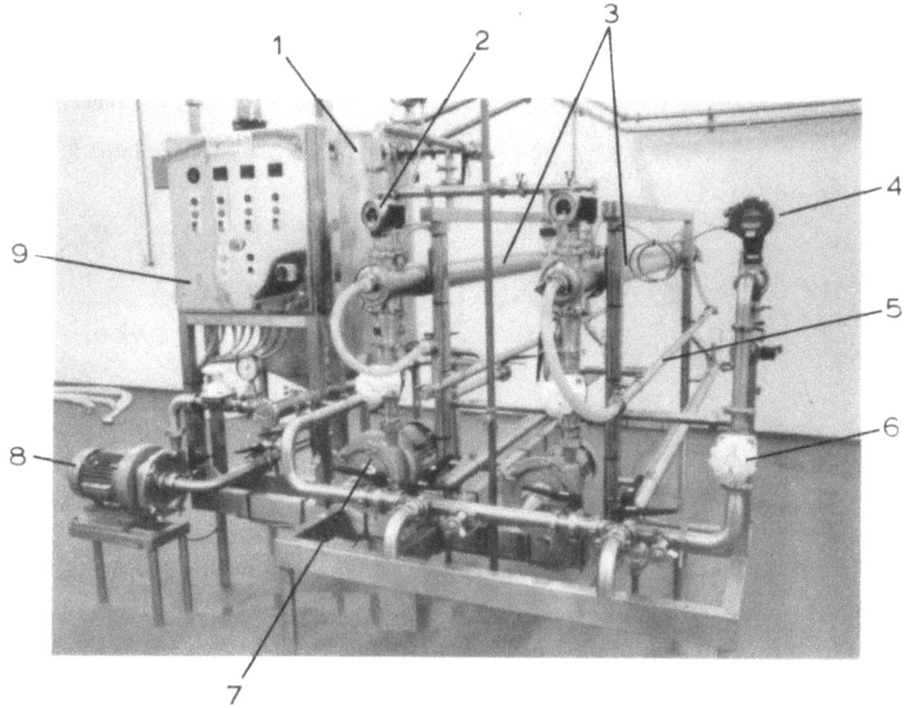

Fig. 19. An illustration of an ultrafiltration plant for the manufacture of Labneh/concentrated/Greek style yoghurt. (1) Feed tank, (2) pressure indicator/switch, (3) UF membrane housings, (4) concentrate flow-meter, (5) permeate manifold, (6) concentrate control valve, (7) recirculation pump, (8) feed pump, (9) control panel. (Reproduced by courtesy of Micro Foods Ltd, Cardiff, UK and Memtech (UK) Ltd, Swansea, UK.)

More recently, Tamime *et al.* (1989*b*) evaluated two different systems for manufacturing concentrated yoghurt by ultrafiltration (see Figs 18 and 19). In this work, the concentration step was applied either before or after fermentation, and if applied to the warm yoghurt, the product was indistinguishable from Labneh made by the traditional cloth-bag method. This apparent similarity was confirmed by examining the products under the scanning and transmission electron microscopes (Tamime *et al.*, 1989*a*—Fig. 20), but the firmness of Labneh made from milk concentrated prior to fermentation was markedly reduced after the 'smoothing' stage as compared with traditional or ultrafiltration Labneh. Post-incubation ultrafiltration has also been

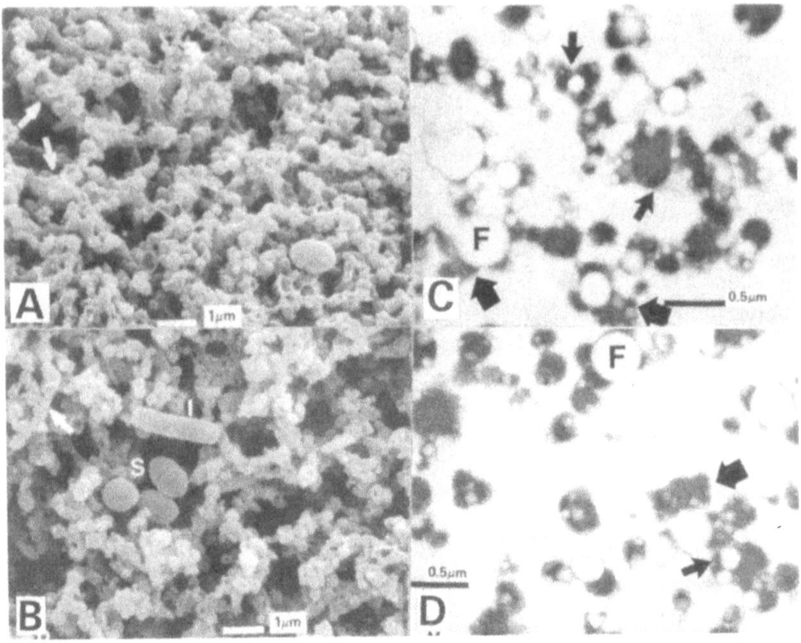

Fig. 20. Illustration of the microstructure of traditional Labneh. (A) SEM of Labneh before 'smoothing', and (B) after passage through a curd homogeniser. The separation of fluffy areas (arrows) is clearly noticeable; I—lactobacilli and S—streptococci. (C) and (D) TEM of the same Labneh samples shown in (A) and (B), respectively. Minute fat globules (small arrows) are embedded in the protein, larger fat globules (F) are surrounded with casein particles (large arrows). The casein particle chains and clusters are larger before smoothening (C) then after it (D). (From Tamime *et al.* (1989*a*), reproduced by courtesy of Scanning Microscopy International, Chicago, USA.)

used successfully for the production of Labneh from the milks of different mammals and cultured with *Lac. acidophilus* and *Bifidobacterium bifidum* (Mahdi, 1990; Mahdi *et al.*, 1990—see Fig. 21). In more recent studies (Tamime *et al.*, 1990, 1991, 1992), ultrafiltration Labneh was processed at different temperatures ranging from 35 to 55°C, and the authors concluded that the best product was achieved by concentrating the yoghurt at 40–45°C.

It is possible also to produce a rather similar product employing a high-protein milk powder and stabiliser to mirror the texture of normal Labneh, but the high cost of the powders has limited the exploitation of this approach.

Fig. 21. SEMs of Labneh made from different types of milks. (A) Residues of fat globule membranes (asterisks) indicate the distribution of fat globules in homogenised sheep's Labneh; compacted protein particles (arrows) formed the walls of a small void space (V) in the matrix. Smoothed goat's Labneh made by the traditional (B) and the ultrafiltration (C) procedures. Compacted casein particle clusters (c) with fat globules embedded in them (G) were found in the traditional product but not in the ultra filtration Labneh (C). Large casein micelles having smooth surfaces (arrows) were observed in both goat's milk products. B—bacteria. (From Tamime *et al.* (1992), reproduced by courtesy of Scanning Microscopy International, Chicago, USA.)

MISCELLANEOUS FERMENTED MILKS

The most important members of this group are the 'health-promoting' products, which are labelled in some countries, e.g. Germany, as 'mild yoghurts', or as 'A/B yoghurts' (milks)/ 'bio-yoghurts' elsewhere. These products may possess the gelled structure of yoghurt or may be more fluid, but they all contain lactic acid bacteria that are normal inhabitants of a healthy, human intestinal tract. Thus, *Lac. acidophilus* (A) and/or *Bif. bifidum* (B) are both essential components of the intestinal microflora and capable of fermenting milk to produce an attractive food item, and the benefits of consuming these types of product have been well documented (Robinson, 1991). Originally, the above cultures, sometimes with *Str. thermophilus* as well, were grown separately and then added to the process milk in the desired combinations/ratios. However, as none of the organisms grow well in normal milk, the tendency now is to employ direct-to-vat starters, and a typical procedure to manufacture an A/B milk product might be as follows (Anon., 1985):

- enrich the protein content of the milk to 3·9–3·9%;
- heat treat strongly, e.g. 95°C for 15 min;
- cool to 37°C, and add deep-frozen cultures of *Lac. acidophilus* (250 g) and *Bif. bifidum* (100 g);

— incubate at 37°C for 16 h, or until the pH reaches 4·2–4·4, and
— cool for mixing and/or blending in other ingredients, and chill to 2–4°C for storage and distribution.

The balance between the two cultures is recommended because

— bifidobacteria produce acetic acid as a product of their normal metabolism, and excessive counts could lead to 'vinegar' taints in the end-product; and
— the mortality rate of this strain of *Lac. acidophilus* is higher than that of *Bif. bifidum* at the pH in question, so that to ensure survival rates above 10×10^5 cfu ml^{-1} after 21 days at 5–7°C, the addition rates must reflect this difference—the target figure for survival is the so-called 'therapeutic minimum' number needed for the product to possess any demonstrable health benefit.

The plant required to handle such products is similar in most respects to a standard yoghurt installation, but because of the overnight incubation, standards of hygiene must be high. It is possible, however, to reduce the incubation time to 3–4 h by simultaneously inoculating the milk with a low percentage of a normal yoghurt starter culture. The product, which can now be referred to as 'A/B yoghurt', has the same 'health-promoting' properties as the A/B milk, and the flavour intensity imparted by the yoghurt culture can be adjusted by manipulation of the levels/strains of *Lac. bulgaricus* and *Str. thermophilus*. However, although this latter approach does offer advantages to the manufacturer, the survival of the *Lac. acidophilus* and *Bif. bifidum*—or other species of bifidobacteria, may have to be monitored to ensure that the expected counts are present at the end of 21 days. Thus, not only are some strains of *Lac. bulgaricus* capable of antagonism to other species of bacteria, but there is evidence that this activity may be enhanced by the presence of the type of fruit/flavoured ingredients that might be added to a fermented milk product (Robinson, 1990). Obviously, strain or ingredient selection can be adjusted to take account of this problem, but attention to detail will be essential for success.

REFERENCES

ADMI (1971). American Dry Milk Institute, Bulletin No. 916.
Anon. (1979) Dairy Science Abstracts, **41**, 200.
Anon. (1985). Technical Report. Chr. Hansen's Laboratorium A/S, Denmark.

Anon. (1990). Technical Report. G. C. Hahn and Company, Lubeck, Germany.
Anon. (1991). Technical Report. Chr. Hansen's Laboratorium A/S, Denmark.
Bjerre, P. (1990). In: *Recombination of Milk Powders.* IDF Special Issue No. 9001, International Dairy Federation, Brussels, Belgium, 157.
Bruderer, J. and Schicht, H. H. (1987). *Swiss Food,* **9**(12), 14.
Cottenie, J. (1978). *Cultured Dairy Prod. J.,* **13** (4), 6.
Crawford, R. J. M. (1962). *J. Dairy Engng,* **79**, 4.
Dannenberg, F. and Kessler, H. G. (1988*a*). *Milchwissenschaft,* **43**, 632.
Dannenberg, F. and Kessler, H. G. (1988*b*). *Milchwissenschaft,* **43**, 700.
Davies, F. L., Shankar, P. A., Brooker, B. E. and Hobbs, D. G. (1978). *J. Dairy Res.,* **45**, 53.
Deeth, H. E. & Tamime, A. Y. (1981). *J. Food Protection,* **44**, 78.
Driessen, F. M. (1984). In: *Fermented Milks.* IDF Bulletin No. 179, International Dairy Federation, Brussels, Belgium, 107.
Driessen, F. M. and Loones, A. (1990). *Proc. XXIII Int. Dairy Congress,* **3**, 1937.
Foley, J. and Mulcahy, A. J. (1989). *Irish J. Food Sci. Techn.,* **13**, 43.
Glover, F. A. (1971). *J. Dairy Res.,* **38**, 373.
Glover, F. A. (1985). *Ultrafiltration and Reverse Osmosis for the Dairy Industry.* Technical Bulletin No. 5, National Institute for Research in Dairying, Reading, Berkshire, UK, p. 141.
Glover, F. A., Skudder, P. J., Stothart, P. H. and Evans, E. W. (1978). *J. Dairy Res.,* **45**, 291.
Haque, Z. and Kinsella, J. E. (1988). *J. Dairy Res.,* **55**, 67.
Lehmann, H. R., Dolle, E. & Bucker, H. (1991). In: *Processing Lines for the Production of Soft Cheese,* Technical & Scientific Documentation No. 8, Westfalia Separator AG, Oelde, Germany.
Lewis, M. J. (1993). In: *Modern Dairy Technology* (Vol. 1, 2 edn), ed. R. K. Robinson. Elsevier Science Publishers, London, UK, p. 1.
Luck, H. and Gavron, H. (1990). In: *Dairy Microbiology* (Vol. 2, 2nd edn), ed. R. K. Robinson. Elsevier Science Publishers, London, UK, p. 345.
Mahdi, H. A. (1990). Some aspects of production and quality control of strained yoghurt (Labneh), PhD thesis, University of Stratchclyde, Glasgow, UK.
Mahdi, H. A., Tamime, A. Y. and Davies, G. (1990). *Egypt. J. Dairy Sci.,* **18**, 345.
Merilainen, V. T. and Dellaglio, F. (1990). *Proc. XXIII Int. Dairy Congress,* **3**, 1917.
Mottar, J., Bassier, A., Joniau, M. and Baert, J. (1989). *J. Dairy Sci.,* **72**, 2247.
Nielsen, V. H. (1972) American Dairy Review, **34**(2) 26.
Nielsen, V. H. (197). *Cultured Dairy Prod. J.,* **11**(1), 12.
Norling, A. (1979). *Cultured Dairy Prod. J.,* **14**(1), 24.
Piersma, H. and Steenbergen, A. E. (1973). *Officieel Orgaan FNZ,* **65**, 94.
Puhan, Z. and Zambrini, A. V. (1990). *Proc. XXIII Int. Dairy Congress,* **3**, 1907.
Robinson, R. K. (1981) *Dairy Industries Int.,* **46**(3), 31.
Robinson, R. K. (1982) *Dairy Industries Int.,* **47**(12), 19.
Robinson, R. K. (1989) *Dairy Industries Int.,* **54**(7), 23.
Robinson, R. K. (1990) *South African J. Dairy Sci.,* **22**(2), 43.
Robinson, R. K. (1991). In: *Therapeutic Properties of Fermented Milks.* Elsevier Science Publishers, London, UK, p. 185.

Robinson, R. K. and Tamime, A. Y. (1986). In: *Developments in Food Proteins* (Vol. 4), ed. B. J. F. Hudson. Elsevier Science Publishers, London, UK, p. 1.

Robinson, R. K. and Tamime, A. Y. (1990). In: *Dairy Microbiology* (Vol 2, 2nd edition), ed. R. K. Robinson. Elsevier Science Publishers, London, UK, p. 291.

Romney, A. J. D. (1990). *CIP:Cleaning in Place.* Society of Dairy Technology, Cambridge, UK, p. 224.

Samuelsson, E. G. and Ulrich, P. (1982). *Proc. XXI Int. Dairy Congress*, **1**, 288.

Sojllema, A. (1988). *Neth. Milk Dairy J.*, **42**, 365.

Steenbergen, A. E. (1971). *Officieel Orgaan FNZ*, **63**, 164.

Tamime, A. Y. (1990). In: *Dairy Microbiology* (Vol. 2, 2nd edn), ed. R. K. Robinson. Elsevier Science Publishers, London, UK, p. 131.

Tamime, A. Y. and Greig, R. I. W. (1979). *Dairy Industries Int.*, **44**(9), 8.

Tamime, A. Y. and Robinson, R. K. (1988*b*). In: *Technology of Manufacture of Thermophilic Fermented Milks.* IDF Bulletin No. 227, 82. International Dairy Federation, Brussels, Belgium, p. 82.

Tamime, A. Y. and Robinson, R. K. (1985). *Yoghurt—Science and Technology.* Pergamon Press, Oxford, UK.

Tamime, A. Y. & Kirkegaard, J. (1991). In: *Feta and Related Cheeses*, ed. R. K. Robinson and A. Y. Tamime. Ellis Horwood, London, UK, p. 70.

Tamime, A. Y. and Robinson, R. K. (1988*a*). *J. Dairy Res.*, **55**, 281.

Tamime, A. Y., Davies, G., Chehade, A. S. and Mahdi, H. A. (1989*a*). *J. Soc. Dairy Technol.*, **42**, 35.

Tamime, A. Y., Kalab, M. and Davies, G. (1989*a*). *Food Microstruct.*, **8**, 125.

Tamime, A. Y., Kalab, M., Davies, G. and Mahdi, H. A. (1990). *Food Struct.*, **10**, 37.

Tamime, A. Y., Davies, G., Chehade, A. S. and Mahdi, H. A. (1991). *J. Soc. Dairy Technol.*, **44**, 99.

Tamime, A. Y., Kalab, M. and Davies, G. (1992). *Food Struct.*, **10**, 345.

Waes, G. (1987). *Milchwissenschaft*, **42**, 146.

Wilcek, A. (1990). In: *Recombination of Milk Powders.* IDF Special Issue No. 9001. International Dairy Federation, Brussels, Belgium, p. 135.

Chapter 2

Modern Cheesemaking: Hard Cheeses

A. Y. Tamime

SAC – Auchincruive, The National College for Food, Land and Environmental Studies, Food Science and Technology Department, Ayr, UK

THE HISTORY AND ORIGIN(S) OF CHEESEMAKING

Cheesemaking is one of the oldest methods practised by man for the preservation of a highly perishable and nutritional foodstuff, e.g. milk, into a product which is not likely to deteriorate. The exact origin(s) of cheesemaking is difficult to establish, but from definite archaeological evidence, cheese was produced around 6000–7000 BC. According to Pederson (1979), the continued existence of man was primarily influenced, some 10–15 thousand years ago, by the change in his way of life from being a 'food gatherer' to a 'food producer'. It is possible that this transition was gradual and occurred at different times in different parts of the world.

Animals, such as the cow, goat, sheep or buffalo, had been domesticated at that time, and the milk was utilised as food. However, some of man's recorded civilisations, e.g. the Sumarians and the Babylonians in Mesopotamia, the Egyptians in north-east Africa, and the Indians in Asia, were well advanced in husbandry methods, and in the production of fermented food products including cheese and yoghurt.

The oldest methods of preserving food known to mankind are as follows: concentration, drying, fermentation and salting, and cheese is a fermented dairy product which is, in addition, concentrated and salted. It is safe to assume that cheese was first produced in the eastern part of the Mediterranean, and a summary of some cheese varieties with the date first noted is shown in Table I.

It is possible that modern cheesemaking could have evolved in two main stages: first, the manufacture of sour milk products, e.g. Leben (Laban), Ayran or yoghurt, and secondly, by partial separation of the

TABLE I
The archaeological references to cheese and their names with the date first recorded[a]

Year	*Cheese variety*
9000 BC	Rock drawings in the Sahara desert illustrate cow worshipping and milking.
7000–6000 BC	Bread and cheese have been identified as the staple food of early civilisations in the 'fertile crescent' which is situated along the eastern region of the Mediterranean (Lebanon and Syria) and Iraq (between the Tigris and the Euphrates rivers).
4000 BC	Detailed records of the Egyptian civilisation appear to illustrate great developments in husbandry and dairy processing.
3000 BC	Cheese has been found in the tomb of Hories-Aha (Sumarian civilisation).
	The act of killing the cow for meat has been regarded sinful by the Vedic Hymns of India; instead, milk and other dairy products have been used as food.
2000 BC	Reference to cheese can be found in the Babylonian records.
1800 BC	Woven reed baskets have been used in Asia to separate the curds from the whey (at present the same process is used to manufacture Surati Panir and Decca curds); similar baskets have been discovered in Dorset dated to the same period which indicate cheesemaking in England may have taken place prior to the Roman occupation.
1500 BC	Evidence of cheese has been referred to in Biblical times.
1200–0 BC	The Greeks and the Romans have been reputed to eat cheese and drink wine during their banquets.
	Homer (1184 BC) wrote about cheese which was manufactured from the milk of the sheep and the goat (at present Feta and Halloumi cheeses are widely produced in Greece, Cyprus and Bulgaria).
	Herodotus (484–408 BC) referred to Scythian cheese made from mare's milk, and Aristotle (384–322 BC) reported the use of mare's and asses' milks for the manufacture of Phrygian cheese.
	Varro (116–27 BC) noted the differences in the nutritional properties of cheeses, e.g. laxative effect, and the nourishing qualities in cheeses made from cow's milk (most), sheep (intermediate) and goat (least).
	Rameses tomb (100 BC) has scenes of goats and milk stored in skin bags.
AD 0–100	Columella (50 AD) in his De Re Rustica reported in detail the process of cheesemaking and pointed out that hygiene in milk production was essential during the production stages.
	Pliney (23–79 AD) wrote about sour milk cheeses which may have been the ancestors of present day 'pickled' cheeses produced in the eastern Mediterranean.

TABLE I (*cont.*)

AD 200–300	The Roman emperor Diocletian enforced maximum prices for cheese, e.g. Lunar which later came to be known as Parmesan.	
AD 400	The Romans brought the art of cheesemaking to Britain.	
AD 879	Gorgonzola	The movement of trade and tribesmen from the Middle East through the Balkan peninsula, the Mediterranean basin, or to India in the east helped to spread the art of cheesemaking and the development of different varieties known today (see text). However, monasteries and other establishments at different dates helped to maintain proper records of the cheesemaking process where the identity of the modern varieties have been preserved.
AD 1000	Schabzieger	
AD 1070	Roquefort	
AD 1174	Maroilles	
AD 1178	Schwangenkäse	
AD 1200	Grana/Parmesan	
AD 1282	Taleggio	
AD 1288	Gruyère	
AD 1500	Cheddar	
AD 1579	Parmesan	
AD 1622	Emmental	
AD 1688	Dunlop	
AD 1697	Gouda	
AD 1783	Gloucester	
AD 1785	Stilton	
AD 1791	Camembert	
AD 1800	Limburg	
AD 1861	St. Paulin	

[a]Data compiled from Marth (1953), Crawford (1960), Davis (1965, 1981), Pederson (1979), Kosikowski (1982) and Scott (1986).

whey and the addition of salt, e.g. yoghurt cheese, concentrated yoghurt or soft cheese. In sub-tropical climatic conditions, e.g. the Middle East, milk sours very rapidly in a few hours after milking, due to the high ambient temperature and the presence of microorganisms in the milk. These bacteria may have originated from the animal, the hands of the milker, the surfaces of utensils used to hold the milk or the environment. These organisms can produce two different types of fermentation: first, a non-lactic fermentation which is brought about by microorganisms other than lactic acid bacteria, and the product is normally stale, insipid or of bad taste when consumed; secondly, a fermentation produced by the so-called lactic acid bacteria that gives a more desirable product which is pleasant to eat.

Traditionally, the containers used for carrying or storing milk were made from animal skins or stomachs. If the milk was left undisturbed, clotting of the milk occurred due to developed acidity as a result of bacterial activity, and possibly due to the presence of clotting enzymes which originated from the stomachs. A soft coagulum was formed, and some of the liquid phase of milk (whey) was absorbed into the skin, or seeped through and was lost by evaporation. The coagulum was concentrated further by hand-squeezing and sun drying. This product was found to have better keeping quality than the original milk, due to higher concentration of lactic acid that limited or prevented the growth of bacteria capable of producing severe taints, etc. However, a longer shelf-life was achieved by preserving the concentrated curd in a salt solution (brine), which also improved its palatability. This fermented dairy product was later known as 'pickled cheese', which is still manufactured in parts of the Middle East, for example, cheeses such as Feta, Domiati, Akkawi or Halloumi.

As pickled cheese became popular in the Eastern Mediterranean region, its popularity spread to western countries, e.g. in some parts of Europe via tradesmen. It is possible that, as 'pickled cheese' became an acceptable dairy product in Europe, efforts were made to learn how to manufacture the cheese locally. With the establishment of dairies in Europe, i.e. in a comparatively colder climate than the Middle East, the preservation of cheese in brine was replaced by partial brining, e.g. Dutch cheese varieties, or by dry salting, because the curd produced was much drier. It could be argued that because local manipulation of the cheesemaking methods had taken place, the soft cheese had evolved into a new kind of cheese, i.e. the semi-hard cheese varieties. Furthermore, the production of even drier curd cheese resulted in the production of different types, e.g. hard-pressed varieties, which can be stored for a long period of time at ambient temperature. Typical examples of such products are Cheddar cheese and the other British territorial varieties.

CHEESE NAMES AND NOMENCLATURE

Throughout the world the names applied to cheese could almost reach 2000 (Scott, 1986). In the US (USDA, 1978) more than 800 cheese varieties have been described, but some of these cheeses have different local names and are practically the same; thus, a more accurate list would include no more than 400 varieties. This confirms the view of Davis (1965). More recently, the International Dairy Federation (IDF, 1981) has produced a catalogue in collaboration with its National Committees describing 510 different cheeses. According to Davis (1965), the exact origin of cheese names can be attributed to the following:

— region and/or towns of cheese manufacture;
— religious institutions;

— type of milk (cow, goat, sheep or buffalo);
— borrowed or made-up names;
— shape, appearance, type of cheese, method of ripening, or the use of additives.

National and international bodies have been involved for the past few decades in preparing or improving the existing specifications for cheese. Different standards exist in each country, and the majority of regulations take into account the following factors: chemical composition, method of manufacture, coagulation of milk, species of starter bacteria and/or method of ripening. These standards are essential on a national level both to the practical cheesemaker and the consumer; however, on the international level, it facilitates marketing of cheese between different countries.

It is clear that the above criteria are used to discuss the definition and classification of cheese, and it is appropriate to consider, therefore, some of the manufacturing procedures available, and to assess the relevance of these different processes in relation to the end-product.

DEFINITION OF CHEESE

The word 'cheese' in the English language is presumably derived from different sources. For example, in Urdi it (chiz) means the 'perfect thing', and 'cheese' could have come into the English language by this Anglo-Indian route, or from the Old English (cese), or from Latin (caseus). Cheese in other languages is also derived from Latin, e.g. Irish (cais), German (kase), Dutch (kaas), Spanish (queso) and Portuguese (quijo). In some languages, i.e. French and Italian (fromage and formaggio, respectively), the word cheese is derived from the Latin word 'forma' meaning shape or form.

In the UK, The Food and Drugs Act of 1955 provides a detailed description of the cheese regulations; these regulations may differ slightly, or come into operation at different times in Scotland, England and Wales and Northern Ireland. For example, The Cheese (Scotland) Regulations (1966/98-S.8), (1967/93-S.8), (1979/108-S.4), (1974/1337-S.115) and (1984/847-S.84) provide the definition of cheese, and a summary of these regulations is shown in Table II. A list of 29 varieties of cheese has been provided (e.g. Sage Derby, Mozzarella, Pecorino, Romano and Feta) and it includes detailed chemical analyses, e.g. the permitted level of fat (expressed as minimum per cent of milk fat in dry matter (FDM)), and the

TABLE II
A summary of cheese regulations in the UK

Type of cheese	*Fat in dry matter* (%)	*Milk fat* (%)	*Maximum water* (%)
1. Hard cheese[a]			
Full fat	⩾48	—	48
Medium fat	<48–>10	—	48
Skimmed milk	⩽10	—	48
2. Soft cheese			
Full fat	—	⩾20	60
Medium fat	—	<20– >10	70
Low fat	—	<10– >2	80
Skimmed milk	—	⩽2	80
3. Cream cheese			
Cream	—	⩾45	—
Double cream	—	⩾65	—
4. Whey cheese			
Full fat	⩾33	—	—
Whey	<33–>10	—	—
Skimmed whey	⩽10	—	—
5. Processed cheese[a]			
Full fat	⩾48	—	48
Medium fat	<48–>10	—	48
Skimmed milk	⩽10	—	48
6. Cheese spread	—	⩾20	60

[a]After 1 January 1973 any hard or processed cheese may have an alternative description to the above regulation, e.g.
x % fat in dry matter (minimum);
y % moisture (maximum);
z % milk fat (minimum).
However, for exact specification of FDM and moisture content in cheese refer to the cheese variety list in the regulations or Tables VII, VIII and IX.

moisture content (expressed as maximum per cent of water). Some other analytical qualities in cheese which are of importance, but not covered in these regulations, are the salt percentage (in some instances it is calculated as per cent salt in water) and the percentage of moisture in fat-free cheese (MFFC).

Other ingredients, which are sometimes used during the manufacture of cheese and processed cheese, are milk powders including caseinates, colour additives, emulsifiers and stabilisers, preservatives and enzymes. The relevant regulations and/or recommendations may differ from one country to another. At present the regulatory bodies within the European Economic Community (EEC) are in the process of standardising these regulations in all member states, and a comparative study of cheese laws and regulations within the EEC regarding the compositional specifications has been reviewed by Pappas (1988).

CLASSIFICATION OF CHEESE

The classification of cheese may vary from one country to another, and different systems have been used which may include one or more of the following technical aspects:

(a) type of milk;
(b) shape and weight of cheese;
(c) type of rind;
(d) method of coagulation;
(e) consistency of cheese;
(f) fat content; and/or
(g) method of preparation and maturation.

Milk from different species of mammals is also used in cheesemaking, and the world production figures of these types of milk are shown in Table III. However, the data of world production of cheese (Table IV) illustrate that, in the 1980s, around 85–88% of cheese was manufactured in North America, Europe and Oceania, which are the major producers of cow's milk. It is possible that the word 'cheese' could be reserved for the product manufactured from cow's milk, and that the name of a cheese produced from the milk of another species of mammal should imply the type used, e.g. buffalo's, sheep's or goat's cheeses. A typical example is the labelling of some French soft-cheese varieties produced from different milks.

The specifications regarding the shape and weight of cheese could provide some useful information, but it would appear that the data is of limited value to a cheese scientist or technologist, because most of the cheeses could be produced in different shapes or weights. However, the type of rind does provide some technical knowledge regarding the starter culture, particularly in respect of the soft-cheese varieties, e.g. surface mould or slime and blue-vein cheeses.

The coagulation of milk can be achieved by one of the following methods: first, the use of acids (e.g. mainly lactic acid by the starter

TABLE III

Production figures of milk of different species of mammals in different parts of the world (million tonnes)[a]

	Cow				Buffalo				Sheep				Goat			
	1970	*1980*	*1985*	*1989*	*1970*	*1980*	*1985*	*1989*	*1970*	*1980*	*1985*	*1989*	*1970*	*1980*	*1985*	*1989*
Africa	9·8	10·3	11·5	13·8	1·0	1·2	1·4	1·4	0·7	0·7	1·2	1·5	1·5	1·4	1·6	2·1
America (North and Central)	66·8	76·5	83·7	86·3	—	—	—	—	—	—	—	—	0·2	0·3	0·3	0·4
America (South)	18·0	23·0	26·7	30·0	—	—	—	—	0·02	0·03	0·04	0·04	0·1	0·1	0·1	0·2
Asia	27·8	37·0	43·3	49·9	23·1	25·8	32·7	38·4	3·0	3·4	3·6	3·4	3·2	3·6	3·5	3·8
Europe[b]	230·0	267·9	282·5	280·1	0·05	0·09	0·09	0·1	3·0	3·4	3·8	3·8	2·1	1·9	1·9	2·1
Oceania	13·5	12·4	14·1	14·0	—	—	—	—	—	—	—	—	—	—	—	—
World	366·3	427·1	461·8	474·1	24·15	27·09	34·19	39·9	6·27	7·73	8·64	8·75	7·1	7·3	7·4	8·6

[a] After FAO (1972, 1982*a*, 1988, 1990*a*).

[b] Data include production figures in the former USSR.

TABLE IV
World production figures of all types of cheeses (thousand tonnes)[a]

	1970	*1980*	*1985*	*1989*
Africa	246	364	437	472
America (North and Central)	1 557	2 482	3 159	3 488
America (South)	341	463	422	525
Asia	1 917	672	681	744
Europe[b]	4 258	7 116	8 125	8 928
Oceania	177	254	278	319
World	8 496	11 351	13 102	14 476

[a]After FAO (1972, 1982*a*, 1988, 1990*a*).
[b]Data include production figures in the former USSR.

organism for the production of Sauermilchkase, Fromage Frais and Quark), second, the addition of a coagulant (e.g. rennet only for the manufacture of Domiati cheese) and third, a combination of acid and a coagulant which is employed for the production of most cheeses. It is apparent, therefore, that such an approach to the classification of cheese would be rather limited.

A scheme for the classification of cheese, which is accepted by both cheesemakers and scientists, is based mainly on the method of manufacture and the chemical analyses of the product. In view of the scheme proposed by the FAO/WHO (1978), some existing cheese regulations, and the IDF (1981), a generalised method for the classification of cheese is illustrated in Table V. It can be observed that such an approach is applicable to all varieties, and an illustrated example regarding the description of cheese in terms of these categories (*I* to *V*) could be as shown in Table VI.

It is impossible to describe in detail all known cheese varieties and, in view of the scheme of classification illustrated in Table V, it was decided that some of these varieties (very hard, hard and semi-hard) would be discussed in this chapter, and the semi-soft and soft/fresh cheeses, including Stilton and Feta would be dealt with in Chapter 3 (see also Robinson and Tamime, 1991). Over the past few decades, the biochemistry of cheese has been reviewed extensively in text books (Davies and Law, 1984; Brochu *et al.*, 1985; Luquet, 1985; Eck, 1987; Fox, 1987; Cross and Overby, 1988; Robinson, 1990), and the subject will not be discussed in this present chapter.

TABLE V
Classification of cheese[a]

Category								
I *Consistency*		*II* *Fat content*		*III* *Moisture content* (%)		*IV* *Scalding temperature* (°C)		*V* *Method of ripening*
Firmness	*MFFC*[b](%)	*Designation*	*FDM*[c](%)					
1. Very hard	<51	High	>60	Very high	55–80	High	55	Starter bacteria → No gas holes; → With gas holes
2. Hard	49–56	Full	45–60	High	45–55	Medium	40	
3. Semi-hard	54–63	Medium	25–45	Medium	34–45	Low	35	Mould → Surface (white); → Internal (blue); → Surface (white); → + Internal (blue)
4. Semi-soft	61–49	Low	10–25	Low	<34	No scald	30	Miscellaneous (surface slime)
5. Soft/fresh	>67	Skim	<10					Unripened

[a]Adapted from FAO/WHO (1978), Kosikowski (1982) and Scott (1986).
[b]MFFC: moisture in fat free cheese.
[c]FDM: fat in dry matter.

TABLE VI

Cheese type	Category				
	I MFFC	II FDM	III Moisture	IV Scald	V Ripening
Parmesan	Very hard	Medium	Low	High	Starter (no gas holes)
Cheddar	Hard	Full	Medium	Medium	Starter (no gas holes)
Cheshire	Hard	Full	Medium	Low	Starter (no gas holes)
Gouda	Semi-hard	Full	Medium	Low	Starter (no gas holes)
Roquefort	Semi-soft	Full	High	No scald	Mould (internal)
Cottage	Soft/fresh	Low	Very high	High	Unripened
Brick	Semi-hard	Full	Medium	Low	Miscellaneous
Emmental	Hard	Full	Medium	High	Starter (with gas holes)

CHEESE SPECIFICATIONS AND STANDARDS

The primary objectives of cheese standards in any one country are to protect the health of the consumer, to produce a quality product, and to describe precisely individual cheeses so as to help ensure fair practices in international trade. The FAO/WHO Codex Alimentarius Commission, which consists of 122 member countries, has been established to provide standards for the major milk products including cheese. The recommended international standards for 25 cheeses and their acceptance by 19 government bodies have been published by FAO/WHO (1972), and the latest report (FAO/WHO, 1984) provides an up-to-date specification for 35 cheeses.

More recently the 29 National Committees of the International Dairy Federation (IDF, 1981) made available information regarding 510 cheese varieties produced in their countries. A selection of cheese specifications and/or standards (very hard, hard and semi-hard cheeses) are illustrated in Tables VII, VIII, and IX). In general, the compositional quality of an individual variety of cheese is rather similar around the world; but differences still exist in different countries. For example, Cheddar cheese (Table VIII) the FDM (min. %), moisture (max. %) and MFFC (mean %) ranged between 48 and 50, 37 and 39, and 50 and 56, respectively. The majority of Cheddar cheese is, therefore, manufactured in different parts of the world to comply with these specifications, but such standards are somewhat limited, because the specifications do not provide the cheesemaker with adequate parameters to produce a 'quality' Cheddar. Under commercial practice, for example, for a long holding, mature Cheddar or cheese intended for export, i.e. from New Zealand and Australia, the chemical composition of such cheese ought to consist of 50% FDM (min.), 37% moisture (max.) and 53–55% MFFC.

WORLD PRODUCTION AND MARKETING OF CHEESE

In 1989 the world production figure for cheese was 14·48 million tonnes, and Table IV shows the trend of cheese production in various continents. It can be observed that, since 1970, cheese production has increased by around 5·9 million tonnes, and 88% of the cheese has been produced in Europe, North America and Oceania. It is difficult to obtain an exact breakdown of all the cheese varieties produced in the world; however, Eck (1984) reported that, in 1981, hard cheese types (e.g. Cheddar,

Emmental and Gruyère) and the semi-hard cheeses, i.e. Gouda and Edam, made up about 60% of total world output. In the 1980s, it is probable that the production trend of these cheese varieties followed a similar pattern, and for the following reasons: first, these cheese varieties are well established and have gained consumer acceptability in the majority of countries trading with cheese; secondly, hard and semi-hard cheeses require a few months of maturation when compared with soft and semi-soft varieties and, as a consequence, these cheeses are more suitable for the international trade because of their long keeping quality; and thirdly, agressive advertisement by the industry to the consumer in most industrialised countries has helped to increase cheese consumption.

In recent reports (IDF, 1979, 1982a; Schelhaas, 1982), the structure of the international cheese trade, i.e. imports and exports, has been analysed. Over the past two decades, the cheese imports have been dominated by North America, Western Europe and Japan. However, this trend is changing, and between 1975 and 1980, cheese imports to these regions have been reduced from 68% to 56% of total world cheese imports, and during the same period, imports to Iran have increased more than five-fold, i.e. from 2 to 11 per cent. According to FAO figures for 1980 (IDF, 1982b), the total world cheese imports were 1.37 million tonnes, and the 10 major importing countries (e.g. Belgium/Luxemburg, France, Germany, Iran, Italy, Japan, Netherlands, Saudi Arabia, UK and USA) accounted for 77% of the total world imports. The most significant varieties of cheese imported are similar to the types reported by Eck (1984) plus Feta (to Iran and Saudi Arabia) and, to a lesser degree, some semi-soft and soft/fresh cheeses.

In 1980, the 10 largest cheese exporting countries (e.g. Australia, Austria, Denmark, France, Finland, Germany, Ireland, Netherlands and New Zealand) accounted for 86·7% of total exports (IDF, 1982a). The cheeses mentioned above are the most significant types exported, including some processed cheese.

In 1988 and 89, the world dairy situation was analysed by Sliter (1989, 1990a, 1990b), Osteras (1989, 1990) and Christiansen (1991). Cheese and butter production are still recognised as the major products for world milk utilisation. They represented 25–26% and 36–37% respectively as compared with 29–30% for liquid milk processing (i.e. including low fat milk, flavoured milks, fermented milks and cream) during the same period. Total world milk supplies have been increasing gradually during the past decade (see Table III), and for this period a comparative analysis of the world cheese market (i.e. imports and exports) is shown in Tables X and XI.

TABLE VII
Specifications of very hard cheese varieties[a]

Cheese variety	Country	Raw material	Description of cheese		Weight (kg)	Chemical analysis (%)		
			Interior	Exterior		FDM[b] (min)	Moisture (max)	MFFC[c] (mean)
Parmesan	Australia		IO[d]	HDR[e]	15–20	32	32	43·7
	Canada		NO[f]	HDR	2·27–22·7	32	32	41·1
	Japan	Cow's milk	NO	SDR[g]	15	35	30	41·1
	New Zealand		IO		10	32	32	43·0
Parmeasao	Brazil		IO		4–20	35	30	46·2
*Parmiggiano Regiano ● ○	Italy	Cow's milk/Cn[h]	SRO[i]/NO		30	32	33	54·1
*Grana Padano ● ○	Italy	Cow's milk/Cn	SRO/NO		24–40	32	34	52·3
Montasio	Belgium		IO	HDR	7	35	30	46·7
Romano	Australia	Cow's milk	IO		3–6	38	35	48·6
	New Zealand		IO		10	38	34	54·5
	Canada	Cow's/goat's milk	NO		5·4–22·7	37·8	34	45·3
Bra	Canada	Cow's milk	NO		0·9–5·4	40·5	30	48·6
*Pecorino Siciliano ● ○	Italy	Sheep's milk/Cn	SRO/NO	SRSS[j]	4–12	40	33	56·9
*Pecorino Romano ● ○	Italy	Sheep's milk	NO	SDR	8–20	36	33	53·2

*Fløtemysost ●	Norway	Cow's milk/ Cn/La[k]/Lg[l]	NO	NR[m]	0·25–4	33	20	26·2
*Geitost ●	Norway	Sheep's milk/ Cn/La/Lg	NO	NR	0·2–4	33	20	26·2
*Getost	Sweden	Sheep's milk/ Cn/La/Lg	NO	NR	0·2–0·5	30	19	—
*Gudbrands dalsost ●	Norway	Sheep's milk/ Cn/La/Lg	NO	NR	0·2–4	35	20	25·7

*Statutory standards.
●Cheese originally produced in country indicated.
○Cheese with international or national protection of origin.
[a]Data compiled from IDF (1981) and Holland *et al.* (1989).
[b]FDM: fat in dry matter.
[c]MFFC: moisture in fat free cheese.
[d]IO: irregular opening.
[e]HDR: hard dry rind.
[f]NO: no opening.
[g]SDR: soft dry rind.
[h]Cn: casein.
[i]SRO: small round opening.
[j]SRSS: soft rind with smeary surface.
[k]La: lactalbumin.
[l]Lg: lactoglobulin.
[m]NR: no rind.

TABLE VIII

Specifications of very hard cheese varieties[a,b]

Cheese variety	Country	Description of cheese			Chemical analysis (%)		
		Interior	Exterior	Weight (kg)	FDM[b] (min)	Moisture (max)	MFFC[c] (mean)
*Emmentaler	Austria			60		38	51·9
*Emmenthaler	Denmark			—		40	53·0
*Emmental	France	LRO[c]	HDR	45–130		38	52·7
*Emmentaler/	Switzerland			60–130		38	52·5
Emmental ● ○					45		
Gruyère	Canada			18·5		38	52·8
*Gruyère	France			20–45		38	52·7
Gruyère	Poland	MSRO[d]	HRSS[e]	30		40	55·7
*Gruyère Greyèrzer Gruviera ● ○	Switzerland			20–45		38	52·3
*Cheddar	Australia	IO/NO	HDR/NR	4·5–36	50	38	53·4
	Canada		NR	8–18·5	50·8	39	56·5
	Denmark		NR	—	50	38	50–54
*	France	NO	HDR	30–35	50	39	56·0
	Ireland		HDR	—	48	36	52·9
*	New Zealand		HDR/NR	2·25–36	50	37	55·1
* ●	UK		HDR/NR	4–30	48	39	55·2

Cheshire[f]		Australia	IO	NR	19–20		44	57·1
		Ireland	NO	SDR	—		44	60·3
		New Zealand	IO	SDR	20		44	55·5
*	●	UK	IO	HDR/NR	4–20	48	44	60·2
*Derby	●	UK	NO	HDR	4–15		42	58·2
*Double Gloucester	●	UK	NO	HDR	4–20		44	60·2
*Leicester	●	UK	NO	HDR	4–15		42	58·2
*Wensleydale	●	UK	NO	HDR/NR	4–10		46	62·1

[a] Data compiled from IDF (1981).
[b] For abbreviations see footnotes to Table VII.
[c] LRO: large round opening.
[d] MSRO: medium sized round opening.
[e] HRSS: hard rind with smeary surface.
[f] Cheshire cheese (Ireland and New Zealand) is referred to as semi-hard variety.

TABLE IX

Specifications of some semi-hard cheese varieties (all produced from full-fat cow's milk)[a,b]

Cheese variety	*Country*	*Description of cheese*			*Chemical analysis (%)*		
		Interior	*Exterior*	*Weight (kg)*	*FDM (min)*	*Moisture (max)*	*MFFC (mean)*
Caerphilly	Ireland	NO	SDR	—	48	46	62·1
* ●	UK	NO	SRWM[c]	4·5	48	46	62·1
Edam	Australia	SRO	HDR	3	40	48	54·5
	Canada	SRO	SRP[d]/NR	18	40·7	46	59·0
	France	SRO/NO	HDR/SRP	1·7–2·5	40	48	60·5
*	Netherlands†	SRO/NO	HDR/NR	0·8–>6	40	45–47	57–59
	New Zealand	SRO	HDR	5	40	45	56
Gouda	Australia	MSRO	NDR/NR	5–10	48	45	54·1
	Canada	SRO	SRP	1·4–18·2	49·1	43	59·7
	France	SRO/NO	HDR/SDR	4–5	48	45	61·1
* ●	Netherlands†	SRO/SDR/NR	HDR/SDR/NR	0·18–>6	48	41·5–45·5	57·6–61·1
	New Zealand	SRO	HDR	10	46	45	56–58
*Fynbo	Denmark†	MSRO	HDR	—	30–45	46–51	58–62
*Tybo ●	Denmark†	MSRO	HDR	—	30–45	46–54	58–62
*St. Paulin	Belgium	NO	SDR	1·75	45	52	65·3
	Canada	NO	SRSS/SRP	0·45–2·26	50	46	63
*	France	NO	SRSS	1–3·2	40	56	66·7
	Norway	SRO/NO	SDR	1·5	45	50	62·8

[a] Data compiled from IDF (1981).
[b] For abbreviations see footnotes to Tables VII and VIII.
[c] SRWM: soft rind with while mould.
[d] SRP: soft rind with paraffin.
† Illustrate different specification within a cheese variety.

TABLE X
World market (import figures) of all types of cheese in different continents

	1980		*1988*	
	Tonnes[a]	*US$*[b]	*Tonnes*[a]	*US$*[b]
Africa	48·9	117·0	70·9	187·0
Algeria	11·1	23·5	8·5	11·0
Egypt	14·2	25·2	32·3	90·2
Libya	11·8	30·2	16·0	36·0
America (N&C)	154·0	440·0	173·5	559·8
Canada	20·5	63·8	21·8	96·1
USA	105·5	307·4	113·4	359·9
America (s)	18·9	44·4	9·8	24·2
Argentina	6·0	14·3	0·2	0·6
Brazil	2·1	5·8	6·4	14·0
Venezuela	7·1	14·6	1·2	3·7
Asia	257·2	516·7	325·9	653·2
Iran	62·0	120·0	80·0	117·0
Iraq	20·0	37·0	13·02	6·0
Japan	74·7	135·7	114·4	242·1
Lebabon	14·0	28·0	8·1	17·0
Saudi Arabia	39·0	95·2	52·4	118·4
Syria	10·1	22·1	1·2	2·7
Europe	873·5	3009·9	1278·3	5237·9
Bel./Lux.	98·5	371·0	121·5	522·5
France	61·2	223·8	98·6	444·8
W. Germany	235·1	876·6	316·1	1426·9
Italy	216·4	730·8	299·9	1213·5
Netherlands	28·4	93·3	54·1	202·3
Spain	20·7	57·4	33·8	124·7
Switzerland	20·1	78·8	24·1	128·4
UK	116·3	395·4	198·2	729·5
Oceania	14·4	41·1	22·1	70·7
Australia	12·0	31·1	19·0	56·5
World	1366·2	4169·1	1880·5	6732·8

Note: [a]Thousand
[b]Million

The above data illustrates the major cheese importing countries in each continent.

Data compiled from FAO (1982b, 1990b).

The data suggest that:- (a) In 1988, the total world cheese imports were 1·88 million tonnes, and the major importing countries were still the same as reported above for 1980, including Egypt; (b) cheese imports to

TABLE XI
World market (export figures) of all types of cheese in different continents

	1980		1988	
	Tonnes[a]	*US$*[b]	*Tonnes*[a]	*US$*[b]
Africa	0·3	1·0	1·4	4·9
America (N&C)	12·4	38·2	35·6	82·7
Canada	3·1	9·2	10·2	33·2
USA	6·2	21·3	25·0	48·2
America (s)	14·1	42·3	16·9	42·0
Argentina	4·1	13·3	9·9	25·8
Colombia	6·8	20·3	NA	NA
Uruguay	3·0	7·8	6·7	15·6
Asia	6·4	16·5	8·8	21·9
Israel	1·0	2·0	1·0	2·1
Lebanon	0·8	1·5	NA	NA
Saudi Arabia	0·5	1·0	0·4	1·1
Turkey	0·7	1·5	3·6	7·2
U.A. Emirates	1·1	2·3	0·2	0·3
Europe	1255·6	3819·8	1671·0	6092·1
Austria	41·0	111·1	36·4	115·0
Bel./Lux.	31·3	108·1	54·5	220·5
Bulgaria	25·9	68·0	23·4	54·1
Denmark	173·1	405·9	211·9	527·4
Finland	46·0	97·7	31·0	75·2
France	231·9	879·8	298·3	1239·5
W. Germany	215·4	614·1	315·7	1108·5
Italy	37·8	121·2	61·7	332·2
Netherlands	275·3	802·3	400·8	1514·1
Switzerland	63·1	289·6	59·9	375·6
Oceania	130·4	209·3	177·3	320·6
Australia	61·2	105·1	68·5	136·2
New Zealand	69·2	104·2	108·9	184·5
World	1419·2	4127·1	1911·0	6564·2

Note: [a]Thousand
[b]Million

The above data illustrates the major cheese exporting countries in each continent.
NA—Not Available

Data compiled from FAO (1982b, 1990b).

these same countries accounted for 79% of total world imports, which does not suggest a significant change in the world cheese market over the past decade; (c) in 1988, the largest cheese exporting countries, i.e. in excess of 20 thousand tonnes (e.g. Australia, Austria, Belgium/Luxemburg, Bulgaria, Denmark, Finland, France, Germany, Ireland, Italy, Netherlands, New Zealand, Norway, Switzerland, UK and USA) accounted for 91.3 per cent of total exports (see Table XI); and (d) it is most likely that the cheeses mentioned above (Eck, 1984), including Feta and processed cheese, are still the most significant types for export (for further information see Anon., 1988a).

Sliter (1990a, 1990b) reported that the world demand for cheese will continue to expand for the next decade at a rate in excess of growth in population, in particular in North America; however, the market for low-fat cheeses will expand at a faster rate than that for full-fat cheeses. Nevertheless, the factors that will have the ultimate impact upon the world's dairy industry are: (a) the current round of GATT negotiations, (b) the EEC after 1992, (c) the impact of re-unification of Germany, (d) the 1990 US Farm Bill and (e) the changes in the economies (i.e. from 'central' to 'market') in Eastern Europe. In addition, the impact of quotas in different countries should not be overlooked (see Raun, 1990; IDF, 1991a).

CHEESE CONSUMPTION

The consumption of different varieties of cheese varies from one country to another, and Table XII shows the *per capita* annual consumption of all cheeses in some European countries. It is evident that there is a steady and uniform increase in the consumption of cheese. Such a trend can be used to classify these countries into two groups: first, countries of low *per capita* cheese consumption (e.g. <10 kg/head), and secondly, countries of high *per capita* cheese consumption (e.g. >10 kg/head). The reason(s) for such differences in cheese consumed in these countries is highly complex, and could be due, in part, to the following aspects:

Eating Habits and Personal Preference

In some countries (Germany, France, Iceland, Iran, Israel and Poland), soft/fresh cheeses constitute a large proportion of total cheese consumption, and food habits, once formed, are difficult to break. However, there is a significant correlation between cheese and wine consumption, and no correlation between cheese and milk consumption. For example, in France, Italy and Germany (cheese consumption is high, but milk consumption is low) the *per capita* annual wine consumption in 1981 was

TABLE XII
Per capita annual consumption for all cheese (kg/head)

Country	*Year*			
	1970	*1980*	*1985*	*1989*
Australia	3·7	6·6	8·1	9·2
Austria	5·8	8·2	9·9	10·9
Belgium	8·8	13·4	14·5	16·5
Canada	6·3	8·8	10·8	15·3
Chile	NA	1·7	3·0	3·5
Czechoslovakia	5·5	9·0	10·8	13·0
Denmark	9·5	9·6	11·3	14·2
Finland	4·5	7·9	10·1	12·9
France	14·0	18·4	20·7	17·3
Germany	9·9	13·7	15·8	18·1
Iceland	8·8	14·0	15·5	16·5
Ireland	2·5	3·3	4·8	5·3
Israel	10·3	13·2	15·3	16·2
Italy	NA	14·2	16·7	17·8
Japan	0·4	0·8	0·6	1·2
Luxemburg	8·3	9·6	10·1	12·9
Netherlands	8·4	13·1	13·8	14·8
New Zealand	4·3	8·9	8·8	7·9
Norway	8·8	12·4	12·9	13·2
Poland	7·5	11·5	3·6	NA
South Africa	1·2	1·3	1·2	1·7
Spain	3·0	3·9	4·9	5·3
Sweden	9·1	13·7	15·1	15·5
Switzerland	10·0	13·4	13·7	15·7
UK	5·4	5·7	7·1	8·1
USA	7·6	10·1	11·6	12·3
USSR	4·0	4·8	6·6	6·6

NA: not available
After IDF (1982a, 1987a, 1991b).

92, 87 and 25 litres head^{-1} respectively. In Ireland and the United Kingdom (cheese consumption is low, but milk consumption is high), the wine consumption during the same period was 3 and 6 litres head^{-1} respectively (Anon., 1982a). Furthermore, in France and Germany, everyday wine was, historically, in plentiful supply and cheap, while milk had very poor keeping quality; hence the habit of drinking wine was established. By comparison, in the United Kingdom (until recently), wine was relatively expensive, while milk was cheap and in plentiful supply; thus milk became the preferred drink.

Uses of Cheese

In Scandinavia, Netherlands and France, cheese is consumed at any meal including breakfast, while in the United Kingdom cheese is not normally eaten during the morning meal. Certain dishes, for example fast foods like hamburgers and pizzas, are major contributors to the increased cheese consumption in the United States of America, and most likely in Western Europe where these types of foods are becoming increasingly popular. It may be that the low cheese consumption in the UK could be mainly attributed to the fact that cheese is normally served after the dessert (i.e. less cheese is eaten), compared with the custom of most countries where cheese is consumed after the main dish before the dessert (i.e. more cheese could be eaten).

Miscellaneous Factors

Factors, such as climatic conditions, availability of different cheese varieties on the market and economic standards, can affect the level and type of cheese consumption; health or dietetic concerns may increase fresh and/or low-fat cheese consumption; advertisements and other promotional activities could increase consumption; a multitude of other factors, which differ from one country to another, can also affect consumer reaction, and a comprehensive study of such factors in some IDF member countries has been recently published (IDF, 1982a; Sliter, 1989, 1990a, 1990b). In addition, it is of interest to note that most of the great religions in the world do not forbid the consumption of cheese. However, certain religious practices may have a negative effect on consumption of cheese, e.g. 'Kosher' or 'Hallal' products, but the use of the right ingredients and adoption of approved manufacturing practices will overcome these religious limitations.

NUTRIENTS IN CHEESE

The nutritional properties of cheese are excellent, and are well reviewed by Renner (1983, 1986, 1988) and Scott (1989). The major constituents of some cheeses discussed in this chapter are shown in Table XIII. The nutrients present in cheese are protein, fat, vitamins, minerals and salts. The level of these constituents in cheese may vary, due mainly to the quality or type of milk used and the variety of cheese produced. In general very hard, hard

TABLE XIII

Promimate composition, energy values, inorganic constituents and vitamins in some cheese varieties (figures are quantities 100 g^{-1} of cheese)

	Parmesan[a]	*Emmental*[b]	*Gruyére*[c]	*Cheddar*[d]	*Gouda*[c]	*Edam*[c]
Water	18·4	35·7	35·0	36·0	40·1	43·8
Protein	39·4	28·2	7·2	25·2	24·02	6·0
Fat (g)	32·7	29·7	33·3	34·4	31·0	25·4
Carbohydrates	Tr	Tr	Tr	Tr	Tr	Tr
Cholesterol (mg)	100	90	100	100	100	80
Energy (*kcal*)	452	382	409	412	375	333
Vitamins (μg)						
Vitamin A[e]	345	320	(325)	325	245	175
Vitamin D	(0·25)	N	(0·25)	0·26	(0·24)	(0·19)
Vitamin E	700	440	(580)	530	530	480
Thiamin	30	50	30	30	30	30
Riboflavin	440	350	390	400	300	350
Niacin	120	100	40	70	50	70
Vitamin B_6	130	90	110	100	80	90
Vitamin B_{12}	1·9	2·0	1·6	1·1	1·7	2·1
Folate	12	20	12	33	43	40
Pantothenate	430	400	350	360	320	380
Biotin	3·3	3·0	1·5	3·0	1·4	1·8

Minerals (mg)						
Sodium	1090	450	670	670	910	1020
Potassium	110	89	99	77	91	97
Calcium	1200	970	950	720	740	770
Magmesium	45	35	37	25	38	39
Phosphorous	810	590	610	490	490	530
Iron	1·1	0·3	0·3	0·3	0·1	0·4
Copper	0·33	1·3	0·13	0·03	Tr	0·05
Zinc	5·3	4·4	2·3	2·3	1·8	2·2
Sulphur	250	200	200	230	N	N
Chloride	1820	690	1040	1030	1440	1570
Manganese	0·1	Tr	Tr	Tr	Tr	Tr

Note: [a] Average of 10 samples (block and powder).
[b] Data compiled from published sources.
[c] Average of 10 samples.
[d] Weighted average of cheese samples from five countries.
[e] Retinol content.
( ) Estimated values; Tr-trace; N-not reported.

Data compiled from Holland *et al.* (1989)

and semi-hard cheeses contain high levels of protein (mainly casein) which is a rich source of the essential amino acids required by man. The energy in dairy products is provided by the fat and carbohydrate fractions; the latter component is very low in matured cheese, because most of the lactose is lost in the whey during processing, or has been utilised by the starter culture bacteria for the production of lactic acid. The high figures for lactose reported in the literature ~3% could be attributed to the type of cheese, or to the fact that the cheese was analysed at a relatively 'young' age, i.e. a few days after manufacture. Indeed, some patients who show symptoms of lactose intolerance when they consume liquid milk can eat cheese without having any allergic response.

The energy (calorific value) in cheese is mainly derived from the fat, but medical opinion may still regard the fat in cheese as a potential cause of coronary heart disease, and physicians may advise their patients not to eat cheese without taking into account the likely impact. For example, in the United States of America (cheese consumption in the 1970s was low, and was eqivalent to 25 g day^{-1}), the average recommended daily intake of cholesterol is 500 mg, and hence cheese would only contribute 26 mg of cholesterol day^{-1}, i.e. not significant (Speckmann, 1979). In addition, the cholesterol content of some cheeses (see Table XIII) is extremely low.

Cheese is considered to be a good source of certain vitamins, but is deficient in ascorbic acid (Vitamin C) which is lost during the manufacturing stages. A varied diet can supplement such a deficiency if cheese is consumed with vegetables. Although up to 90% of the water soluble B vitamins are lost in the whey (Renner, 1988; Scott, 1989), the concentration of these vitamins in the cheese is still quite high (see Table XIII), and able to contribute significantly to the supply of these vitamins in the human diet.

Cheese contains appreciable quantities of minerals (e.g. calcium and phosphorus) which are essential for teeth and bone formation, and pregnant women. According to Tunick (1987), the level of calcium in 28 g of cheese, together with the contribution to the US Recommended Dietary Allowance (RDA) for calcium, is as follows:

Product	*Calcium (mg)*	*RDA (%)*
Parmesan	336	42
Swiss	272	34
Cheddar	204	26
Gouda	198	25
Edam	207	26

The salt level, i.e. sodium chloride, in cheese varies with the type of cheese produced, method of salting employed (dry or brining) and/or the amount used. Salt is used as a preservative and flavour enhancer, but recently cheese has been considered by the medical profession as a contributing factor to high blood pressure due to its 'high' salt content. Some trials have been carried out to produce Cheddar cheese low in salt, or with a mixture of sodium/potassium chloride (1:1 molar basis) (Lindsay *et al.*, 1982; Green, 1986; Schroeder *et al.*, 1988). The results showed that the mean consumer preference was towards 'salty' cheese, and some bitterness was observed in cheese containing a 1·5% salt mixture. However, a French patent (Lefier *et al.*, 1990) describes a process for the manufacture of low-sodium cheeses where magnesium salts are added to the curd rather than sodium chloride alone. The sodium and the magnesium contents in the cheese were $\leqslant$50 mg and $\leqslant$80 mg, respectively. The industrial production of low-salt Edam cheese has been reported by Prokopek *et al.* (1990b). Up-to-date information regarding the reduction of salt in dairy products including cheese has been reported by Reddy and Marth (1991).

Milk proteins undergo hydrolysis (i.e. liberation of amino acids, pepetides and soluble nitrogen during the maturation of cheese) by proteinase enzymes that originate from the starter culture organisms, coagulant retained in the curd, indigenous milk enzymes that survived the heat treatment of the milk and any non-starter bacteria. Thus, the rate of protein hydrolysis differs in relation to the cheese variety, and the role of such proteinases in the hydrolysis of the various casein fraction have been reported by Fox (1989). Peptide formation and the release of free fatty acids in cheese have been recently reviewed by Law (1987).

Free amino acids undergo decarboxylation during the maturation of cheese to produce amines (e.g. cadaverine, histamine, phenyl-ethylamine, putrescine, tryptamine and tyramine) (Renner, 1983), and the amine concentration can vary depending on the duration of maturation and the microbial flora of the cheese variety (see Table XIV). Amines can potentially affect blood pressure (i.e. hypertensive or hypotensive effect) and may be a health risk to those lacking the mono- or di-amine oxidase enzymes (Scott, 1989). For most people, the risk in eating cheese is very negligible, and for further information on amines in cheese refer to the reports published by Taylor (1985) and Chang *et al.* (1985).

TABLE XIV
Variation in the concentration (mg 100 g^{-1}) of some amines in different cheeses

Cheese	Tyramine	Histamine
Parmesan	0·4–29	0–58
Emmental Gruyére	3–73	1–94
Cheddar	35–109	4–27
Gouda Edam	0–60	0–90

Data compiled from Scherz and Kloos (1981) and Scott (1989).

CHEESEMAKING PROCESS

Although there are many varieties of cheese, the basic concepts of cheesemaking are similar. In principle, milk from any species of mammal could be used as the main raw ingredient, plus the addition of starter culture and coagulant. Hence, a recipe is used for the manufacture of a specific variety of cheese, and the basic stages of manufacture may include the following:

(a) milk handling, storage and further processing;
(b) starter cultures;
(c) formation of the coagulum, cutting and scalding;
(d) handling of the curd after de-wheying; and
(e) miscellaneous treatments which include milling, salting, pressing of the curd (incidentally, such treatments may not occur in the same order, depending on the type of cheese produced or the mechanised system employed).

Milk Handling, Storage and Processing

Raw milk, which is normally cooled on the farm, is received at the dairy in bulk tankers. A wide range of microorganisms can be present in milk (Gilmour and Rowe, 1990), and the microbial count reflects the on-farm hygiene of milk production and storage (IDF, 1980; Bramley and McKinnon, 1990). In general, the presence of high counts of psychrotrophic and thermoduric microorganisms can affect the quality of the cheese. Therefore, the handling methods of the milk at the dairy, prior to cheesemaking, have developed over the years so as to contribute positively towards the manufacture of 'quality' cheese. The overall practice of milk handling may include (i) metering or weighing, (ii) filtering to remove some contaminants (e.g. straw, hairs, soil, etc.), (iii) cooling to $<5°C$ using a plate cooler and (iv) storing in a silo.

Psychrotrophic bacteria can grow and survive at ordinary refrigeration temperature, and the peptide hydrolase and lipase-producing species can affect both the milk-clotting mechanism and the flavour of the cheese. Law *et al.* (1976) reported that cheese milk containing more than 10^7 colony forming units (cfu) ml^{-1} resulted in Cheddar cheese which was rancid after 4 months, and the cheese had a distinctive 'soapy' flavour. Some of the methods which have been used to prolong the storage of raw milk and control psychrotrophic proliferation include (i) thermisation of milk, (ii) deep cooling of milk to 2°C, (iii) a combination of (i) and (ii), (iv) addition of carbon dioxide, or (v) the use of lactic acid bacteria (Zall, 1980; King and Mabbitt, 1982; Al-Darwash, 1983; Banks *et al.*, 1986, 1988; Rashed *et al.*, 1986; Manap, 1988; Banks, 1990; Muir, 1990; Griffiths *et al.*, 1991).

Before further processing, the cheese milk is filtered in order to remove certain contaminants, such as cellular material, straw, hairs or soil, etc. Cloth filters are the universal system employed, but centrifugal separation and/or bactofugation is used where the presence of spore-forming organisms (*Clostridium* spp.) in milk can lead to product loss due to the development of 'late blowing' and butyric acid fermentation in the cheese. Such faults have been identified as a serious problem in brined cheeses, such as Emmental, Grana Padona, Provolone, Gouda and Danbo. The methods, which have been used in the cheese industry to minimise clostridial activity in cheese, are as follows.

(a) *Bactofugation*—The bactofugate, i.e. the separated fraction, amounts to 2–3% of the total volume of milk, and it contains both the undesirable microorganisms and the fat content of the milk (Scott, 1986). Since the heat treatment of the cheese milk is limited to 72°C for 15 s, the bactofugate is sterilised by live steam injection at 130–140°C for a few seconds, and after cooling, is added back to the pasteurised cheese milk (Berg *et al.*, 1989).
(b) *Addition of nitrate*—An alternative approach to bactofugate sterilisation is the addition of sodium or potassium nitrate ($NaNO_3$ or KNO_3, respectively) to the cheese milk in order to inhibit the

growth of gas-forming butyric acid bacteria. Kosikowski (1982) and Robertson (1987) reported that $NaNO_3$ was added at a rate of 50–90 g 455 litres^{-1} or 5–20 g 100 litres^{-1} of milk, respectively, and that Gouda and Edam cheeses may contain 23 mg $NaNO_3$(45·5 g of cheese)$^{-1}$. However, Scott (1986) reported that the addition of $NaNO_3$ (known as saltpetre) to the milk can cause colour defects in the cheese, due to the reaction of tyrosine with the nitrite (i.e. after the reduction of nitrate), and hence may be limited in its use. The addition of $NaNO_3$ is controlled by statutory standards in some countries due to the reported carcinogenic effect of its reaction products.

(c) *Addition of lysozyme*—This enzyme (muramidase) is present in milk and egg white (~0·5%) and commercial preparations of lysozyme have been used successfully to control clostridia and butyric acid fermentation in certain European cheese varieties. Most of the added lysozyme (~80–90%) becomes bound to the cheese curd, where it decomposes the cell walls of clostridia and other Gram-positive bacteria. Published data (IDF, 1987*b*) regarding the amount of lysozyme added to inhibit clostridia in the cheese milk without affecting the activity of starter cultures were: 500 units ml^{-1} (Netherlands and Germany), 25 or 30 mg litre^{-1} (Italy and France, respectively) and 1–2 g 100 litres^{-1} (Denmark). Recently, Dutch workers (Venema *et al.*, 1990—European Patent 0 380 823 AI) have successfully achieved the production of a lysozyme-type enzyme by genetic manipulation of certain lactic acid bacteria. Such a development could overcome the prohibited addition of lysozyme to the cheese milk in certain countries. Up-to-date scientific data regarding the control methods employed (i.e. (a), (b) and (c) mentioned above) to control butyric acid fermentation in cheese have been recently published by the International Dairy Federation (IDF, 1990*a*).

(d) *Microfiltration/Bactocatch*—The regulatory status of nitrate or lysozyme in cheese milk may place them as food additives in some countries, and may limit their application in cheesemaking, especially where regulations stipulate that the residual level of nitrate or lysozyme in the cheese has to be very low. Recently, a method, which is claimed to produce a nitrate-free and very low bacterial count, including clostridia, in cheese milk has been developed by Alfa-Laval Filtration Systems A/S in Denmark, and is known as

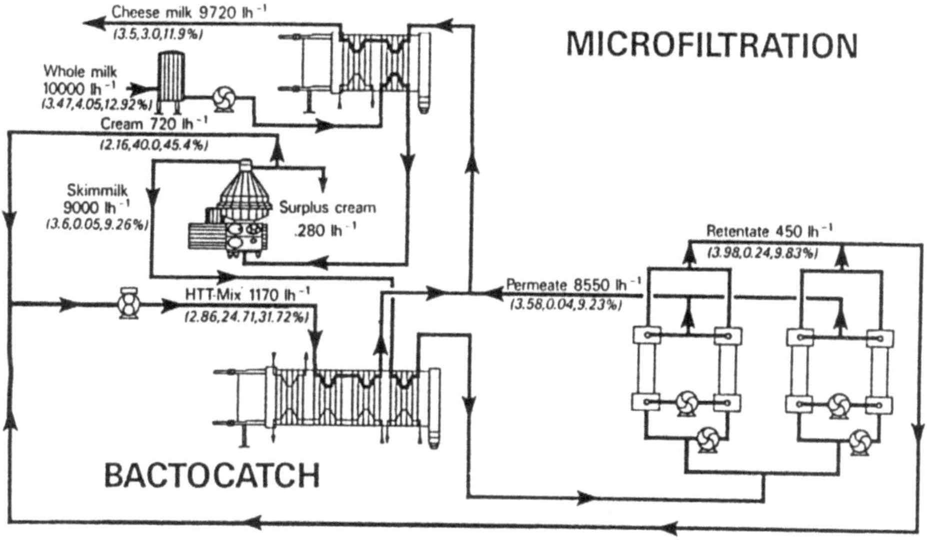

Fig. 1. Flow chart illustrating microfiltration/bactocatch treatment of cheese milk. Figures in parenthesies represent protein, fat and total solids contents, respectively. (Reproduced by courtesy of Alfa-Laval Filtration Systems A/S, Aarhus, Denmark.)

Microfiltration/Bactocatch (Fig. 1). According to Anon. (1987*a*), Malmberg and Holm (1988) and Meersohn (1989) the cheese milk is processed as follows.

- — Pre-warm the milk to 40–50°C and separate the cream.
- — Microfilter the skim-milk at 50°C and concentrate by microfiltration (MF) to 10- or 20-fold. The MF unit is fitted with a large pore size membrane (0·8–1·4 μm in diameter), so that all the milk solids permeate through, but not microorganisms and spores.
- — Standardise the retentate (~5% of the original volume of the skim-milk) with cream to the level of desired fat in the cheese milk, and heat at 110–130°C for 4 s in a plate heat exchanger. This latter process is known as high-temperature treatment (HTT).
- — Cool the HTT standardised retentate–cream mixture, blend with MF permeate (see Fig. 1), pasteurise at 72°C for 15 s, cool to 30°C and finally transfer to the cheese vat.

In processing the cheese milk by this method, the advantages claimed by the manufacturer are as follows: first, ~12% of the cheese milk is processed at high temperature and, as a consequence, the coagulation of the milk proteins or the taste of the cheese are not affected, second, >99·5% of microorganisms and spores present in the milk are inactivated or removed, and third, the running time of Microfiltration/Bactocatch is similar to the pasteuriser (i.e. 8–10 h) before cleaning is required,

The running cost per annum of such a process for cheese milk has been reported by Meersohn (1989). However, Fig. 1 illustrates the mass balance of separation–MF–standardisation and pasteurisation of the cheese milk which can be used, on a daily basis, for the production of cheese with similar chemical composition throughout the year. The suitability of microfiltered milk for the manufacture of Greve cheese (i.e. Swedish hard variety) has been reported by Lidberg and Bredahl (1991).

Seasonal variation in the chemical composition of milk has been shown to occur (Harding and Royal, 1974; Muir *et al.*, 1978; Walstra and Jenness, 1984; see Table XV), and such variations can ultimately affect the yield and compositional quality of, for example, Cheddar cheese (Lelievre and Gilles, 1982; Banks *et al.*, 1984*a*, *b*; Banks and Tamime 1987). The standardisation of the Cheddar cheese milk to a casein to fat ratio of 0·7 is ideal (Kosikowski, 1982), in order to optimise the yield of cheese and obtain the highest retention of fat in the curd ~92·5% (Banks and Tamime, 1987).

Over the past decade, the introduction of new agricultural policies in the EEC has influenced milk production and, as a consequence, its composition. Other factors, which may have influenced the compositional quality of milk, include (i) changing the calving pattern, (ii) changes in the feeding management of cows in order to comply with the quotas, and (iii) changes in the breed of cow.

Thus, changes in the casein to fat ratio occurred in milk supplies during this period, and the pattern of change in the Scottish Milk Marketing Board (SMMB) region where 96% of Scottish Cheddar cheese is produced (J. Russel, pers. comm.) is shown in Fig. 2. Standardisation of cheese milk is highly recommended to occur between October and April, and is essential to optimise the yield of cheese. A similar observation was reported by Banks (1990) for England and Wales, and for ther EEC countries (G. Smith, pers. comm.).

TABLE XV
Seasonal variation in silo raw milk composition (%)[a,b]

	Month	Total solids	Fat	Solids-non-fat[c]	Protein	Casein	Casein number	Lactose[d]	Ash[e]	Calcium
1982	October	13·03	4·00	9·03	3·45	2·58	74·78	4·73	0·85	0·12
	November	12·65	3·95	8·70	3·20	2·45	76·56	4·67	0·84	0·12
	December	12·60	3·95	8·65	3·13	2·31	78·80	4·70	0·83	0·12
1983	January	12·60	3·90	8·70	3·09	2·31	74·76	4·79	0·82	0·12
	February	12·56	3·80	8·76	3·05	2·28	74·75	4·87	0·84	0·12
	March	12·32	3·90	8·42	3·03	2·27	74·92	4·62	0·82	0·12
	April	12·21	3·85	8·36	2·91	2·24	76·98	4·65	0·80	0·12
	May	12·31	3·60	8·71	3·20	2·38	74·38	4·68	0·84	0·12
	June	12·54	3·53	9·01	3·36	2·51	74·70	4·83	0·84	0·12
	July	12·54	3·73	8·81	3·31	2·50	75·68	4·67	0·84	0·12
	August	12·58	3·79	8·79	3·32	2·49	74·10	4·66	0·82	0·12
	September	12·73	3·83	8·90	3·34	2·50	74·10	4·76	0·81	0·12

[a]After Tamime (unpublished data).
[b]The values are average of two samples per month. Each milk sample was analysed in duplicate.
[c]Solids-not-fat was calculated by subtracting the fat value from total solids.
[d]Lactose was calculated by difference.
[e]The ash level was rather high due to the potassium dichromate ($K_2Cr_2O_7$) which was added to the milk sample for preservation.

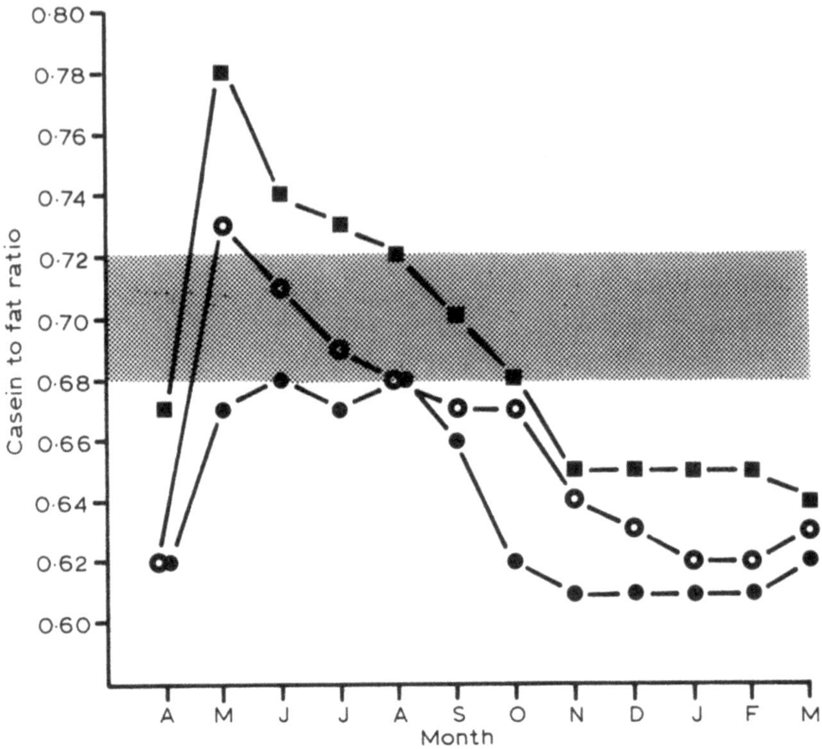

Fig. 2. Seasonal changes in the casein to fat ratio in milk in the Scottish Milk Marketing Board region. (■) 1980–81, (○) 1984–85, (●) 1990–91, shaded area is the optimum ratio. Data compiled from G. Smith (pers. comm.).

As mentioned earlier, the breed of cow can influence the chemical composition of milk, and this can ultimately affect the casein-to-fat ratio (Fig. 2). Over the past decade, the pattern of change of breeds of dairy cow in the SMMB region (expressed as a percentage) is as follows:

	1981	*1984*	*1990*
Ayrshire	31·2	24·2	13·5
Friesian*	64·9	69·5	72·6
Holstein*	3·5	5·8	13·1
Channel Islands	0·3	0·3	0·5
Others	0·2	0·2	0·1

*Data include same breed of cow/Ayrshire cross (G. Smith, pers. comm.).

The theoretical approach to milk standardisation is mainly dependent on the following: first, the type of equipment used and the efficiency of separation obtained, and secondly, the control system used, and illustrations of some of the cheese milk standardisation systems, which could be used in large dairies, have been recently reported by Tamime and Robinson (1985), Wilbrink (1985), Hellstrom (1986), Scott (1986) and Tamime and Kirkegaard (1991). Incidentally, the process of Microfiltration/Bactocatch of cheese milk (Fig. 1) can also be used for standardisation of the casein and fat contents.

Certain additives, such as calcium chloride and colouring matter, are added to the cheese milk after pasteurisation. The former compound is important in order to provide the appropriate balance between the soluble and colloidal calcium in milk which leads to successful coagulation. Al-Obaidi (1980) has shown that a calcium concentration in milk of 140–160 mg 100 ml^{-1} is ideal, and Scott (1986) has reported that no more than 0·02% calcium chloride is needed for a satisfactory coagulation because larger doses can destabilise some of the casein fractions. The most popular colouring matter used in the dairy industry is the water-soluble annatto, which is extracted from the fruit of *Bixa orellana*. The rate of annatto addition per 450 litres may be in the following ranges:

(a) <35 ml for slightly coloured cheeses;
(b) 35–55 ml for medium coloured cheese; or
(c) 115–230 ml for highly coloured cheese.

The annatto is normally added to the cheese milk with the starter culture, or alternatively, some cheesemakers add the colouring matter 15 min before the addition of the coagulant.

Swiss cheeses manufactured in stainless-steel vats rather than the traditional copper kettles have been criticised as lacking a typical cheese flavour(s). However, in Finland, the defect has been minimised by the addition of 15 parts per million (ppm) of copper sulphate to the cheese milk, which is equivalent to the amount of copper absorbed by the milk when using the copper kettles (Lampert, 1975). The role of copper sulphate, with respect to the flavour of the cheese, could be associated with slight fat lipolysis and/or activation of certain enzymes which are important during the maturation of the cheese.

Starter Cultures

The role of starter cultures in cheesemaking can be summarised as follows. First, the organisms produce D, L or DL-lactic acid, as a result of lactose fermentation which is important during the coagulation and texturising of the curd. Secondly, the production of flavour compounds and, in some instances, gas, e.g. Swiss cheeses. Thirdly, the starter culture enzymes (peptide hydrolases and lipases) play a major role during the maturation of the very hard, hard and semi-hard cheeses. Fourthly, the lactic acid may contribute towards the flavour of the cheese, and the acidic condition prevents the growth of pathogens and many spoilage organisms.

The classification of dairy starter cultures has been reported by Garvie (1984) and Tamime (1990), and the type of cultures employed during the manufacture of the cheese varieties mentioned above are the following:

(a) Mesophilic lactic starters: *Lactococcus lactis* subsp. *lactis*, *Lactococcus lactis* biovar. *diacetylactis**, *Lactococcus lactis* subsp. *cremoris* and *Leuconostoc mesenteroides* subsp. *cremoris** (* these cultures are used as aroma-producing bacteria).

(b) Thermophilic lactic starters: *Streptococcus salivarius* subsp. *thermophilus*, *Lactobacillus delbrueckii* subsp. *bulgaricus*, *Lactobacillus helveticus*, *Lactobacillus casei* subsp. *casei* and *Lactobacillus plantarum* (these cultures are only used during the manufacture of high-scald cheese, e.g. Swiss varieties).

(c) Miscellaneous bacteria: *Propionibacterium freudenreichii* subsp. *freudenreichii*, *globosum* or *shermanii* (these organisms are used in conjunction with thermophilic lactic cultures mainly for their ability to produce large gas holes in the cheese during the maturation period).

These sub-species of *Propionibacterium* are differentiated on the following basis:

	Ability to reduce nitrate	*Acid from lactose in milk*
P. freudenreichii	+	−
P. globosum	+	+
P. shermanii	−	+

Furthermore, these microorganisms are sometimes used under the synonym of *Propionibacterium casei*, because of their significance in the dairy industry (Moore and Holdeman, 1974).

Starter cultures may be preserved in one of the following forms, i.e. liquid, concentrated freeze-dried or concentrated frozen at −196°C. The latter two types can be used for direct-to-vat-inoculation (DVI) for the production of the bulk starter, or added directly to the cheese milk. The rate of acid development using an active liquid culture *vis-à-vis* DVI during the manufacture of Cheddar cheese is shown in Fig. 3.

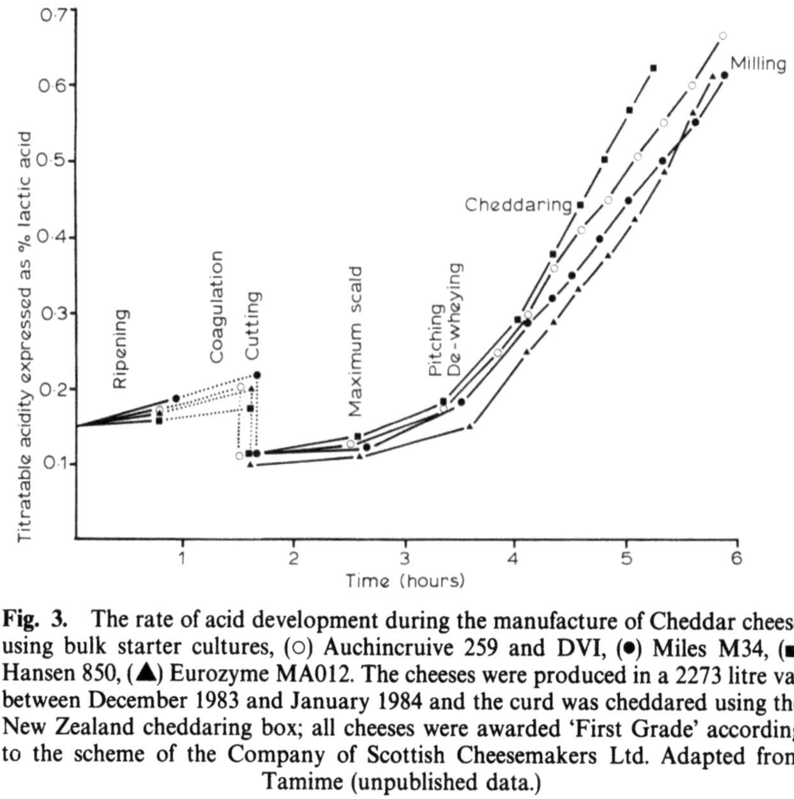

Fig. 3. The rate of acid development during the manufacture of Cheddar cheese using bulk starter cultures, (○) Auchincruive 259 and DVI, (●) Miles M34, (■) Hansen 850, (▲) Eurozyme MA012. The cheeses were produced in a 2273 litre vat between December 1983 and January 1984 and the curd was cheddared using the New Zealand cheddaring box; all cheeses were awarded 'First Grade' according to the scheme of the Company of Scottish Cheesemakers Ltd. Adapted from Tamime (unpublished data.)

The production of the bulk starter in a dairy is an important feature of a successful cheesemaking process. Over the past few decades, starter culture technology has advanced towards the ultimate objective of producing a pure and active culture, and a description of some of the available systems, including illustrations of the equipment, has been recently provided by Tamime and Robinson (1985) and Tamime (1990).

Coagulants, Coagulum Formation and Cutting of the Coagulum

Coagulants used in the dairy industry can be classified into the following groups:

(a) animal (calf, lamb, kid, cow, pig);
(b) microbial (*Mucor meihei, Mucor pusillus, Endothia parasitica* and *Bacillus subtilis*);
(c) plant (*Carica papaya* and *Ficus carica*); and
(d) recombinant or genetically engineered chymosin (*Kluyveromyces marxianus* var. *lactis, Escherichia coli* K-12, *Lactobacillus delbrueckii* subsp. *lactis*).

These enzymes are known as peptide hydrolases (proteinases), because of their activity on the casein micelle during the coagulation process of the milk. Rennet, which consists of chymosin (80%) and pepsin (20%), is extracted from the fourth stomach of a young calf, and this enzyme is widely used in the cheese industry. The ratio of pepsin to chymosin in coagulants increases as the donor animal ages, and practically 100% pepsin preparations are obtained from the stomachs of cows and pigs. Other rennet coagulants can be used, but due to the broad specific proteolytic activity of these enzymes, the finished cheese may have texture and flavour faults. Some aspects of the application of coagulants in the dairy industry and the use of recombinant/genetically engineered chymosin have been published elsewhere (IDF, 1985, 1990*a*, *b*; Koch *et al.*, 1986; Hicks *et al.*, 1988; Bines *et al.*, 1989; Chamba, 1989; Gripon, 1989; Reimerdes, 1990; Prokopek *et al.*, 1990*a*, LaGrange and Goering, 1991; Morris and Anderson, 1991).

The hydrolytic activity of chymosin on the casein micelle can be divided into three phases. The primary phase is the destabilisation of the κ-casein by enzymatic hydrolysis of the susceptible amino-acid bond at the 105–106/phenylalanine–methionine linkage. The result is the formation of two components, i.e. *para*-κ-casein (residues 1–105) and the casinomacropeptide (residues 106–169) (Galesloot, 1958; Tamime *et al.*, 1991). The former fraction is insoluble, highly hydrophobic (1310 cal mol^{-1} (5·5 kJ mol^{-1}) residue) and, in the presence of divalent ions (mainly calcium, magnesium, phosphates and/or citrates), aggregates to form the coagulum which is an integral stage of cheese production. However, the caseinomacropeptide fraction is soluble, not markedly hydrophobic (1082 cal mol^{-1} (4·54 kJ mol^{-1}) residue), and is lost in the whey after cutting (Dalgleish, 1982*a*, 1987). Two aspects are still unclear regarding the specific activity of the enzyme: first, the mechanism of cleavage by the enzyme at the Phe_{105}–Met_{106}, and secondly, the exact nature and kinetics of coagulum formation. A vast literature exists on the possible action of these enzymes in milk, and it is beyond the scope of this publication to review this topic in detail; however, comprehensive data in this field have been published by Cheeseman (1981), Fox and Morrissey (1981), Dalgleish (1982*a*, *b*, 1987, 1990), Fox (1981, 1982*a*, 1984, 1986) Fox and Mulvihill (1983), and McMahon and Brown (1984).

According to Davis (1965) and Fox (1982*b*) around 90% of the coagulant is lost in the whey, and the rest is retained in the curd. Hence, the secondary phase of enzyme activity is partial hydrolysis of the protein during the cheesemaking process, and the tertiary phase, where proteolysis plays a major role during the maturation period, is significant in flavour and texture development (Fox, 1989). The latter two phases may not be important in Swiss cheese, because the whey–curd mixture is heated above 50°C, i.e. inactivates the coagulant. However, the role of non-starter lactobacilli and the enzymatic activities of mesophilic starter organisms have been reviewed in detail by Kamaly and Marth (1989), Khalid and Marth (1990) and Peterson and Marshall (1990).

The amount of coagulant (standard calf rennet) added to milk is 28·4 ml 114 $litres^{-1}$, and the coagulation process is normally carried out at 30°C. It is common practice to dilute the coagulant with water, i.e. three to four times its volume, in order to ensure homogeneous distribution of the enzyme in the milk. Factors affecting the efficiency of the coagulant have been published by Tofte-Jespersen and Dinesen (1979), and the usual duration of agitation of the milk–coagulant mixture is 5 min.

After the coagulum has achieved the desired firmness, the curd is cut. Stainless-steel knives (vertical and horizontal) are widely used for cutting the coagulum, but recently designed cheese vats employ only vertical knives. The knives are also used as an agitation mechanism, so that the curd is cut with the sharp side of the blades, the stirring of the curd–whey mixture is carried out with the blunt side facing forward with respect to rotation. However, stainless-steel wires, which are known as a 'cheese harp', are still used during the manufacture of Swiss cheeses employing the traditional process, and the sequence of cutting is a figure-of-eight shape.

The duration of cutting the coagulum is around 20 min, depending on the size of the curd particle required, type of cheese manufactured and/or the size of the vat. For example, 'fine' or 'coarse' size curd paticles are used for the manufacture of hard and semi-hard cheese varieties, respectively. The coagulum is cut at a very low speed, and the speed is progressively increased to avoid fat and casein losses in the whey which could influence the yield of cheese.

During the manufacture of certain cheese varieties, e.g. Gouda, Edam, Fynbo and Tybo cheeses, 25–30% of the whey is removed and replaced by warm water. The primary objective is to reduce the lactose content in the whey and the production of acid in the cheese.

Maximum Scald and Pitching

After the coagulum has been cut, two phases, i.e. whey and curd particles, become apparent in the cheese vat. The treatment of the curd–whey mixture at this stage varies in relation to the type of cheese produced, and a summary of such treaments, including the scalding temperatures employed, is shown in Table XVI. On completion of cutting, the application of heat and/or the partial removal of the whey, physical and chemical changes in the curd and whey phases start to occur. Walstra *et al.* (1987) and Pearse and Mackinlay (1989) have reviewed extensively all the possible factors that affect the syneresis of rennet milk gels, and the relevant changes (see also Czulak, 1981; Borcherds, 1983, 1985; Grant, 1987) can be summarised as follows:

(a) A semi-permeable membrane is formed across the outer layer of the curd particle and, as the curd is stirred and the temperature is raised, the casein network continues to alter in a manner which results in the formation of a semi-rigid framework, i.e. fibrous connections. Incidentally, the milk solids retained in the curd particle constitute the main cheese components.
(b) Due to the combined action of the application of heat and the development of the network of casein micelles, syneresis (i.e. expulsion of whey entrapped in the curd particle) becomes evident, and the curd particles shrink in size.
(c) Most of the lactic acid is produced within the curd particle, because the starter bacteria are embedded in the casein framework. The excess lactic acid permeates through the membrane to the whey, so assisting in further shrinkage of the curd particles. Furthermore, the level of acid is higher inside the curd particle than in the whey, but not the pH, due to the buffering effect of the milk solids.

(d) An equilibrium between the lactose and mineral salts in the curd particles and the whey becomes evident due to the following: first, as the lactose is utilised by the starter bacteria, the level is reduced, and lactose from the whey permeates through the membrane to the curd particle; and secondly, due to the development of lactic acid inside the curd particle, some of the divalent ions (i.e. bound to the casein micelles) become free and permeate to the whey.

(e) The scalding temperature may range between >31°C and <55°C and, at high temperatures and acidity, the rate of syneresis from the curd particle will be reduced due to the plasticising of the curd, e.g. Swiss and Italian cheeses.

(f) Partial removal of the whey and the addition of water, for example, during the manufacture of Dutch cheeses, retards syneresis, perhaps due to a lowering of the level of lactose in the whey, hence upsetting the lactose equilibrium between the curd and the whey.

(g) The continuous stirring of the curd–whey mixture during the scalding and the pitching periods exerts a physical force or pressure which expels moisture from the curd particles.

(h) High acidity at rennetting, scalding the curd–whey mixture very quickly, and cutting the coagulum when it is too firm can intefere with the rate of syneresis or moisture retention in the curd particle.

De-wheying and Curd Handling

When acid development in the curd has reached the desired level, whey is drained off in order to allow the texturising of the curd. Different treatments are employed in relation to the type of cheese produced, and some typical examples are illustrated in Table XVI. In brief, the consistency of cheese curds can be classified as given in Table XVII, and it is evident that the temperature during scalding, level of acid in the cheese, method of handling the coagulum, amount of pressure applied at the pressing stage and/or the amount of moisture retained play major roles in deciding the body characteristics of the final cheese.

Milling, Salting/Brining and Pressing

Milling of the texturised/matted curd must be carried out at the right acidity, e.g. 0·5–0·7% lactic acid in Cheddar cheese; these acidities at milling are normally recommended when using DVI or bulk starter cultures, respectively. Passing the mellowed curd through the mill (chip

TABLE XVI

Differences in the treatment of the curd during the manufacture of certain cheese varieties[a]

Variety of cheese	*Size of cut coagulum*	*Temperaure of scalding and other treatments*
		I. High scalding temperature
Grana Padona	3 mm	Scald the curd–whey mixture to 58°C; de-whey and press the curd followed by brining and dry salting.
Gruyère and Emmental	10 mm	After cutting, stir for 30 min and scald to 52–54°C in around 30 min; pitch for further 30–60 min, remove curd from whey and press; salt the cheese by brining and dry salting.
		II. Medium scalding temperaure
Parmesan	3–4 mm	Raise temperature from 32 to 42°C in 30 min; settle curd and de-whey at 0·2% lactic acid, place the matted curd in moulds and press; float cheese in brine and dry salting.
Cheddar	6–8 mm	Raise temperature from 30 to 39–40°C in 45 min and pitch for the same period; de-whey when the acidity of the pressed curd is 0·2% lactic acid; form curd into 20 cm blocks, pile and press; mill at 0·65–0·68% lactic acid, salt at a rate of 2½–3% (estimated w/w), fill in moulds (20 kg) and press overnight.
Dunlop	8 mm	Similar to Cheddar but scalding temperature is 36–37°C and less salt is added; the cheese (20 kg) is pressed overnight.
		III. Low scalding temperature
Cheshire	> 10 mm	Cut the coagulum coarse and scald to 32–34·5°C in 45–50 min and draw off the whey when acidity reaches up to 0·23% lactic acid; cut the curd into 7·5 cm blocks, turn over and cut by hand every 20 min (*do not* pile the blocks of curd on top of each other); mill when the acidity reaches 0·65–0·7% lactic acid using a peg mill and salt at a rate of 2%; press cheese overnight.

Wensleydale	>10 mm	Cut the coagulum coarse and for the first 15–20 min do not apply any heat; scald the whey–curd mixture to 35°C in 15–20 min followed by idle stirring for 15–20 min; de-whey when acidity reaches 0·2% lactic acid, from into blocks and turn several times until acidity reaches 0·55% lactic acid; cut the blocks in half, mill at 0·6% lactic acid and salt at a rate of 1·8%; small size cheese is pressed for a few hours.
Lancashire	>10 mm	This cheese is produced in a similar way to Wensleydale but the curd is formed into 15 cm blocks and turned continuously until the acidity reaches 0·5% lactic acid; cut the blocks into smaller pieces and mill at 0·6% lactic acid; salt at a rate of 2% and press overnight.
Caerphilly	5 mm	Cut the coagulum and scald to 35°C in 15–20 min, stir and de-whey when acidity reaches 0·2% lactic acid; form blocks and pile along the side of vat; mill at ~0·3% lactic acid, salt at a rate of 1%, press overnight and place in a brine bath up to 24 h.
Gouda/Fynbo	>12 mm	Cut coagulum, stir and de-whey 25–30% of the volume and replace with water; scald to 35°C and drain whey when curd is consolidated, press, brine and store (curd making to pressing 60 min, pressing 90 min, brining 4–5 days); Fynbo cheese is scalded to a slightly higher temperature with brining for 2 days only.
		IV. No scalding
Edam	>12 mm	Similar to Gouda but major differences are as follows: remove only 20–25% of whey, scald to 30–31°C, curd making to pressing 50 min; pressing time 75 min and brining for 65 h.

[a]Data compiled from Anon. (1959), Davis (1976), Vries and Ginkel (1980), Kosikowski (1982), Scott (1986), Nieuwoudt (1987) and J. H. Dijkstra (pers. comm.).

TABLE XVII

Nature of curd particles	*Description of texture*	*Cheese variety*
Granular	Plasticised	Gruyère and Emmental
Texturised (in some instances it is granular, e.g. American Cheddar)	Close	Cheddar and Dunlop
Slightly texturised	Coarse/Crumbly	Cheshire
Granular	Elastic	Gouda, Edam and Fynbo

or peg type) merely cuts it into smaller pieces, so increasing its surface area so that the salt can be applied more evenly. In some instances, i.e. Cheddar cheese produced in a Tebel Crockatt system, the curd is normally salted prior to milling.

Salting of the curd is achieved in two different ways: first, the addition of dry salt, and secondly, brining in which the cheese, after pressing, is immersed in a brine solution. The function of salting has been studied in detail by Guinee and Fox (1987), and can be described as follows:

— the salt acts as a preservative and flavour enhancer in the cheese;
— it helps to reduce the metabolic activity of the starter culture bacteria, and assists in the liberation of their enzymes which play a major role in flavour development in the cheese;
— salting (dry or wet) helps to reduce the moisture content in the curd as a result of differences in osmotic pressure (for example, Cheddar cheese curd at milling may contain $<40\%$ moisture and after pressing $\geqslant 35\%$ moisture; 60% of the added salt is retained in the cheese and 40% is lost in the pressed whey and on the equipment (Davis, 1965). In Gouda and Edam cheeses, the moisture contents after pressing are 46·5% and 51·5%, respectively, and after brining, the moisture levels in the cheeses are reduced to 41·8% and 44·7%, respectively (J. H. Dijkstra, pers. comm.);
— salt suppresses the growth of undesirable microorganisms in the cheese;
— it helps to alter the physico-chemical characteristics of the curd; and
— dry salting of the curd at high temperature can increase fat losses in the whey which can affect the yield, and the cheese may become greasy after pressing.

The duration of salting may vary from 15–20 min (i.e. dry salting of Cheddar), to a few hours or days, for example the brining of Dutch, Swiss and Italian cheeses.

Pressing of the salted curd in moulds assists in removing more moisture from the curd, and helps to form the final shape of the cheese. The pressing of the cheese is influenced by many factors, such as the following:

(a) the amount of pressure applied is dependent on the cheese variety; cheeses with low moisture require high pressures to consolidate the curd to the desired structure;
(b) the pressure should be applied gradually otherwise whey retention in the cheese will be higher due to premature rind formation; this leads to higher retention of lactose and results in a lower pH in the cheese;
(c) the duration of pressure is dependent on the type of cheese and the pressing system employed (see the section entitled Mechanised cheesemaking);
(d) pressing the curd at high temperature increases fautly losses in the whey, reduces the rate of exudation of the whey due to rapid rind formation, and the fat is in the 'liquid' state which can impede the fusion of the milled curd;
(e) open texture in a cheese could be associated with faulty pressing equipment, or the curd being pressed at too low a temperature; and
(f) during pressing, better fusion of the curd particles and faster removal of the whey is achieved with smaller curd particles than with larger pieces.

Miscellaneous Handling and Storage

The preparation of the 'green' cheese, after pressing, for bulk storage and maturation is an important aspect of production, and one of two different approaches is usually adopted. The conventional process involves the following stages of operation:

— scalding and re-pressing of the cheese to form a harder rind;
— drying;
— bandaging/dressing; and
— waxing.

The second method is to package the cheese in a barrier film which is impermeable to oxygen and moisture, or permeable to carbon dioxide.

The maturation of the cheese takes place in a controlled atmosphere, e.g. temperature and humidity, and both conditions are relevant to cheeses packaged in the conventional method. The relative humidity is maintained around 85% otherwise the cheese becomes very dry due to the evaporation of moisture. If cheese is packed in a barrier pack, i.e. moisture-proof, no evaporation can occur, and only the temperature requires to be controlled. The temperature of maturation of cheese is dependent upon the variety, and while, for example, Cheddar cheese and most of the British territorial varieties are stored at 10°C, Gouda and Edam are held at 13°C and Swiss cheeses up to 20°C.

Turning of the conventionally packaged cheese in the store is essential during the early stages of maturation in order to overcome certain potential faults, e.g. deformaton of shape, rotting of the cheese due to condensation between the cheese and the shelf and/or aid even distribution of the moisture content throughout the block of cheese.

MECHANISED CHEESEMAKING

The period from 1960 could be considered as one of the most significant as regards progress that has been achieved in the development and application of mechanised equipment for the manufacture of cheese. According to King (1966), the primary objectives of mechanisation of the cheesemaking process can be summarised as follows:

— increase productivity in a given size of factory;
— reduce the cost of manufacture, i.e. by labour savings;
— improve the working conditions and avoid heavy manual work; and
— improve, if possible, the quality of the cheese.

At present, there are many different types of cheesemaking systems available on the market for the manufacture of the very hard, hard and semi-hard varieties (Berg, 1990). The reason(s) for diversification can be mainly attributed to technological advancements achieved in the manufacture of certain cheeses (i.e. Cheddar, Emmental or Edam), and to the development of different systems to produce the same variety of cheese, e.g. Cheddar. Nevetheless, Crawford (1976) has pointed out that 'significant in the planning of mechanisation is the establishment of well-defined stages of the cheesemaking process, as follows':

Stage 1: curd production;
Stage 2: de-wheying, texture forming and/or handling operations;

Stage 3: milling, salting/brining and/or mould filling; and
Stage 4: pressing and packaging.

Employing this approach, the recent developments in mechanised cheesemaking will be discussed.

Curd Production

Traditionally, the cheese vat (open type) was utilised for all the operations of cheesemaking, i.e. starter addition to milling and salting, with the exception of the two-level system where curd handling (e.g. cheddaring, milling and salting) took place in coolers or finishing vats. However, the major developments in cheese vats have included

(a) enclosed vat design (vertical and horizontal);
(b) cutting/stirring devices; and
(c) optional fittings.

Some types of cheese vat, which had been installed in cheese factories in the UK during the 1970s, were described by Crawford (1976). Other cheese vats, which have been widely used in different parts of the world, were reported by Scott (1986). Some recently developed cheese vats include the following types:

Damrow 'Double O' vat

The Damrow Company of the US manufactures the 'Double O' cheese vat, and it is marketed in Europe by Gadan A/S in Denmark. This type of vertical cheese vat resembles an open figure-of-eight, and it can be used for the production of curd for the majority of cheese varieties. The movement and mixing of the curd from one section of the vat to another is illustrated in Fig. 4, and the capacity of the vat ranges from 1000 to 35 000 litres.

The overall specifications of the vat are as follows:

— the vat is totally enclosed;
— double blades operate in one direction for stirring and reverse for cutting (Fig. 5 shows the blade assemblies);
— the sequence of operation of the vat, which could be fully or semi-automatic, was described in detail by Crawford (1976); and
— the construction of the vat offers improved cheese yields, production of consistent, high-quality cheese, and minimises fat and curd 'fines' losses in the whey (see Hansen, 1986*a*).

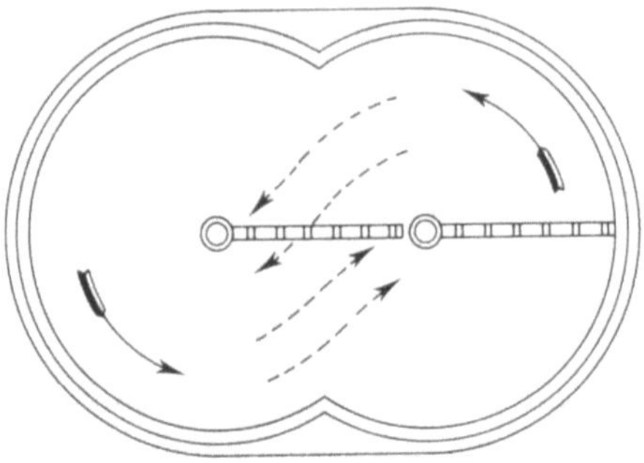

Fig. 4. Horizontal cross-section of the Damrow 'Double O' vat illustrating the direction of the cutting/stirring devices and the movement of the curd. (Bold arrows: direction of rotation; dashed arrows: curd/whey mixture movement towards centre of vat). (Reproduced by courtesy of Gadan A/S, Them, Denmark.)

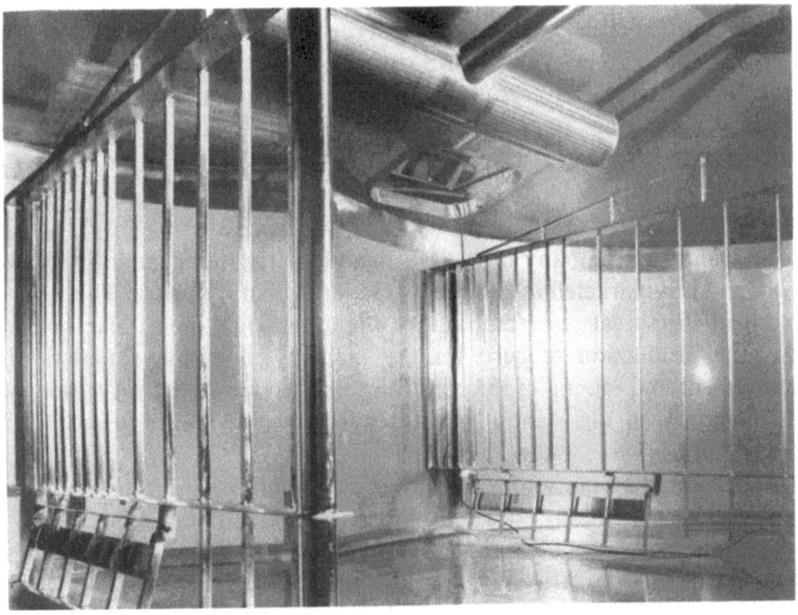

Fig. 5. Inside view of the Damrow 'Double O' vat. (Notice the vertical structure of the cutting/stirring blade assemblies attached to the top-mounted motor and the horizontal strainer is only required to replace some of the whey with water, e.g. during the manufacture of Dutch cheeses). (Reproduced by courtesy of Gadan A/S, Them, Denmark.)

APV–Pasilac cheesemaking vat type OCH

The OCH cheesemaking vat (previously known as CT 2000) is an enclosed horizontal type of tank which is manufactured by APV-Pasilac AS in Denmark. The capacity of this vat ranges from 6700 to 20 000 litres, and an overall schematic view is shown in Fig. 6. The cheese vat is bottom-filled to minimise frothing, and only 46% of the volume is utilised during production. The vat itself is provided with a heating jacket on the lower part of the shell, and a level gauge is fitted to give an exact measure of milk. A horizontal centre shaft, which is supplied with separate cutting/stirring tools (see Fig. 6), operates with a pendulous movement through 180°. Immediately after the addition/stirring of the milk coagulant, the shaft assembly is positioned vertically by turning the stirring paddles up and the wire cutters down into the milk. The reverse position of the shaft is achieved by turning the stirring paddles

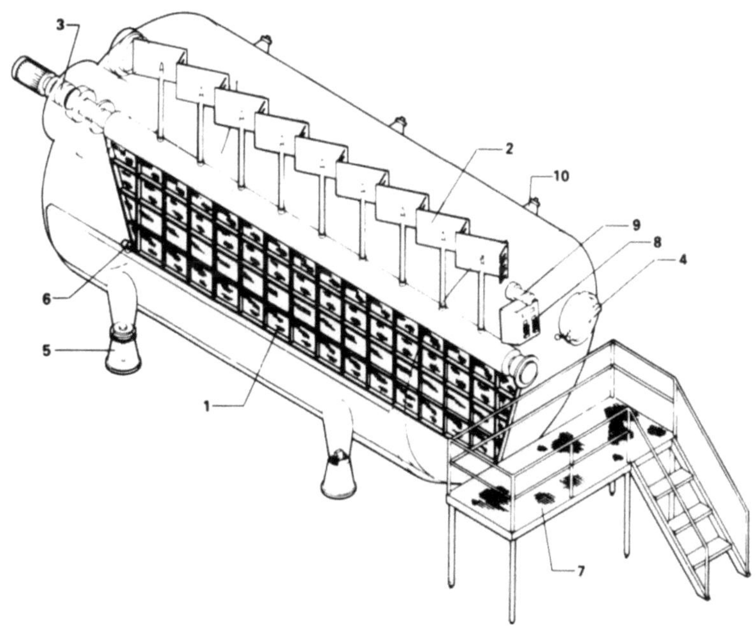

Fig. 6. Schematic illustration of the APV-Pasilac Cheesemaking Vat type OCH. (1) Cutting device, (2) stirring tools, (3) motor and drive, (4) manway, (5) support fitted with pneumatic bellows, (6) level gauge, (7) service platform, (8) control panel, (9) tank light, and (10) spray ball. (Reproduced by courtesy of APV-Pasilac AS, Silkeborg, Denmark.)

down into the curd–whey mixture after the coagulum has been cut. Both the speed of agitation, and the angle of deflection of the stirring/cutting device can be controlled to permit gentle handling of coagulum (Nielsen, 1990*a*).

The OCH vat is mounted on four pyramidal or conical-shaped feet. The front pair are fitted with swivel joints, and the rear pair are fitted with pneumatic cylinders for tipping the vat when pumping the curd–whey mixture to the de-wheying and texture formation units.

Eleven such vats were installed at the Golden Cheese Company in California, USA. The capacity of each vat is 20 000 litres and they are used for the production of Cheddar, Monterey Jack and Colby cheeses (Hansen, 1986*b*).

APV-Pasilac cheesemaking vat type OCT

This vertical and enclosed cheese vat is manufactured by APV-Pasilac AS in Denmark and was originally known as the CT 3113. The bottom of the vat is in the shape of a figure-of-eight, and is insulated (see Fig. 7). The vat bottom and shell are equipped with a jacket for indirect heating, and all parts of the cheesemaking vat are made out of stainless-steel. The vat is designed for cleaning-in-place (CIP), and is supported by three pairs of legs with adjustable, ball-type feet. The vat bottom is designed as two, flat, 7° cones, each with centrally located outlets which ensures optimal emptying of the curd residues.

The vat is manufactured in 14 sizes and shapes ranging from 2000 to 26 000 litres capacity. At the vat end, there are two pipe connections which allow removal of 25 or 50% of the whey, respectively. However, the agitation/cutting mechanism consists of two vertical and synchronised drive shafts, each designed with a one-sided frame with vertical knives. A horizontally movable agitator plate is fitted at the base (rear) of the frame. During the cutting of the coagulum, the shafts rotate in the cutting direction and the agitator plate will be positioned horizontally so acting as a horizontal knife. By reversing the rotation of the shaft, the liquid pressure will move the agitator plate into an angled position to ensure proper mixing of the curd and whey.

The tank top is constructed as two flat cones, and this design provides strength and stable support for the over-head motor and gears. The manway hole is fitted with a ventilation grid and safety switch. The standard vat is equipped with a control unit so that the operator controls the various functions, i.e. cutting or agitation, and the speed. However,

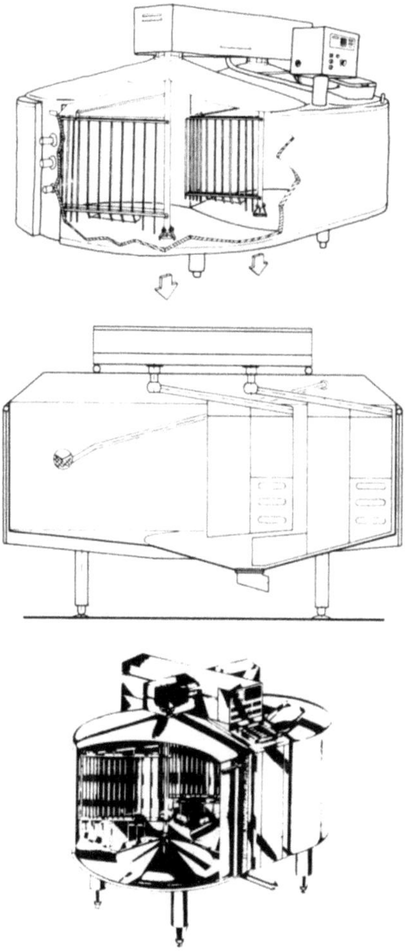

Fig. 7. Schematic illustrations of different cheese vats in the shape of the figure-of-eight. (Reproduced by courtesy of APV-Pasilac AS, Silkeborg, Denmark; Schwarte-Werk, GmbH, Ahlen, Germany; Gadan A/S, Them, Denmark.)

additional equipment, which could be installed in the vat, includes the following:

— semi-automatic or fully automatic process control;
— digital recording for contents and temperature;
— digital speed indicator; and
— level-controlled whey strainer.

It is possible, however, that an alternative vat design could be manufactured, i.e. flat bottom, with a slope of 4% inclination towards one outlet.

Schwarte cheese maker tank

This vertical, enclosed or open-top tank in a figure-of-eight is manufactured by Schwarte-Werk GmbH in Germany, and is suitable for the manufacture of a wide varieties of cheeses. The cheese maker tank is manufactured in different sizes ranging from 500 to 25 000 litres (Fig. 7). The construction of this tank is triple-walled, and the heating system (steam or hot water) and the cooling system are installed between the middle and the inner walls, so providing a large surface area and rapid heating and cooling times. On-site operation of 2 × 12 000 litre vats at Milchhof Osnabruck cheese factory was reported by Hansen (1987*a*).

The inner side of the cheese maker tank is made of stainless chrome–nickel steel; however, for Swiss type cheeses, the tank is manufactured in copper. The enclosed version of this tank is recommended, especially for automatic process control and CIP. This tank can be supplied with (i) manual control, (ii) semi-automatic control, where the individual stages of operation are initiated by hand and, thereafter, the operation becomes automatic, and (iii) fully automatic control of the entire cheese production process. Furthermore, the removal of whey can be achieved by siphoning or by pumping. Whey extraction from the tank is possible when the agitator is stationary or running.

Gadan multicurd cheese vat

This cylindrical stainless-steel cheese vat is constructed in a figure-of-eight shape, and it is manufactured by Gadan A/S in Denmark (Fig. 7). The vat is manufactured in all sizes ranging from 1000 to 25 000 litres. According to the supplier, the features of the Multicurd cheese vat are as follows:

- self-emptying through one centrally placed outlet;
- effective stirring and cutting of the entire product volume;
- heavy, stable construction entirely of stainless-steel;
- resistant to pressure and vacuum;
- good re-start qualities after curd settling;
- minimum loss of fat and fines due to advanced agitator design;
- vertical construction in up-to-date design;
- CIP cleanable; and
- adapted for manual or fully automatic process control.

ASTA cheesemaking tank

This is a vertical and cylindrical cheesemaking tank (type SKF) which was developed by ASTA-Eismann GmbH in Germany (Anon., 1987*c*). The tank is manufactured in different sizes ranging from 500 to 12 000 litres. The cutting/stirring mechanism is located centrally in the tank and rotates anti-clockwise and clockwise, respectively. The tank is installed with a tilt of 15° towards the outlet valve which causes limited rotation of the curd during the cutting stage. The outlet valve (i.e. disc type) is rather large and welded to the bottom of the tank in such a way that 'dead pockets' do not occur.

Emptying of the tank is very effective because (i) of the tilting arrangement available, and (ii) the domed-bottom design which minimises the vortex effect and air suction. The cheesemaking tank can also be adapted for whey removal at different levels, and it can be cleaned by CIP.

Tebel-MKT cheese vats

These stainless-steel cheese vats are manufactured by Tebel-MKT in Holland and Finland. The Finnish company was previously known as MKT Tehtaat OY, and the Emmental/Swiss cheese equipment is manufactured in Finland. Primarily, three main models of cheese vat are available: (i) Tebel-MKT-S (cylindrical), (ii) Tebel-MKT-K (vertical in a figure-of-eight), and (iii) Tebel-MKT-O (cylindrical and horizontal which is used for the production of cottage cheese), and Fig. 8 illustrates the overall construction of these vats. Both vats (MKT-S and K) are designed with a driving mechanism which is top-mounted, while MKT-O vat is

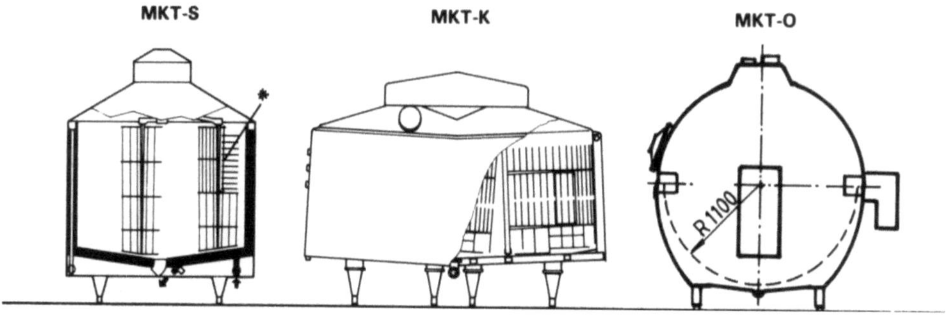

Fig. 8. Schematic illustrations of different cheese vats suitable for the manufacture of Dutch, Swiss and Cottage cheeses. (Reproduced by courtesy of Tebel-MKT BV, Jarvenpaa, Finland.)

fitted with a horizontal shaft for cutting/stirring of the curd. The capacities of the MKT-S, K and O are up to 20 000, 25 000 and 18 000 litres, respectively. The emptying procedures of the MKT-S and K are easy due to the conical design of the bottoms of the vats. Partial removal of the whey can be achieved during agitation of the curd–whey mixture through a special, fixed, sucking device. The outer cutting mechanism (Fig. 8) is designed as a 'gate', i.e. it swivels as the whey-sucking device is lowered to follow the level of the whey in the vat. Illustrations of these vats in cheese factories have been reported by Hansen (1987*b*, 1988*a*, 1989*a*).

APV Swiss cheese vat

This is a cylindrical cheese vat with a spherical bottom, and the interior section of the vat is normally made out of copper and the outer shell of resistant stainless-steel. The vat is manufactured by APV-Rosista AG in Switzerland and is suitable for the production of Swiss cheeses. The vat is fitted with specially designed agitator and cutting mechanisms (see Fig. 9).

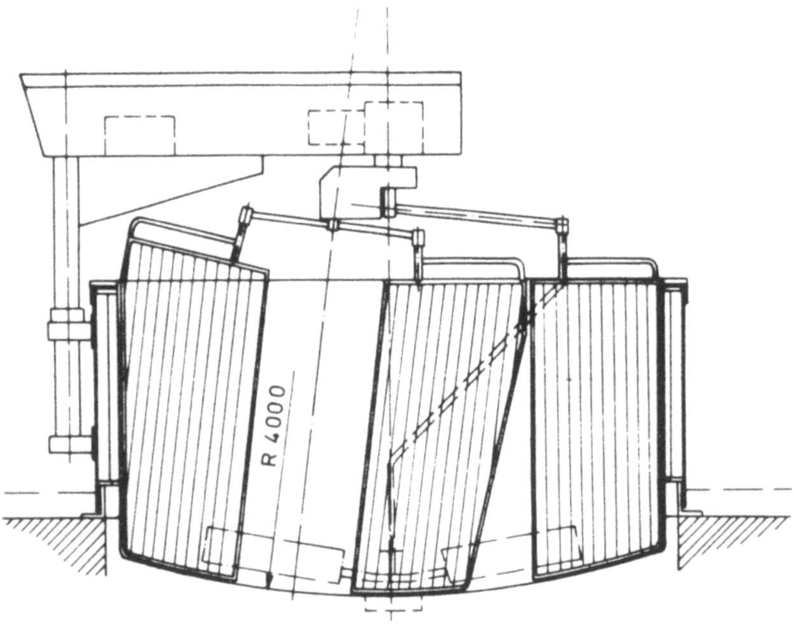

Fig. 9. Schematic illustration of Swiss cheese vat. Notice the special design of the harp cutter and the agitator mechanism of this cylindrical cheese vat. (Reproduced by courtesy of APV-Rosista AG, Worb, Switzerland.)

The latter part consists of two cheese harps, and the exterior one is fitted with a spherical agitator. Upon rotation, the entire coagulum is cut into uniform pieces. Other specifications of the APV Swiss cheese vat include the following:

(a) the vat is fitted with an electronic device to control the speed of the agitator;
(b) the curd–whey mixture can be emptied through a pneumatic discharge valve situated on the bottom of the vat, or by a suction device over the edge of the vat;
(c) the heating of the vat contents can be achieved by circulating warm water in the jacket (i.e. via two combined bottom and side wall circuits) or with steam.

A modification of the above vat is the oval-shape type, in which the bottom is flat; it has a lower filling height, but the same capacity. This cheese vat is tipped pneumatically for discharging the curd–whey mixture through a bottom valve, either gravimetrically or by pump. Further information regarding dairy equipment, including cheese vats, made by APV–Rosista AG has been reported by Anon. (1987*b*).

Tebel OST-III and IV

This horizontal and enclosed cheese vat is manufactured by Tebel BV in Holland, and is suitable for the production of different cheeses. For example, the OST-CH is developed for the manufacture of Cheddar cheese, and does not contain an advanced whey strainer; this latter equipment is required for some types of cheese, e.g. Gouda and Edam. The tank is manufactured in different sizes ranging from 3000 to 30 000 litres capacity, and the milk occupies 80–85% of the volume of the tank (Fig. 10(A)). The volume of the tank is increased by extending the length of the tank; the diameter, height of liquid, peripheral equipment and drive-unit remain unchanged (see also Damerow, 1988; Hansen, 1985*a*).

The tank is supported by four legs, and inclined at a slope of 1:40 towards the valve outlet which is 10·16 cm in diameter; the same valve is used for bottom filling of the vat.

The cutting and stirring device (see Fig. 10(A)) is divided into sections and welded onto the drive shaft. During the cutting of the coagulum, the shaft rotates clockwise (when viewed from the manhole) through 360°, and anticlockwise during stirring, i.e. blunt side of the knives. In order to prevent the coagulum from rotating in the tank, the cutting operation is

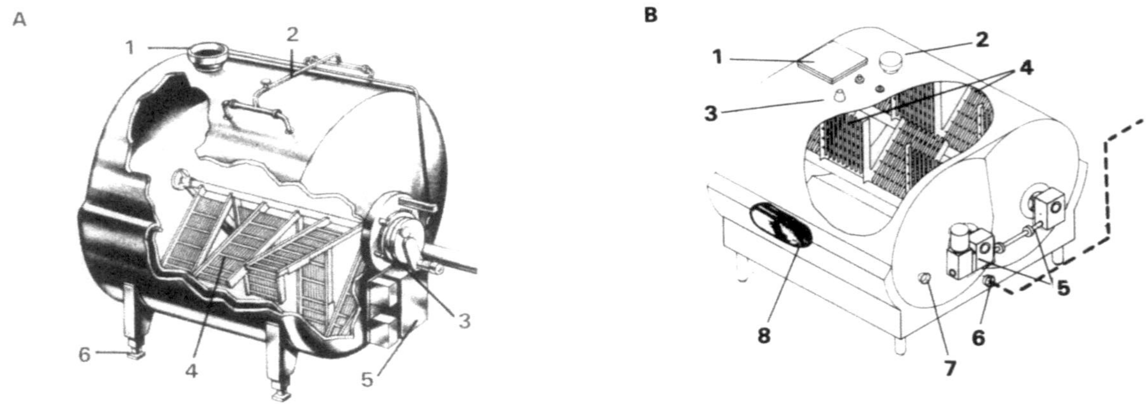

Fig. 10. Schematic illustrations of horizontal cheese vats. (A) Tebel OST-IV; (1) vent, (2) CIP manifold, (3) main drive shaft, (4) combined cutter/stirrer frames, (5) frequency controller, and (6) adjustable legs. (Reproduced by courtesy of Tebel BV, Leeuwarden, The Netherlands). (B) Scherping twin-shaft horizontal vat; (1) manway, (2) vent, (3) light, CIP inlet and coagulant injection filling, (4) reversible cutting/stirring paddles, (5) dual shaft drive, (6) milk inlet and curd/whey outlet, (7) whey pre-draw connection, and (8) steam or hot-water jacket with sparge pipes. (Reproduced by courtesy of Scherping Systems, Minnesota, USA.)

carried out intermittently, i.e. stops twice per revolution and only 80–85% of the volume of the tank is utilised.

A permanent whey drainer can be fitted to the tank in one of its end walls and at a pre-set level. For example, if the strainer is situated at the 60% tank level, the amount of whey removed is equivalent to 40%. Incidentally, such a strainer is only required for the production of Dutch and related cheeses.

Scherping horizontal twin-shaft cheese vat

This horizontal twin-shaft and enclosed cheese vat (HCV) is manufactured by Scherping Systems in the US, and is suitable for the production of wide range of cheeses. The vat is manufactured in six sizes ranging from 13 500 to 31 500 litres capacity (Fig. 10(B)). The volume is increased by extending the length of the vat, and the other dimensions remain unchanged, i.e. the construction principle could be considered as 'Double O' horizontal tank—similar to two Tebel OST tanks joined together.

The vat is supported by four legs, and inclined at a slope of 1·9 cm per 30·5 cm towards the valve outlet, which is also used for bottom filling and curd–whey emptying of the vat (Fig. 10(B)).

Some of the characteristics of this vat can be summarised as follows:

- improvements in the consistency of the size of the curd cut can be achieved, due to the gentle speed of cutting and limited movement of curd in the vat;
- minimises curd damage in the vat and hence increases cheese yield;
- the vat is fitted with a coagulant injection system and proper agitation, which provides uniform setting of the coagulum and minimises fat losses into the whey during the cutting stage;
- cooking is achieved using steam or hot water in the jacket, and high cook temperatures, e.g. Swiss cheese, can be reached in 30 min without creating hot spots or burnt curd;
- the cut/stir operations of the vat are controlled by a programmable control centre.

Performance of New-Style Cheese Vats

Crawford (1976) has reported that 'it is very difficult to obtain factual and unbiased information on the peformance of new equipment, and this is as true of the new race of cheese vats as it is of other items of machinery.... The efficient function of the cheese vat and proper control by the cheesemaker over the first stage of the process is vital to the following stages and ultimate quality and yield of product'. However, the OST-III and IV, 'Double O', Schwarte and APV-Pasilac vats have been evaluated at the Netherlands Dairy Reseach Institute (Vries, 1979; Vries and Ginkel, 1980, 1983, 1984; Ginkel *et al.*, 1987) for the production of Gouda cheese, and their overall conclusions are as follows:

Tebel OST III

(a) The fat and the curd 'fines' losses in the whey were the same and slightly lower, respectively, compared with the figures commonly obtained at the Institute.
(b) The curd particle size was too fine, but the manufactured cheese had a satisfactory composition.
(c) The tank had no influence on the quality of the cheese, i.e. microbiological analysis at 14 days and organoleptic evaluation after 6 weeks.
(d) The curd-making tank had an adequate CIP system, as confirmed by bacteriological tests.

Damrow 'Double O'

(a) The fat and curd 'fines' losses in the whey were lower than the figures commonly obtained at the Institute.
(b) A good curd size was obtained when the renneting temperature was increased from 30·5 to 31·0°C.
(c) Curd production and/or handling in the vat could be properly controlled, and the composition of the cheese was good; at grading, i.e. 6 weeks after production, the quality of the cheese was good.
(d) After CIP cleaning, the bacteriological results were good.

The effects of speed and duration of cutting, in the Damrow 'Double O' under factory conditions in New Zealand using mechanised Cheddar cheesemaking, on curd particle size and yield have been recently reported by Johnston *et al.* (1991).

Tebel OST IV

(a) The fat and curd 'fines' losses in the whey were lower compared with the figures normally obtained at the Institute.
(b) The difference between the calculated and the determined fat content of the second whey tended to be a little too high.

(c) The curd particle size was somewhat fine for the manufacture of Gouda cheese, but did not affect the curd 'fines' losses in the first whey.
(d) Other comments reported were similar to those mentioned for Tebel OST III, i.e. (c) and (d).

Schwarte

(a) No lumps of curd were formed in the vat during production of the coagulum or during the transfer of the curd–whey mixture to the strainer vat.
(b) Losses of fat in the whey (i.e. 5·75% of the fat in the cheese milk) and curd 'fines' were lower than the figures usually obtained at the Institute.
(c) The difference between determined and calculated fat content in the second whey was slightly higher than the Institute standards, despite changing the stirring assembly and reducing the speed.
(d) The curd particle size was good.
(e) The cheesemaking process in the tank was properly controlled, and hence cheese of good composition was produced.
(f) At grading of the cheese, the quality of the product was normal.
(g) After CIP cleaning, the bacteriological results were good.

APV – Pasilac type 3113

(a) The fat and 'fines' in the whey were lower than the figures usually obtained at the Institute.
(b) The difference between the determined and calculated fat in the second whey was higher during the test period.
(c) The distribution of the curd particle size easily met the normal standard.
(d) Other comments reported were similar to those mentioned for Schwarte, i.e. (e), (f) and (g).

De-wheying, Texture Forming and/or Handling Operations

The advent of mechanisation in the cheese industry has brought about some changes in the traditional process, and in most cases, the main improvements have involved the following stages of cheesemaking: first, the de-wheying, i.e. the separation of the whey from the curd particles, and second, the texturising/handling of the curd. However, the latter aspect is of great importance, because the mechanised cheese systems have been designed for the manufacture of one or more types of cheese, and hence the available systems will be reviewed in relation to the variety of cheese.

Swiss Cheese Varieties

Traditional process

At the pitching stage, a heavy-duty cheese-cloth, which is fastened onto a metal frame, is used for the de-wheying of the curd. The curd and whey are vigorously agitated in a circular motion, and the swirling effect is stopped by the stirring equipment which causes the curd to collect in the middle of the vat. At that moment, the cheese-cloth is inserted into the vat beneath the curd, and pulled from one side to the other side. A pulley is used to lift the curd out of the copper kettle into the cheese mould. The cheese is pressed overnight in order to remove the excess moisture and to matt the curd particles. Normally, the curd weighs 100 kg which is the equivalent to the yield from each cheese vat needed to produce one cheese 'wheel'.

Damrow Swiss cheese vat

This type of mechanised vat (Fig. 11) is used for de-wheying and pressing the curd. The sequence of operations is as follows: the curd and whey are pumped from the conventional cheese vat through a curd distributor to fill the curd evenly in the Swiss cheese vat. The curd is covered with a sanitary, re-usable woven plastic mesh (2) and the press plate is lowered into position (3). The rate of whey drainage is controlled by the cheesemaker at the whey outlet (4). Upon completion of pressing, the press plate is removed, together with the plastic mesh, and the block cutting mechanism is positioned at the curd unloading door. The bottom belt (8) continuously moves the pressed cheese to the unloading door for cutting into blocks. Finally, the cheese blocks are transferred to the brining tank.

The Tebel-MKT system

This mechanised system has been developed in Finland for the production of 'block'- and 'wheel'-shaped Emmental (Kiuru, 1976; Hansen, 1984*a*). For the manufacture of block cheeses, i.e. 84 kg in weight, the curd–whey mixture is pumped from the Tebel-MKT cheese vat to a de-wheying and pressing vat through a nozzle tube (see Fig. 12(A)). The pressing vat is fitted with sieve plates along its sides to assist whey

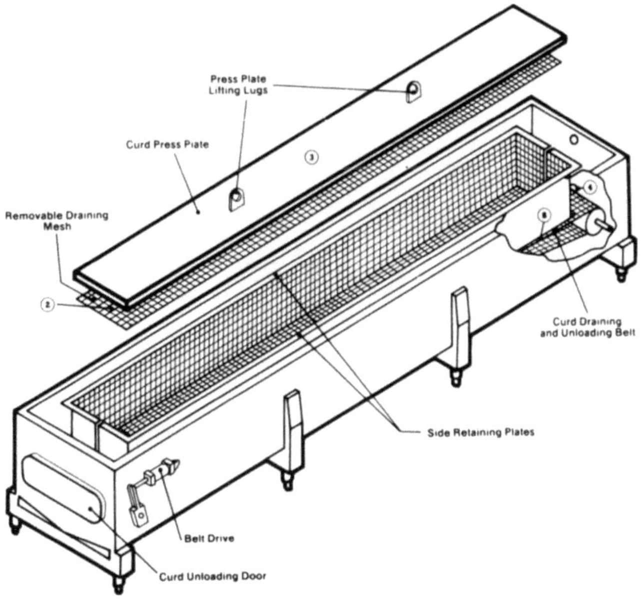

Fig. 11. Schematic illustration of the Damrow Swiss cheese vat. (Reproduced by courtesy of Gadan A/S, Them, Denmark.)

drainage and, in order to increase the removal of whey, the curd is covered with a woven, plastic sheet and press lid. After the pressing stage, the side sieve plates are unhinged, and the pressed curd is cut to the desired size, turned and placed over a conveyor to be transported to the brining basin.

More recently this method of production has been modified so that the curd and whey is delivered directly to the moulds through an octopus-like distributor (Fig. 12(B)). The accuracy of mould filling is $\pm 2{\cdot}0$ kg per 84 kg block of cheese. The curd is pre-pressed for 15 min, and afterwards the moulds are transferred to the tunnel press (see Fig. 28).

The production of 'wheel'-shaped Emmental is achieved using a pressing vat (similar to the type mentioned above), which is furnished with cylindrical moulds formed as closed girdles made out of perforated stainless-steel. The moulds are fitted with a filling tray onto which the curd–whey mixture is delivered. The curd in each mould is covered with a woven, plastic sheet and a press lid. The latter unit is fitted with a pressing cylinder actuated by compressed air at low pressure. After the pressing stage, a hydraulic lift is used to empty the moulds, and the cheese is transferred to the brining section.

(A)

(B)

Fig. 12. A view of the Tebel-MKT cheese equipment for the production (A) block-shaped, and (B) wheel-shaped Emmental cheese in Finland. (Notice the MKT-S cheese vats in the background and some of the pressing vats are being emptied, pressed or CIP; some of the cylindrical moulds are being un-assembled and the 'octopus' curd distributor is moved onto the next pressing vat). (Reproduced by courtesy of Tebel-MKT BV, Jarvenpaa, Finland.)

Miscellaneous equipment

Hansen (1975) reported on a special pressing vat, which was designed by Gadan A/S in Denmark for the manufacture of a 450 kg 'block'-shaped Emmental. The pressing table consisted of three parts of perforated stainless-steel (frame, lower and upper sheet) of the following dimensions: 1·12 m wide × 2·8 m long × 0·3 m high. After the de-wheying stage, the table is titled at an angle of 15°, and the curd is turned once to remove the excess whey before pressing commences. The 450 kg cheese is later brined and portioned to 10 × 45 kg blocks, dried and finally packed in Cryovac® bags (registered trademark of W. R. Grace Ltd).

Cheddar Cheese and Related Varieties

The main systems available for the manufacture of Cheddar cheese are as follows:

(a) Bell-Siro/Cheesemaker 2;
(b) Cheddarmaster;
(c) Lacto-O-Matic;
(d) Tebel-Crockatt;
(e) Alf-O-Matic;
(f) Damrow Drainage/Matting Conveyor (DMC);
(g) Scherping Cheese Curd Drainage Conveyor (CCDM); and
(h) Wincanton Draining and Cheddaring Conveyor (DCC).

Up to the 1970s, the mechanised systems employed for Cheddar cheese (i.e. (a)–(d) above) were widely used in the UK, Australia, New Zealand, and North America. A comprehensive review of such systems, including illustrations, has been published by Crawford (1976) and Scott (1986) and hence only the last four systems will be discussed here.

Alf-O-Matic

This machine, which has been developed and manufactured by Alfa-Laval A/B, includes a number of different models (i.e. different number of belts ranging from one to four), that can be used for the manufacture of separate varieties of cheese. For example, the four-belt Alf-O-Matic is used for the production of cheddar cheese (see Fig. 13), and in principle, the unit consists of the following:

- a fixed screen for de-wheying, and conveyor number (1) is for further drainage and moisture control in the curd;
- conveyors number (2) and (3) are for curd matting/fusing and cheddaring, respectively; and
- conveyor number (4) is for salting and mellowing of the milled curd.

The conveyors are maufactured from perforated stainless-steel and are mounted on top of each other; the whole unit is enclosed within a stainless-steel casing. The Alf-O-Matic is equipped for CIP, and the specifications of the four-belt model are as given in Table XVIII. The usual sequence of operations is as follows.

(a) The curd–whey mixture is pumped from the cheese vat to the inlet of the Alf-O-Matic and onto a fixed de-wheying screen.
(b) The curd particles are delivered to conveyor (1); further drainage can be achieved by stirring the curd with rotating cylinders covered with pegs. The pumping speed of the curd–whey mixture is controlled by a curd level sensor situated at the beginning of the belt, and the residence time in this section ranges between 5 and 15 min.
(c) Matting and/or fusing of the curd particles take place in conveyor (2), and the curd depth is maintained around 25–30 cm. At the end of this conveyor (e.g. residence time 10–45 min), the curd mattress is inverted through a guided chute onto the cheddaring conveyor (3) where the residence time ranges between 15 and 70 min depending on the type of cheese produced.
(d) At the end of conveyor (3), the cheddared curd passes through a mill; salt mixing and mellowing of the curd take place on conveyor (4). The mellowing stage is assisted by stirring and the residence time in this section ranges between 10 and 40 min.
(e) The salted curd is transferred to the mould-filling machines by an auger mechanism or by air. The latter system is employed with the continuous 'Block Former'.

Damrow draining/matting conveyor (DMC)

The Damrow DMC equipment is a combined unit in which the de-wheying and texture formation of the curd takes place during the

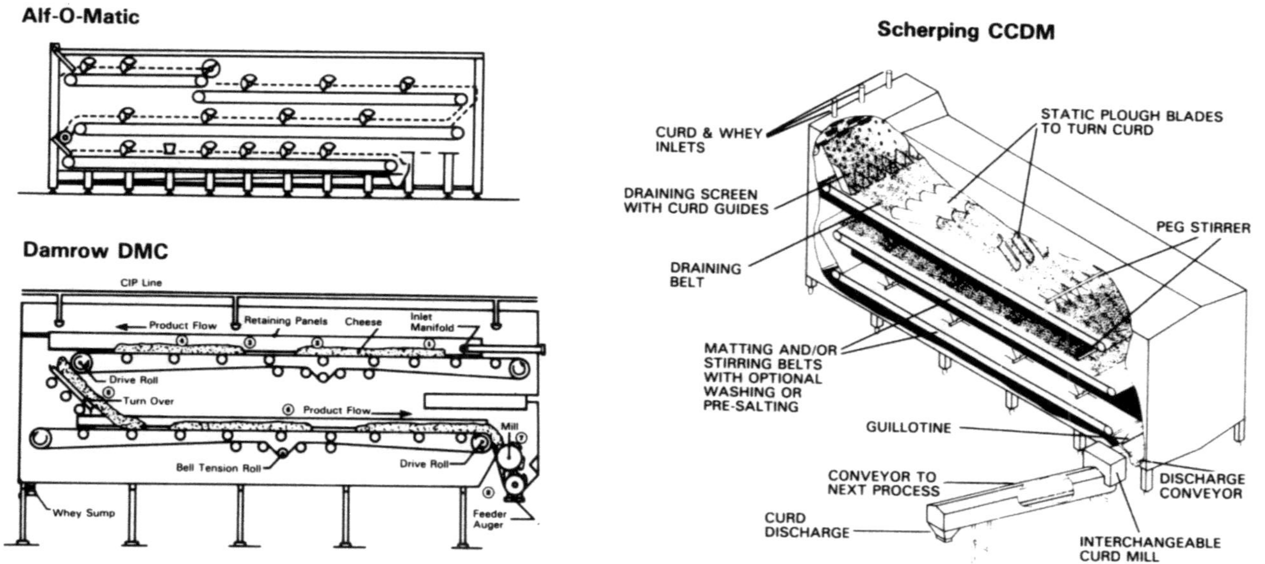

Fig. 13. Schematic illustrations showing different equipment for de-wheying, texture forming and handling operations for the manufacture of Cheddar and related cheeses. (Reproduced by courtesy of Alfa-Laval Cheddar Systems Ltd, Somerset, UK; Gadan A/S, Them, Denmark and Scherping Systems, Minnesota, USA.)

TABLE XVIII

Type	Dimensions (m)			Capacity (kg h^{-1})[b]
	Length[a]	Height	Width[a]	
3000	7·5	5·8	2·5	1350–2000
6000	11·2			2700–3200
12 000	20·5			5400–6400

[a]Excluding the walkways.
[b]Depending on the residence time and treament of the curd.

manufacture of Cheddar cheese. The DMC is totally enclosed within a stainless-steel chamber and, in theory, it consists of three sections, i.e. de-wheying, matting of the curd on moveable mesh belts made out of double-woven plastic made of polypropylene, and a milling unit. The capacity of the DMC may range from 0·1 to 6 t h^{-1}, and the number of plastic belts installed in such a unit is adjusted according to the type of cheese produced. For example, a DMC unit with two belts is used for the manufacture of Cheddar cheese, but other cheese varieties may only require one belt, e.g. blue vein, Feta and other brine-salted cheeses. The sequence of operations of the DMC (see Fig. 13) can be summarised as follows:

— The curd–whey mixture is pumped from the cheese vat to the DMC inlet fitted with a special manifold for even distribution of the curd on top of the first belt.
— As the belt travels forward, the whey drains off through the belt and by the time the curd has reached the end of this belt, matting of the curd has occurred.
— The speed of the belt and the retention guides located along and over the top of the belt ensure control of the width and depth of the curd.
— At the end of each cycle from the cheese vat, the belt continues to move on, e.g. around 120 cm, so as to separate the production of one vat from another.
— The matted curd will turn over automatically when it reaches the end of the top belt, and as it leaves the top belt, the curd is picked up (upside-down) on the lower belt.
— On the lower belt, the matted curd will stretch and flow, thus simulating the traditional process of cheddaring.
— A 'chip' mill is situated at the end of the bottom belt, and the milled curd can be cut to any desired size, e.g. ranging from 13 × 10 mm to 18 × 24 mm.

— The milled curd falls into a hopper, and an auger mechanism transports the curd to the salting unit.
— The DMC is equipped with CIP.

Scherping cheese curd draining conveyor (CCDM)

The Scherping CCDM equipment is another example illustrating a combined unit in which the de-wheying and texture formation, or granular curd handling, takes place during the manufacture of Cheddar, American varieties and other related cheeses. The CCDM is totally enclosed within a stainless-steel chamber, and it consists of three belts mounted on top of each other (Fig. 13). The draining conveyor uses a self-tensioning and tracking belt which is made of a series of moulded polypropylene sections linked together with polypropylene pins. However, the hydrophobic nature of this plastic material makes it easy to clean the belt. This belt is driven by drive sprockets (also made from polypropylene) which are positioned across the entire width of the belt to ensure even distribution of load. The sequence of operation of the CCDM (see Fig. 13) is as follows:

— The curd–whey mixture is pumped from the cheese vat to the three inlets of the CCDM for even distribution of the curd on to a sloped pre-draining screen. At the bottom half of this screen, curd guides are fitted which helps the curd to form 'ribbons' as it flows forward onto the first, moving, draining belt.
— As the belt moves forward, the curd 'ribbons' pass through static plough blades to be turned and stirred using rotary pegs to assist in whey drainage.
— Matting of the curd takes place on the second or third belt depending on what type of cheese is being produced.
— The bottom two draining/matting belts are equipped with rotary peg stirrers which are used for the production of granular curd (e.g. dry stirred Cheddar cheese); thus, a mill is not required and optional equipment, which can be added on this conveyor, includes a spray device to wash the curd and/or a salting system. Incidentally, this method of curd handling is used during the manufacture of Colby and Monteray Jack cheeses.
— At the end of the third conveyor belt, a guillotine is fitted to section the matted curd which is then delivered onto a discharge conveyor that feeds the external curd mill.
— This unit is equipped with CIP.

Wincanton draining cheddaring conveyor (DCC)

Another equipment, which is a combined unit for de-wheying and curd handling, is the Wincanton DCC system (Fig. 14). This equipment is manufactured by Wincanton Engineering Ltd in the UK, and it is suitable for the production of Cheddar and related cheeses. The DCC is a totally enclosed system housed within a stainless-steel body, but the conveyors are made of plastic that complies with the hygienic regulations of the United States Department of Agriculture (USDA).

The DCC equipment consists of two belts mounted on top of each other, and the handling of the curd–whey mixture is as follows.

- The curd–whey mixture is pumped from the cheese vat to the DCC unit via a turbulence trough and weir in order to distribute the product evenly over an inclined draining screen.
- The draining screen is constructed of a 'wedge wire' which is capable of removing 95% of the free whey without damaging the curd granules that roll gently onto the draining belt.
- The draining belt is perforated in order to aid further the release of whey, and it is supported on stainless-steel guides over the full length of the DCC unit. The drive mechanism is situated at the outlet end, and the movement of the conveyor is achieved via a stainless-steel shaft and plastic sprockets. The tension of the conveyor belt is maintained by its own weight and a roller tensioner that adjusts automatically, especially to compensate for thermal expansion during cleaning.
- A number of agitators are located at regular intervals along the conveyor, and these assist in further whey drainage before matting of the curd takes place.
- At the end of the top conveyor belt, the matted curd drops onto the bottom conveyor belt via a chute which helps to turn the curd over.
- The residence time on the draining/matting conveyor belts is adjustable (i.e. manual or remote control).
- At the outlet of the matting conveyor there is a 'chip' mill, auger, rotary valve and a pneumatic conveying system to transfer the milled curd to the salting eqipment(s).

In general, the Wincanton DCC unit is mounted on adjustable legs for ease of installation, and manways/access doors and lights are positioned at intervals along the length of the machine for checking the curd during cheddaring. All drive motors and gearboxes are supplied with vented

Fig. 14. On-site view of the Wincanton Cheesemaking equipment at T. W. Clothier & Sons Ltd, Somerset, UK, which has a capacity of 3·6 tonnes h^{-1}. (A) The DCC unit (right-hand side) and the salting unit (left-hand side)—note the 'swan neck' pipe which connects these machines for transferring the milled curd from the DCC to the salting unit. (B) Curd leaving the de-wheying screen on the top belt of the DCC unit—notice the first set of stirrers. (C) Cheddard curd approaching the curd mill. (D) A view of the milled and salted curd in the salting unit. (Reproduced by courtesy of Wincanton Engineering Ltd, Dorset, UK.)

stainless-steel covers. This DCC machine is cleaned by CIP, and the spray balls and nozzles are located at strategic points for optimum cleaning and sanitisation. The spray devices are fed, via a number of inlet feed pipes, from the main CIP pipe-line attached along the full length of the machine (D. Shaw, pers. comm.).

Finishing coolers

Some degree of mechanisation can be achieved, in small cheese factories, using finishing coolers, and some of these coolers have been described by Davis (1965), Crawford (1976) and Scott (1986). These coolers have been developed in the US by different companies, and some examples are:

(a) Damrow—finishing vat, (e.g. latest design is fully enclosed).
(b) Stoelting—Stoelting table
(c) Kusel—Cheddar rite vat

The finishing coolers facilitate the following stages of cheesemaking, thus releasing the cheese vat for a second re-fill:

- the curd–whey mixture is pumped to the cooler;
- de-wheying can take place through the perforated area at the bottom of the cooler;
- cheddaring and other texture-forming treatments of the curd are carried out along traditional lines;
- a mill is fitted on the cooler at the milling stage, and the blocks of curd are delivered manually into the mill; the curd chips fall back onto the cooler, and salt is applied manually.

The mechanised features of the finishing cooler are as follows:

(a) mixing tools (i.e. operated by overhead-mounted gear) during the de-wheying and mellowing stages;
(b) mechanical unloading devices;
(c) the salted curd is transferred to the mould-filling station by the use of a spiral/screw lift elevator and/or a pneumatic system.

Dutch Cheeses and Related Varieties

During the manufacture of these varieties of cheese, the curd is not 'textured' to any great extent, and the handling of the curd–whey mixture, after leaving the cheese vat, may involve the following stages: first, partial de-wheying and pre-pressing, and second, mould filling and final pressing. Different equipment is available on the market for the production of Gouda, Edam and similar varieties, and some examples are as follows.

Pre-pressing vat

These units are manufactured by different companies, e.g. the Tebel BV and Stork-Friesland in The Netherlands, APV-Pasilac AS and Gadan A/S in Denmark and Schwarte–Werk in Germany, (Hansen, 1985*a*,*b*, 1988b,*c*, 1989*a* Anon.1985*a*, 1989; Skovhauge, 1989; Ostergaard, 1990, 1991; Nielsen, 1990*b*). In principle, these pre-pressing vats are similar in their function, and, in general, they are square-ended. The bottom and the sides are made out of perforated stainless-steel, and the bottom section is designed as a wide-slat conveyor which is progressively moved forward at the end of the pre-pressing stage, so that the matted curd is cut into sections that fit the plastic moulds.

A modified version of this system is the Gadan tunnel pre-pressing vat (Hansen, 1980*a*), which consists of two parts, i.e. a stationary upper section which automatically operates the press, and the mobile section which is the vat unit. The latter unit can be divided into compartments (i.e. known as the bottomless mould) ready to be filled with curd, and the whole vat is moved on rails to the press section.

In some installations, a rotating, perforated drum is used (e.g. Roto-strainer–Tebel BV) on the inlet side of the pre-pressing vat or pre-pressing tower to separate the bulk of the whey from the curd, and the slightly drier curd is then distributed evenly before pressing commences.

'Block'-shaped Gouda cheese (e.g. 500 kg in weight) has been produced in the Republic of Ireland (Hansen, 1981*a*) using a pre-pressing vat system which is 3 m long and 1·12 m wide. After the brining stage, the large block of cheese is portioned into 10x50 kg Gouda sections, and this method of production is somewhat similar to the system reported earlier for the production of 450 kg 'block'-shaped, rindless Emmental cheese.

The newly designed, pre-pressing systems are completely enclosed during the pressing, cutting and cleaning stage except for an opening through which the sectioned cheese block leaves the machine. One example is the APV-Pasilac type OPD, where the drainage belt is made of polypropylene (i.e. lamella belt) which is easy to clean (Ostergaard, 1991). The capacities of such batch, pre-pressing systems range from 6000 to 18 000 litres, and they are equipped with specially designed curd distributors to ensure uniform levelling of the curd in the pre-pressing vats. The APV-Pasilac OCD curd distributor can be supplied as a combined unit for both wet curd distribution for the manufacture of Gouda and related cheeses, and dry curd distribution (e.g. Tilsit and related cheeses), or in a single version, i.e. wet or dry (see also Anon., 1985*b*).

Cheeses, such as Danbo, Gouda and Jarlsberg, can be produced in a continuous, pre-press type OPP as manufactured by APV-Pasiliac AS in Denmark. The curd–whey mixture is delivered to the rear end of the pre-press and distributed onto a perforated conveyor belt, i.e. the full width of the vat; the bulk of the whey is continuously discharged through the bottom valve. The curd is transferred by the belt under an adjustable trimming plate to ensure a constant level of curd. Subsequently, the curd passes under a pre-press belt which gradually increases the pressure. The height and pressing angle of the pressing belt can be adjusted in relation to the drainage belt, thus, controlling the pressing conditions in relation to curd height. The compressed curd forms a cheese block which is gradually raised above the whey level and drained. Finally, the pre-pressing curd is cut, using a two-dimensional guillotine, and turned before transfer to the mould filling, lidding and final pressing sections. This unit is completely enclosed and cleaned by CIP.

Another example of pre-pressing vat is the Tebel DBS Strainer (i.e. open top) which can be manufactured in different sizes to match the production programme and the capacity of the cheesemaking vat. The main part of the DBS strainer is a rectangular-shaped vat which is made from stainless-steel and equipped with an electrically operated draining floor consisting of a series of perforated and hinged plates of stainless-steel. Each plate is provided with slots for fixing the vertical end-plate; thus, making it possible to adjust the length and depth of the curd bed.

The curd–whey mixture is distributed over the draining floor of the DBS strainer in an even layer. Pressing of the curd is carried out by pneumatically operating the pressing plate which, initially, is 500 mm from the sliding door at the front end of the strainer. When the whey draining is completed, the floor is moved forward and the matted curd is cut into bars by knives in front of the strainer door. When the curd has moved forward to a pre-set length, the door moves downwards and acts as a quillotine. Then the pressed curd, which has been cut, moves forward to be placed into the moulds. The pressing plate is automatically lifted when the floor is moved forward, and after cutting the last pre-pressed curd, the knives are pneumatically lifted.

The perforated floor of the DBS strainer is pulled forward and pushed backward by a motor-driven system which is mounted under the vat. All the operations of the strainer can be manually or automatically controlled, including the cleaning.

Pre-pressing tower

The de-wheyed curd is compacted in a tower, and a sliding plate at the base of the column allows the curd to fall into the cheese mould. This system of curd handling was primarily developed for the manufacture of 'round-eye' or granular cheeses, such as Gouda and Edam cheeses, as was de-wheying, pre-pressing and mould-filling unit. However, the same equipment could be used for other varieties of cheese, e.g. Swiss and some Danish types. Furthermore, these machines are highly flexible for the production of 'round', 'rectangular' or 'square'-shaped Dutch cheeses, and are adaptable for continuous and automated production lines. Examples of some pre-pressing towers are Curd Moulding, Curd Dosing, Conomatic-R, Pre-Pressing Tower and Casiomatic (see Fig. 15). Descriptions

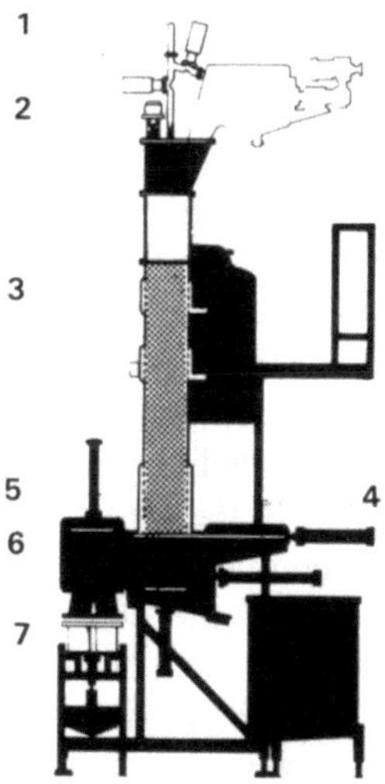

Fig. 15. Schematic illustration of the Casiomatic mark IV. (1) curd–whey inlet, (2) curd level indicator, (3) perforated whey discharge, (4) combined bottom/knife, (5) perforated dosing/pressing plate, (6) sliding holder; (7) cheese mould. Note—a buffer tank is situated between the cheese vat and the Casiomatic which has the following advantages: the cheese vat can be emptied rapidly; the curd–whey mixture is more homogeneous in the buffer tank, and it is cooled which results in less moisture loss from the curd; the elimination of air entrapment from the system; fluctuations in the flow of product are overcome. (Reproduced by courtesy of Tebel BV, Leeuwarden, The Netherlands.)

of such installations in cheese factories have been reported by Hansen (1977*a,b*, 1978*a*, 1979*a,b*, 1983*a*, 1984*b*, 1985*c,d*, 1986*b*, 1987*a,c*).

Curd Recovery Systems

Cheese whey is composed of the following milk components: fat, curd 'fines' (mainly fat and casein), whey proteins, lactose and minerals. The recovery of the fat and the curd 'fines' from the whey can, in large, centralised cheese factories, play a major role in improving the cheese yield and reducing financial losses. In principle, the recovery process is achieved in two stages: first, the removal of the curd 'fines', which are later mixed with the bulk of the cheese curd after the de-wheying stage; and secondly, the separation of the fat from the whey mechanically, with the cream being utilised for the manufacture of whey-butter. Some examples of the equipment which could be employed for the recovery of the curd 'fines' from the whey, are as follows:

Russell fines saver

This type of curd 'fines' separator is manufactured by Russell Finex Ltd in London (UK), and the overall specification of the machine is illustrated in Fig. 16(A). The filter element is made out of two polyester or nylon, mesh sleeves mounted on a stainless-steel basket, which also houses the paddle-blade impellers that connect to the drive shaft of the motor. The whole unit is enclosed within a stainless-steel body. As the curd 'fines' – whey mixture enters the mesh basket, the impellers pump the liquid through the filter, and the separated curd travels up inside the basket to the outlet. By adjusting the tilt of the whole unit, the moisture content of the curd discharge side can be manipulated, i.e. the greater the angle of tilt, the drier the consistency of the curd 'fines' and vice versa. The resultant slurry or curd 'fines' can then be pumped away and distributed over the cheese curd after the de-wheying stage, or alternatively a dry 'fines' can be collected into buckets and manually spread over the cheese curd at regular intervals.

The Russell Fines Saver has an adjustable speed motor with a range of 200 to 1070 rpm, but the efficiency of the curd 'fines' recovery is primarily dependent on the mesh size of the filter element. For example, a unit (40 000 litres h^{-1} throughput) fitted with a 40 μm mesh filter can recover approximately 22·7 kg of curd, but by using a finer mesh filter, e.g. 5 μm, the flow rate drops to 330 litres h^{-1}, and the weight of recovered curd will be approximately doubled (see Anon., 1987*d*).

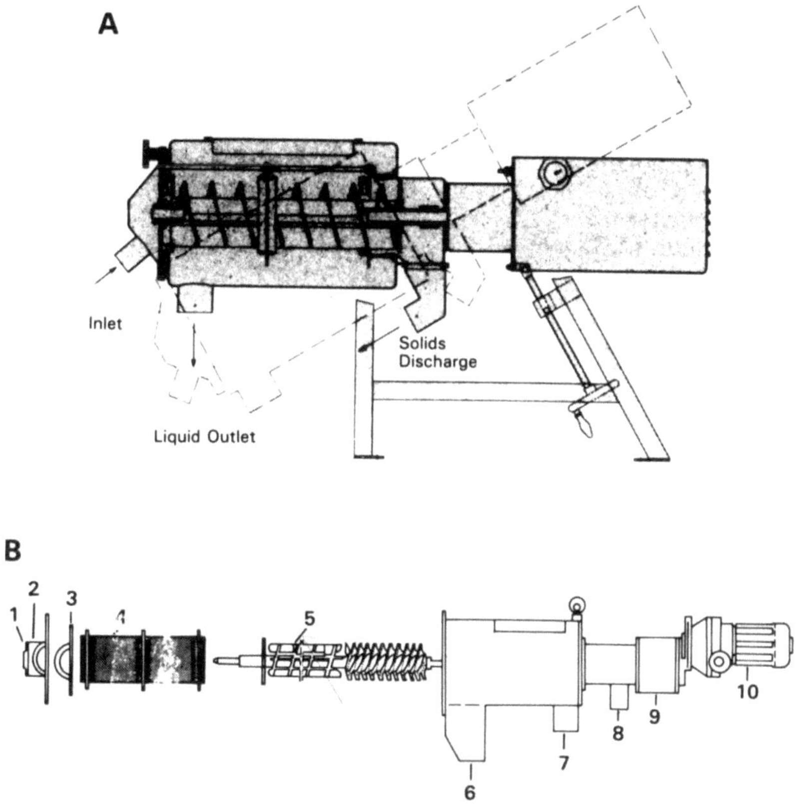

Fig. 16. Schematic illustrations of units for mechanically separating curd fines. (A) Russell Fines Saver. (Reproduced by courtesy of Russell Finex Ltd, London, UK). (B) AZO Fluid Sifter: (1) shaft bearing, (2) end cover, (3) cover for the sieve basket holder, (4) sieve basket, (5) rotor shaft fitted with helical screw, (6) curd 'fines' outlet, (7) whey outlet, (8) curd 'fines'–whey mixture inlet, (9) distance piece, (10) motor with variable drive. (Reproduced by courtesy of Adolf Zimmermann GmbH, Osterburken, Germany.)

AZO Fluid Sifter

This sifter is manufactured by Adolf Zimmermann GmbH, Osterburken in Germany, and an illustration of the unit is given in Fig. 16(B). The rotor is enclosed within a cage on which is mounted a woven, nylon screen (20–40 μm in size), and the whole unit is enclosed in a stainless-steel sheet. The geared motor has a variable speed ranging from 155 to 775 rpm, which helps to create a centrifugal force inside the separation chamber. This force helps to send the liquid through the screen, and the 'fines' are carried by a helical screw to the discharge port (see No. 6 in Fig. 16(B)). The consistency of the discharged solids is controlled by rate of throughput, size of sieve, speed of rotation and slope of the machine. Some examples of the efficiency of solids recovery by the AZO Fluid Sifter are as follows:

Size of sieve (μm)	*Product throughput* (*litres* h^{-1})	*Fines recovery* (%)
20	17 000	84
30	22 000	72
40	28 000	64

However, if a sieve size of 30 μm is used with a product throughout of 22 000 litres h^{-1}, by changing the rotor speed (expressed as percentage of maximum speed) from 40, 50, 60 to 70%, the efficiency of 'fines' recovery would be 78, 71, 70 and 63%, respectively.

Damrow cheese saver II

This cheese curd 'fines' recovery unit is manufactured by the Damrow Company in the US. Recovery is achieved by filtration of the whey, as compared with the other systems which rely on mechanical separation. A schematic drawing of the Cheese Saver II is given in Fig. 17(A). As can be seen, this unit consists of a sloped, nylon screen of 100 μm size enclosed within a stainless-steel frame. The filtration capacity of the Cheese Saver II ranges between 20 and 60 m^3 of whey h^{-1} depending on the model. The recovered curd slurry is collected into a stainess-steel compartment at the bottom of the unit, and then returned to the cheese vat, or distributed over the de-wheyed curd using a positive pump. Rasmussen (pers. comm.) reported the following recovery efficiency of curd solids from whey after a 2 month trial period using whey from a Cheddarmaster drainage belt:

amount of 'fines' in whey	80 mg 100 g whey^{-1}
amount of 'fines' in filtered whey	26 mg 100 g whey^{-1}
amount of recovered 'fines'	54 mg 100 g whey^{-1}

Therefore, a cheese plant having a throughput of 100 000 litres of whey per day will save an amount of dry 'fines' equal to 54 kg (i.e. $(540 \times 100\,000)/(1000 \times 1000)$) this is equivalent to 74 kg ($54 \times 1{\cdot}37$) of

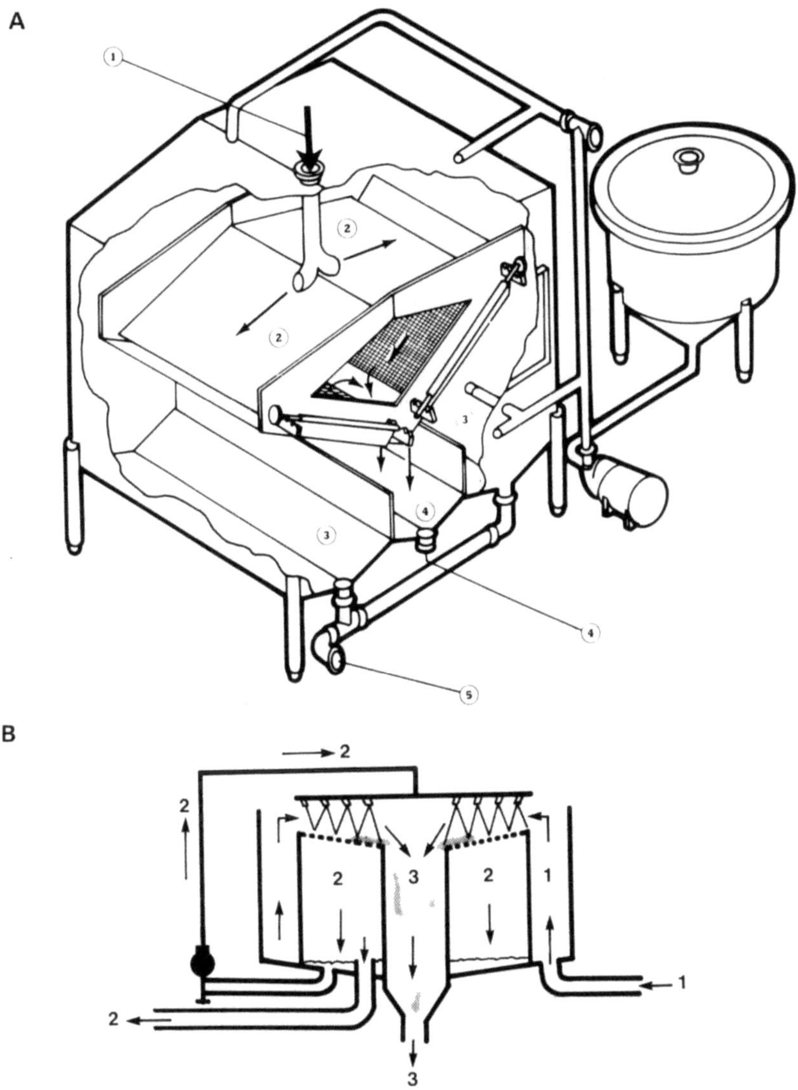

Fig. 17. Schematic illustrations of curd 'fines' filtration units. (A) (1) Curd 'fines'–whey mixture inlet from whey tanks to manifold for distribution, (2) manifold distributes cheese whey over the separation media, (3) separated whey is pumped to storage tank, (4) curd 'fines' slurry is returned to cheese vat or drained curd with a positive pump, and (5) separated whey outlet. (Reproduced by courtesy of Gadan A/S, Them, (Denmark). (B) (1) In-coming unfiltered whey, (2) filtered whey, and (3) curd 'fines' slurry. (Adapted from Lee (1981).)

cheese containing 37% moisture. Taking into consideration the cost of the unit and the price of Cheddar cheese in 1980, the capital outlay of Cheese Save II was realised in 120 days.

AES 'fines' recovery system

Another example of a whey filtration unit, which could be employed for the recovery of curd 'fines' from whey, is the AES system manufactured by Albaney International in the US. The unit is fabricated from stainless-steel and is totally enclosed (Fig. 17(B)). The filtration screen is made out of food-grade polyester available in 100–325 μm mesh size, but it is recommended that the 200 μm mesh be used for curd 'fines' recovery from whey of most hard cheese types. The application of the AES system has been reported by Lee (1981), and, in brief, the filtration sequence is as follows:

- the cheese whey is pumped into the outer plenum (1), and surges over through the stationary mesh screen;
- the curd 'fines' are moved from the nylon screen to the curd slurry section (3) by a continuously rotating shower of filtered whey from section (2); and
- the curd slurry is pumped at a rate of 7·6–11·4 litres min^{-1} to the finishing tables of the cheddaring unit(s).

Milling, Salting/Brining and/or Mould Filling

The methods available to handle the curd a this stage of production depend on the type of cheese produced, but, in general, the handling of the curd can be divided as indicated in Table XIX.

TABLE XIX

Sequence of operation	*Cheese variety*
Mould filling/pressing/brining[a]	Parmesan/Grana Gouda/Edam Emmental/Gruyère
Milling[b]/salting/mould filling/pressing	Cheddar and related British territorials

[a]The Swiss and Italian varieties are rubbed with salt on the surface each day after turning in brine; however, the 'block' shaped Emmental and/or Gruyère are only brined.
[b]The salting of Cheddar is carried out prior to milling when employing the Tebel–Crockatt equipment, or during the manufacture of American Cheddar where the curd is not matted together, and hence milling is not required.

It is evident, however, that the above division is also applicable to the method of salting the cheese curd, i.e. salting for Cheddar and related varieties and brining for Swiss, Italian and Dutch cheeses. In some of the latter cheese varieties (Italian and Swiss), the surface of the pressed curd is rubbed with dry salt in order to produce a harder rind to protect the cheese during the maturation period.

Milling equipment

Most of the mechanised equipment employed for the manufacture of Cheddar cheese includes automated milling and/or salting and mellowing facilities. The Alf-O-Matic, the Damrow DMC and the Wincanton DCC units are fitted with a curd mill at the end of the cheddaring conveyor (i.e. belt numbers 3, 2 and 3, respectively—see Figs 13 and 14). However, the mill in the Scherping CCDM unit is located outside because it may not be required for the manufacture of cheeses from granular curd. The matted curd is cut into curd chips of roughly 1·25–1·85 cm square section and 17·5–20·0 cm in length (Brockwell, 1981). However, the length of the curd chip is influenced by such factors as:

— depth of the matted curd;
— space between the rotating blades;
— speed of rotating cutters;
— and/or design of the mill.

An alternative type of mill, which could be employed during the manufacture of most British varieties, is the peg type; however, a different type of peg mill is used for milling Cheshire cheese curd to give 'granular' pieces which help the finished cheese to attain the crumbly, texture characteristics.

Incidentally, the 'block'-shaped Cheshire cheese is somewhat drier, less acidic and tends to have a closer texture *vis-à-vis* the traditional cheese and, therefore, is more suitable for cutting and pre-packaging into retail portions.

Salting equipment

The salting should be completed in an atmosphere isolated from any humidity in order to avoid the ingress of moisture to the metering device, and thus changes in the accuracy of salting. The salting equipment in the Alf-O-Matic operates as follows:

— warm, dry air transfers the salt from storage to a salt metering belt;
— hot air passing through the valve transfers the salt to a stainless-steel, oscillating distribution tube to be spread onto the cheese curd; the oscillating pipe is driven by an electric, geared motor;

- the amount of salt added is controlled by a floating sensor that determines the depth of the milled curd, and actuates the salt metering device either to increase or decrease the salt dosage in line with the height of the milled curd; and
- salt mixing and mellowing take place on conveyor number 4 (see Fig. 13), where the curd chips are treated in a series of 'resting' and 'stirring' periods. The residence time ranges between 10 and 40 min.

The salting and mellowing systems developed by Damrow Co., Scherping Systems and Wincanton have, by contrast, been established as separate units from the de-wheying/curd handling equipment. Two different systems are available from Damrow Company, i.e. the Salt Retention Unit (SRU) and the Salting and Finishing Vat (SFV) (Fig. 18). In brief, the sequence of operations of the SRU is as follows:

- the curd chips are air conveyed from the DMC or other cheesemaking equipment to a cyclone situated on top of the SRU (1);
- the curd chips fall to the bottom of the cyclone (2) and are delivered onto a weighing belt (3); the air is vented at the top of the cyclone;
- The curd chips enter the mixing drum (4), where the exact amount of salt is added and mixed for a short period; (5) the curd exits onto the mellowing top belt; the principle of using a mixing drum is similar to the Bell-Siro 'Cheesemaker' 3 system (Crawford, 1976);
- mellowing of the salted curd takes place on two belts which are similar in operation to the DMC; and
- at the end of the second belt, the mellowed curd leaves the SRU through a 'star' valve (6), and is air conveyed to a mould or hoop filling station.

The SFV unit could be used for salting untextured curd, e.g. American Cheddar, or conventional cheddared and milled curd. The curd chips are conveyed from the DMC or other cheddaring/milling units to the SFV which is fitted with load cells to accurately weigh the curd. The salt is air-conveyed and added at a rate predeterminded by the cheesemaker. The mellowing of the salted curd is achieved by continuous or intermittent stirring. The salted curd is emptied by reversing the stirrers, and discharges through the bottom, via a 'star' valve and air-conveying system, to the mould-filling station.

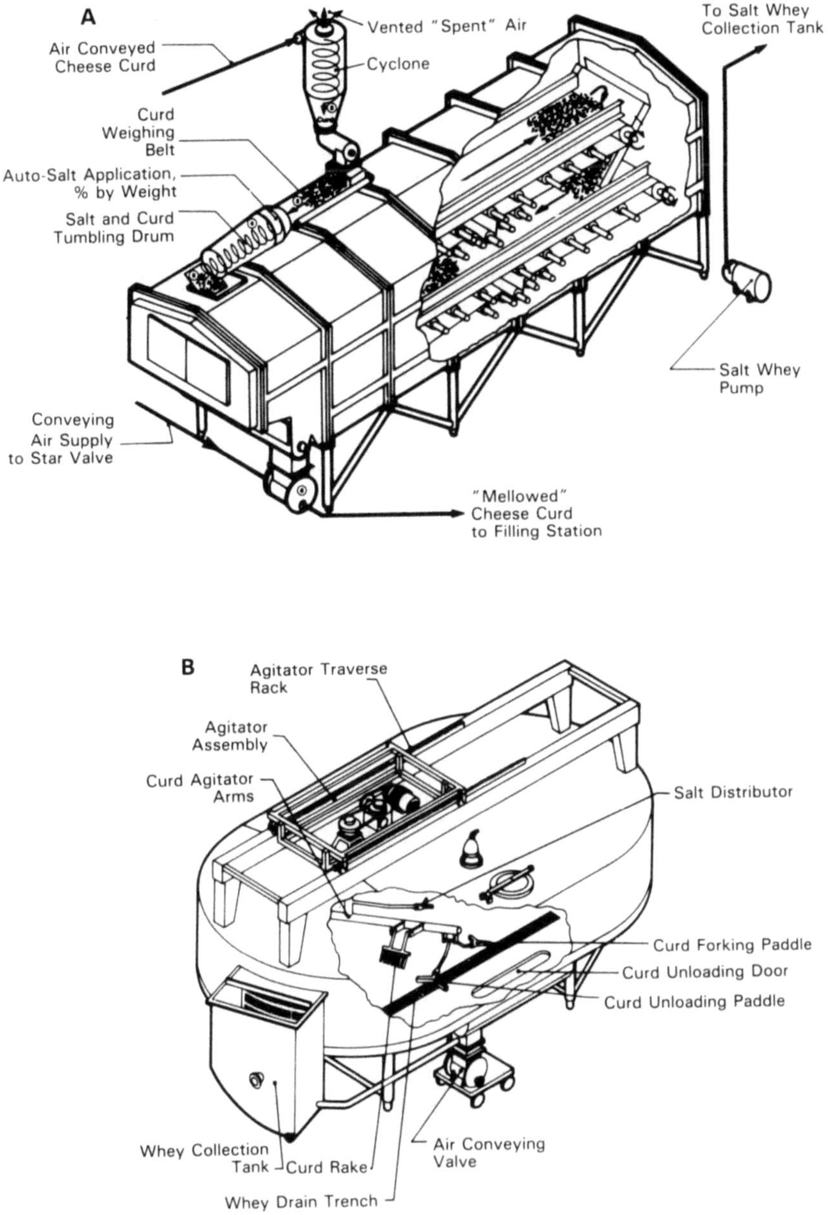

Fig. 18. Damrow enclosed salting equipment. (A) SRU salting unit. (B) SFV salting unit. (Reproduced by courtesy of Gadan A/S, Them, Denmark.)

Both units (i.e. SRU and SFV) are designed for CIP, and the salting stage could be made continuous by using an SRU of the same capacity of the DMC unit, or by installing two or more SFV units (Park, 1979).

The Scherping two-stage, continuous system for the salting and mellowing of milled or graular curd is shown in Fig. 19. The unit consists of two belts, two salt dispensers and peg stirrers. The curd is delivered to the salting chamber on an inclined, self-tensioned and perforated belt which is made of poypropylene. A levelling screw on the first belt disperses the curd depth evenly across the width of the belt, and some syneresis also occurs due to the action of the levelling screw. The first curd sensor measures the depth of the curd, and a proportion of salt is applied through a special spreader, followed by mixing. Subsequently, the second curd sensor actuates the distribution of the remaining amount of salt required in the final cheese. Heated air is used to blow the salt from the hopper to the salt dispensers, where oscillating booms spread the salt onto the curd across the salting belt.

Curd mellowing takes place on the second belt which is regulated by adjusting the speed of the peg stirrers, or by whether they are in the on/off position. The curd residence time during the mellowing period ranges from 10 to 45 min and the unit is designed for CIP.

The Wincanton salting unit consists of a plastic belt housed in a stainless-steel body similar to the Wincanton DCC. The milled curd is transferred to the salting conveyor via a pneumatic conveyed system and, immediately after the inlet of the curd, it is agitated in order to maintain (i) separation of the milled curd or 'chips', and (ii) uniform distribution of the curd across the width of the salting belt. A stainless-steel fork and linear potentiometer measure the height of the curd, and are situated near the curd inlet, i.e. after the first agitator. The output is then transmitted to the salt metering system which comprises of the following:

— a small salt hopper with high/low level sensors;
— a volumetric feeder with variable speed control;
— a rotary valve; and
— a positive displacement blower and appropriate pipework to transfer the salt to the salting conveyor.

In operation, the signal from the curd height sensor is passed to the metering system which is calibrated in order to provide a quantity of salt which will vary according to the height of the milled curd. Consequently, the salt is distributed evenly over the milled curd bed using a swinging arm with an action similar to that of a cheesemaker.

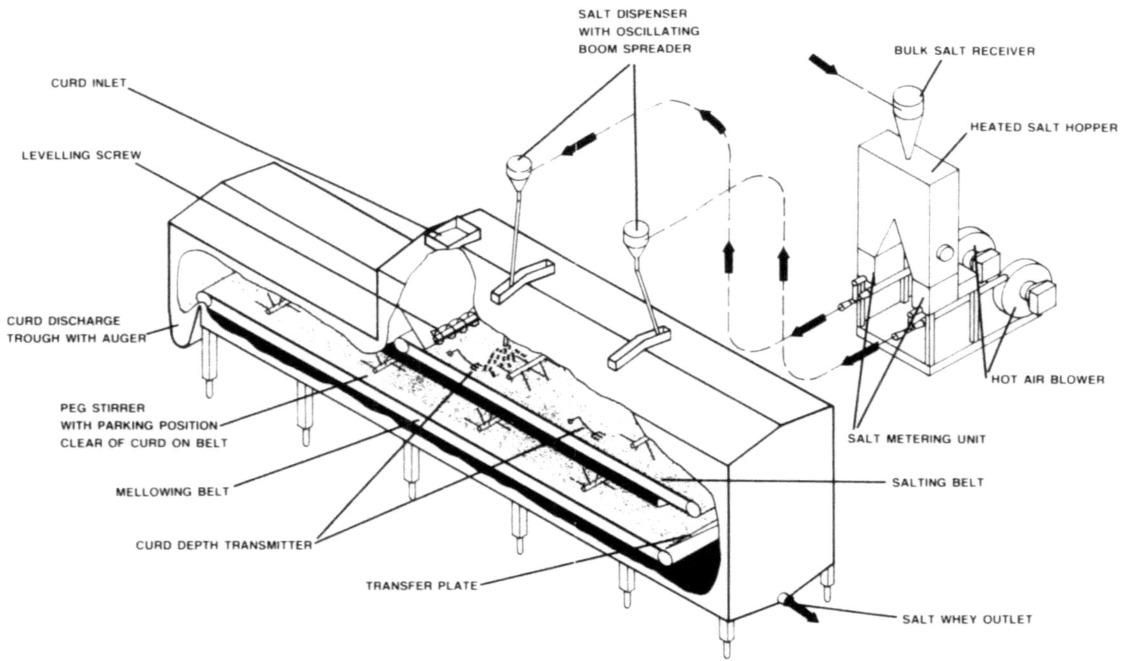

Fig. 19. The Scherping two-stage, continuous salting and mellowing of milled or granular curd. (Reproduced by courtesy of Scherping Systems, Minnesota, USA.)

At further intervals along the salting conveyor, a number of agitators are located in order to achieve proper mixing of the salt and curd. The residence time of the curd is adjustable, and the salting conveyor is cleaned (CIP) in a similar way to the Wincanton DCC equipment, including the provisions mentioned earlier. At the end of the salting conveyor, there is an agitator/auger assembly and a rotary valve to remove the salted curd to the moulding stage, e.g. Wincanton cheese tower or 'Press-n-Fill' systems (see Figs 23 and 29) (D. Shaw, pers. comm.).

Hand-salting of the milled curd is still practised in small cheese factories using coolers, or in mechanised systems employing the Tebel-Crockatt equipment. The 'lawnmower' device in the Tebel strainer performs the cheddaring process, and partly mills the curd. The salt is added manually, and by passing the 'lawnmower' cutters a few times over the table, the curd is cut, turned, mellowed, and finally transferred to a peg mill (Crawford, 1976). Most of the finishing coolers are equipped with stirring devices to mix the salted curd chips, and a specially operated blade to push a portion of the curd to a conveying device leading to the mould-filling station.

In contrast to this method of salting, i.e. of the curd chips, the immersion of pressed cheese in brine is often used, and examples of such varieties are the Italian, Swiss, Dutch and Danish cheeses. In general, there are three different systems which are employed to brine cheeses: first, surface brining, where the cheese floats in brine tanks; secondly, deep brining where the cheese is stacked on racks and lowered into brine tanks, i.e. total immersion in the salt solution; and thirdly, spraying the brine directly onto the cheese (Anon., 1986).

Most brine tanks are made of plastic, in order to minimise the corrosive effects of the salt. In some instances the stainless-steel racks filled with cheese are supported with nylon netting before immersion into the brine (Hansen, 1986*b*). The construction of brining tanks has to take into account the following aspects:

(a) salt dosing equipment, i.e. dry form or in solution, so that the desired concentration of the brine can be maintained;
(b) temperature control mechanism, so that the cheese can be brined at two different temperatures, e.g. at 18–20°C for the first few hours, followed by 14°C during the brining of Danbo cheese (Danish variety) in a Gadan/Brine-O-Matic unit (Hansen, 1979*c*,*d*);

(c) provision of some eccentric movements of the racking unit in the brine tank, in order to overcome the problem of cheese buoyancy (i.e. top side) resting on the same bars during the brining stage;
(d) in large cheese factories, the brining and handling of the cheese is highly automated (Anon., 1985*b*; Bojgaard, 1988; Bogh-Sorensen, 1989), and some examples are illustrated in Figs 20 and 21.

The quality control of the brine is critical because it can influence the quality of the cheese and, according to Tschager (1988), some aspects include the following:

— maintaining the right concentration of salt;
— monitoring for microbiological contamination, e.g. total count, coliforms, yeasts and moulds;
— maintaining the appropriate temperature;
— monitoring the whey content, which leads to the formation of sludge and hence a shift in pH;
— maintaining the calcium content in the region of 0·17–0·24% for semi-hard cheese, which corresponds to the calcium percentage in the cheese; lower calcium concentrations in the brine can affect the firmness of the rind.

The amount of salt retained in the cheese is governed by a multitude of factors, such as the following:

— level or percentage of added salt;
— method of salting (i.e. addition of salt to the milled curd or brining the pressed cheese);
— duration of salting;
— moisture content of the cheese; and
— type of cheese, etc.

According to Scott (1986), the salt content in the cheese may range between 1·5 and 2·5%, but in some instances, the desired salt level may reach 5·0–7·0%, as in Pecorino cheese (an Italian variety), or as low as 0·6% salt in Emmental. Thus, the amount of added salt and/or the concentration of the brine is dependent on the variety of cheese, and Scott

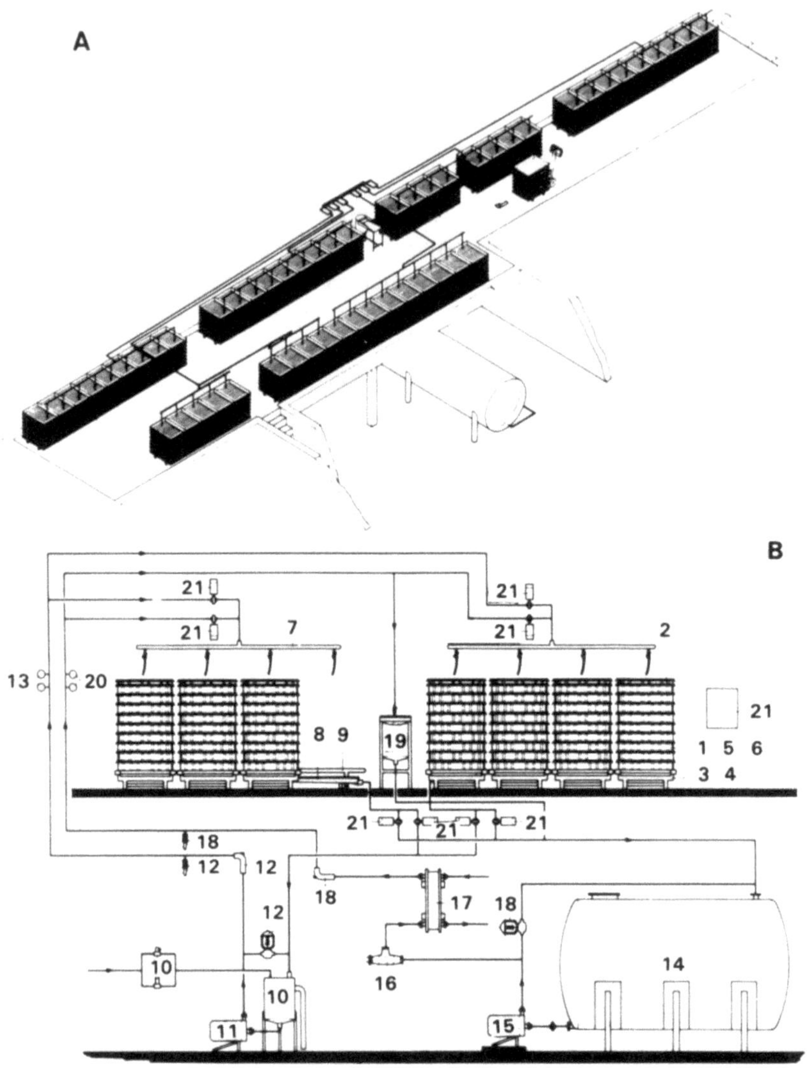

Fig. 20. Gadan cooling/Brining system type KS-600. (A) An overview of the brining system constructed at two floor levels. (**B**) (1) Cooling/brining trays, (2) top trays, (3) bottom trays, (4) bottom frames, (5) perforated plates, (6) cheese nets, (7) distribution pipes, (8) gutters, (9) buffer rods, (10) balance tank with valves and level control, (11) centrifugal pump for water, (12) temperature and pressurising equipment for water, (13) indicating instruments for temperature and pressure on water, (14) brine tank, (15) centrifugal pump for brine, (16) brine filter, (17) plate heat exchanger, (18) temperature and pressurising equipment for brine, (19) salt dosimeter, (20) indicating instruments for temperature and pressure of the brine, and (21) electron controller with flow-forwarding and return valves. (Reproduced by courtesy of Gadan A/S, Them, Denmark.)

Fig. 21. Salting basins for the brining of Gouda cheese in cages at DMV–Campina factory in Born. (Notice the hoist which transfers the cages stacked with Gouda cheeses ready for immersion in brine). (Reproduced by courtesy of Tebel BV, Leeuwarden, The Netherlands.)

(1986) has reported some recommended guidelines. For example, the percentage of added salt in some of the British varieties is given in Table XX.

Caerphilly cheese is the only British variety where the curd is salted at a rate of 1·0%. The cheese is dry salted while it is being turned in the press, and finally immersed in brine (i.e. 18·0% concentration) at 15·6°C for 12–24 h.

TABLE XX

Salting (%)	*Cheese variety*
1·6–1·8	Wensleydale
1·7–1·9	Single and Double Gloucester
1·8–2·0	Cheshire, Cotswold, Derby, Dunlop, Kingston and Leicester
2·5–3·0[a]	Cheddar

[a]According to the experience of the author, 3% salt could be added to the curd chips when using direct-to-vat starter cuture in order to halt acid development during the early stages of maturation, or in instances where the rate of acid increase is fast during the cheddaring stage. Incidentally, the salt content in such cheeses at one day ranged between 1·7 and 1·8%.

Cheese brine is a sodium chloride (NaCl) solution, and its concentration is expressed either as percentage NaCl (w/w) (e.g. saturated brine is 35% NaCl) or °Baumé. The brining conditions of the Swiss, Italian and Dutch cheeses may vary slightly, and some examples are shown in Table XXI.

In practice, the brine should be monitored to maintain the following specifications:

(a) adjust the pH level to 5·2 by the addition of hydrochloric acid;
(b) the microbiological content of the brine should be controlled by the addition of saltpetre (KNO_3) or hypochlorite (e.g. 0·5 litres 1000 litres^{-1} of brine), or by heat treatment of the brine (i.e. by boiling);
(c) the appearance of a reddish discolouration on the surface of the cheese may be attributed to a high KNO_3 content in the brine, and/or to certain *Lactobacillus* spp. which have contaminated the cheese after the brining stage.

Types of moulds and mould-filling equipment

The traditional, multi-piece iron or tinned steel moulds have been replaced by perforated stainless-steel or aluminium alloy, and recently, plastic moulds have become very popular.

In the past, cotton cheese-cloth was normally used to line some types of cheese mould in order to provide a smoother finish to the pressed

TABLE XXI
Some examples of brining specifications of certain cheeses

Cheese variety	*NaCl* (%)	*Temperaure* (°C)	*Duration* (*days*)	*References*
Parmesan/Grana	24	7–10	14–15	Kosikowski (1982)
Emmental	23	10	2–3	Scott (1986)
Gruyère	23	10–13	2–6	Scott (1986)
Edam	22–25	16	2–3	Scott (1986)
	22–25	12–14	3–4	Scott (1986)
	18[a]	13	3	Dijkstra (pers. comm.)
Gouda	20	15	3	Scott (1986)
	18[a]	13	4–5	Dijkstra (pers. comm.)
Reading Yellow	20	15	0·5–1	Scott (1986)

[a]°Baumé (°Bé), e.g. 18, 19 and 20% NaCl is equal to 16·9, 17·8, 18·7°Bé, respectively.

cheese, and to prevent the cheese from sticking to the mould. Thus, the preparation of large numbers of moulds in a cheese factory was a labour-intensive process, and washing, sterilising and drying the mould liners also increased the cost of production. These problems have been overcome by the introduction of perforated, stainless-steel moulds (e.g. Gadan A/S in Denmark) which has made it feasible to dispense with the cheese-cloths, or alternatively make use of disposable, polyethylene liners. The specifications of one such a cheese mould liner manufactured by Smith & Nephew Plastics Ltd (UK) are as follows:

— tear strength 70–110 g;
— burst strength 127·53–156·96 kN m^{-2};
— porosity 40–70 s; and
— gauge 100–150 μm.

Plastic moulds, which are manufactured by Crellin BV (Kadova) and Arend BV (Lauda) in The Netherlands, are widely used for semi-hard cheese varieties (Fig. 22). These moulds are made out of polyethylene, do not require a cheese liner, and are suitable for moulding, washing and refilling. Basically, the moulds are constructed of two pieces, and they are produced in a multitude of shapes. However, the block-type cheese mould, which is suitable for Cheddar cheese, consists of three pieces constructed within a stainless-steel frame to withstand higher pressures. Some of the technical specifications of the Kadova and Lauda cheese moulds have been reported by Anon. (1976, 1980), Hansen (1978*b*, 1981*b*, 1985*e*,*f*, 1986*c*, 1987*d*, 1989*b*) and Dijkhuizen (1981), and some of these plastic cheese moulds can be supplied as multi-cheese moulds.

Mould-filling equipment is primarily designed to deliver a certain amount of curd to a mould, and hence to produce cheeses, after pressing, of roughly the same weight. Such equipment could be incorporated either as part of the curd-handling system, or in a position dependent on the variety of cheese produced. The following are examples:

Emmental—traditional 'wheel'-shaped cheese

The entire curd content from each vat is delivered to a mould after de-wheying, and the cheesemaker has to alter the volume of the milk used in order to overcome the effects of seasonal variation on yield; thus, producing cheeses of roughly the same weight.

(A) (B)

(C) (D)

(E)

Fig. 22. Laude cheese moulds are made of plastic. (A) Cylindrical mould without net enclosure; (B) Gouda mould without net enclosure; (C) block moulds without net enclosure; (D) Cheddar mould without net enclosure; and (E) 'ball'-shaped Edam with net enclosure. (Reproduced by courtesy of Arend BV, AB Ter Apel, The Netherlands.)

Cheddar and related varieties

The sequence of filling the salted, curd chips into moulds is as follows:

(a) tare-off weight of mould;
(b) first stage filling—up to 70% or more of the final weight of the cheese is delivered into the mould;
(c) pre-press the curd;
(d) second stage filling of the mould to the desired weight, e.g. 20 kg; and
(e) transfer cheese moulds to the press.

Two-stage filling ensures that the pressed cheeses are uniform in weight, and Hansen (1981*c*) describes a filling sequence of Cheddar cheese chips into Lauda moulds in a large factory at Waitoa in New Zealand.

A different approach to the filling of Cheddar cheese moulds is the 'Press-n-Fill- Mk V (see Fig. 23), which is a volumetric hoop filler

Fig. 23. The Wincanton Mark V 'Press-n-Fill' Unit. (Note that the cheese moulds in-feed can either be from the left or the right-hand side of this unit). (Reproduced by courtesy of Wincanton Engineering Ltd, Dorset, UK.)

manufactured by Wincanton Engineering Ltd in Dorset (UK). The salted curd is pre-pressed into a small, rectangular chamber of similar dimensions to a 20 kg 'block'-shaped hoop. While the discharge door is in the closed position, a horizontal guillotine cuts the curd, and an ejector-ram pushes out the block (i.e. after the discharge door has opened), and gives the curd a second pre-press in the hoop. During the mould-filling stroke, the mould liners are held in position by a clamp frame which also prevents spillage of the curd. The rate of filling is 3 × 20 kg moulds min^{-1}, and the mould filling can be arranged either from the right-hand or left-hand side.

Semi-hard cheese varieties

The de-wheying and curd handling equipment, e.g. pre-pressing vat or tower, are designed as mould fillers, and the accuracy of filling is very good (see Fig. 24).

The cheese moulds (metal, i.e. aluminium alloy, or plastic, e.g. Lauda and Kadova) can be easily cleaned in a tunnel washing system. A special detergent should be used when cleaning the metal moulds to avoid black discolouration, and the polyethylene type should be cleaned with an alkaline and/or acid detergent at 50–70°C. However, it is recommended that an acid treatment should follow the alkaline wash in order to neutralise the latter compound.

Pressing, Packaging and Storing

Pressing equipment

The very hard, hard and semi-hard varieties of cheeses are pressed in either vertical or horizontal presses under normal atmospheric conditions or under vacuum. The pressing parameters, i.e. application of the right amount of pressure for a specific duration of time, gives the cheese its final shape, produces a cheese with a firm and smooth surface, and helps to lower its moisture content to the desired level. The pressure applied is dependent on the type of cheese, and it can be achieved using one of the following:

- spring-loaded or screw mechanisms;
- hydraulic presses;
- pneumatic systems; or
- vacuum presses.

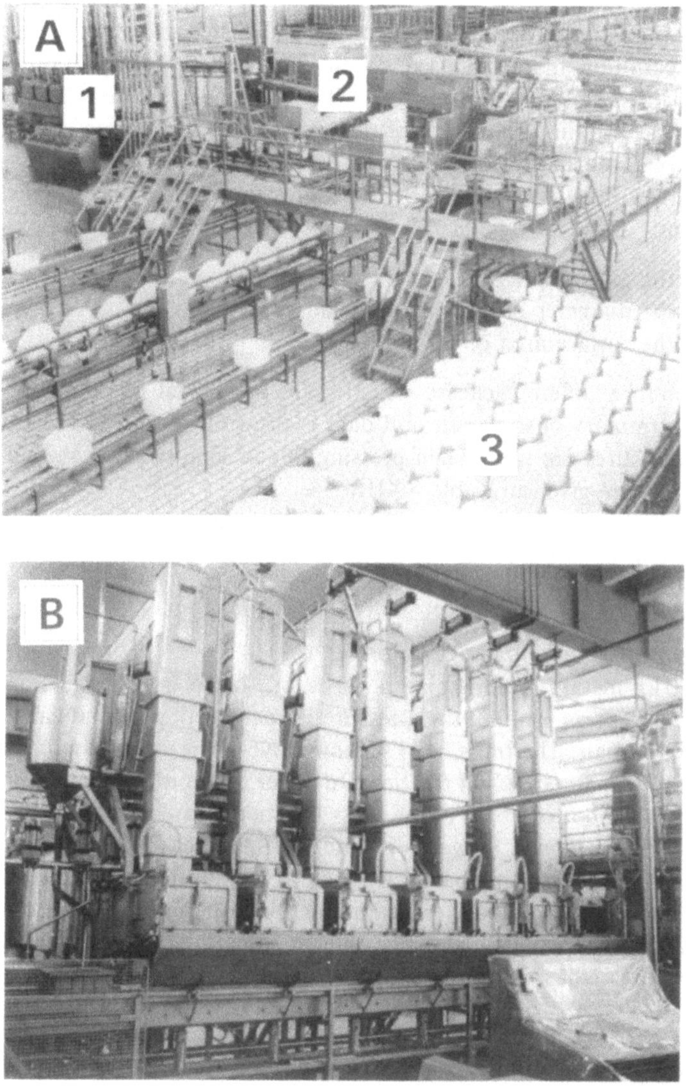

Fig. 24. (A) A Gouda cheese factory DMV-Campina in Born in the Province of Limburg. (1) Six Casiomatics mark III capable of delivering 264 pre-pressed and moulded 12·5 kg Gouda cheeses h^{-1}; (2) mould cleaning equipment; (3) Kadova cheese moulds ready to be conveyed to tunnel presses. (**B**) An illustration of the new generation of Casiomatic mark IV which has been recently delivered to DMV-Melkunie at Rijkevoort. (Reproduced by courtesy of Tebel BV, Leeuwarden, The Netherlands.)

The former type is manually operated, and hence the other systems are more common in mechanised cheese factories. One important aspect, which must not be overlooked, is the difference between the 'air-line' pressure and the 'actual' pressure applied to the cheese; the former pressure is normally lower. Since the pressure is defined as mass per unit area, and the calculated 'actual' pressure on the cheese takes into account:

— the air-line pressure;
— the diameter of the cylinder head; and
— the surface area of the cheese.

Therefore, different cheese varieties are pressed at different pressures, and a summary of some relevant data is shown in Table XXII. However, high pressures are used when pressing 'block'-shaped cheeses, and some examples are given in Table XXIII.

In general, the cheeses are pressed in individual moulds (i.e. in single or multi-row) or in bulk, and examples of some cheese-pressing systems are as follows:

Horizontal 'creeping' or gang press

The 20 kg metal mould is placed on its side between two guiderails, and when the press is full, the pressure is achieved pneumatically using compressed air. The press head section moves forward gradually, and the pressure is maintained all the time. In order to avoid too long a pressing cylinder, the bottom base of the press has metal notches welded to it, and as the press moves forward, it interlocks at certain intervals, which prevents it from moving backwards. This type of press is not mechanised, and the cheese moulds are placed manually in the press. Corrosion of the metal frame is a major problem and, in some factories in the UK, two presses are built on top of each other in order to reduce the floor area required.

Vertical press

This type of press comes in different designs, and the simplest, manually operated type consists of up to four vertical units. Each unit is divided into four sections which can hold, for example, 2 × 20 kg block cheese moulds. The pressing cylinder is mounted on the top, and the base of each section is guided by two vertical guide rails.

The development of a vertical press, which is around 10 m high and suitable for 20 kg block moulds, has been reported by Crawford (1976).

The press is filled and emptied using an automatic stacking and destacking mechanism, and it can hold around 42 moulds. The press is enclosed with plastic sheeting, and is suitable for CIP operation. A similar press (Towerpress) is manufactured by Tebel BV in The Netherlands, and a combined vacuum and tower press suitable for Cheddar cheese is shown in Fig. 25.

A different type of vertical press, i.e. the 'Pallet' press has been reported by Scott (1986), and it has a pressing capacity of 14 t of cheese. The press is divided into two main sections, and each section holds 36 × 20 kg block moulds stacked in six layers. This press is highly mechanised, and can also be used under partial vacuum. Incidentally the 'Pallet' press is not manufactured anymore.

Table, trolley, conveyor, tube and/or tunnel presses

These types of presses have been mainly developed for pressing semi-hard cheese varieties, and the pressure is applied from above (see Fig. 26(A)). These presses consist of a number of rows which can be filled manually or mechanically, and pressing commences whenever a row is filled. For highly mechanised cheese plants, the latter two types of press are widely used (Scott, 1986). In some instances, cooling of the pressing area is recommended during the manufacture of Dutch cheeses and most equipment suppliers provide such facilities. Furthermore, some of these presses are automatically programmed to increase the pressures when required. The economics and efficiency of three of the above presses is illustrated in Fig. 27.

A recent development of the conveyor press is the 'air cushioning' press. The pressing cylinder does not press directly onto the cheese mould, but instead the pressure is applied on to a pressure box containing a tightly packed, pneumatic pressure tube. This tube adjusts any irregularities of the positioning of the lids, and all the cheeses are pressed under the same pressure. This development has facilitated the pressing of cheeses of different sizes at the same time (Hansen, 1984*c*).

Tube and/or tunnel pressing systems, which are completely enclosed, highly mechanised and are designed for CIP cleaning, are manufactured by Gadan A/S in Denmark (Fig. 26(B)). Both pressing systems are suitable for small and large cheese plants for the production of hard and semi-hard cheeses including Danbo. The curd from the pre-pressing vat is cut, portioned into smaller square blocks, and a special gripping device places the cheese in a clean mould. Subsequently, the filled moulds are

TABLE XXII
Pressures applied and treatment of some traditional cheese varieties during pressing[a]

Cheese variety	*Size (cm)* *Diameter × height*	*Weight (kg)*	*Pressing treatment*
Caerphilly	25·0 × 6·0	3·0–4·0	Press at 12 kN m^{-2} for 10–20 min; turn, rub with salt and re-press overnight at 49·8 kN m^{-2}.
Cheddar	30·5 × 35·5	18·0–22·5	First day press at 75 kN m^{-2} for 12–16 h; second day turn, pull cloth, scald (optional) and re-press at 200 kN m^{-2} for 24 h.
Cheshire	30·5 × 25·5	up to 27·2	First day press at 24·9 kN m^{-2} for 2 h, turn and re-press at 49·6 kN m^{-2} overnight; second day increase pressure to 99·6 kN m^{-2}.
Cotswold	Variable	0·45 – 1·0	First day press with 13 kg per 1 kg of cheese; second day turn into fine cloth and press at 24·9 kN m^{-2} for 24 h.
Derby	41·0 × 11·5		First day press at 24·9 kN m^{-2} for 2 h, turn and re-press at 49·8 kN m^{-2} increasing to 74·7 kN m^{-2} overnight; second day turn and press at 124·5–149·4 kN m^{-2} for 24 h.
Dunlop	35·0–40·0 × 20·0	13·6–15·8	Press lightly for 15 min increasing to 50 kN m^{-2} for 3 h; turn and re-press at 150 kN m^{-2} overnight.
Edam	30·0 × 13·5 25·0–11·0	4·0 – 4·5 2·0–2·5	Press cheese in *groups* at 980·0–1470·0 kN m^{-2} for 3 h; turn and re-press at 1470·0–2450·0 kN m^{-2} (number of moulds not given).
Emmental	70·0–100·0 × 13·0–25·0	100·0–110·0	Press at 7·0 kN m^{-2} for 5–15 min, and press overnight at 30·0–60·0 kN m^{-2}

Gloucester double	45·7 × 12·6	18·0–20·0	First day press gently increasing to 24·9 kN m^{-2} for 30 min, and turn and press at 49·8 kN m^{-2} for 16 h; second day turn and press at 74·7 kN m^{-2} for 24 h.
single	38·0 × 7·6	6·8 × 8·0	Press at 34·0 kN m^{-2} for one day increasing to 133·0 kN m^{-2} on the second and third day.
Gouda	24·0–25·0 × 6·5–12·0	up to 20·0	Press lightly increasing to 96·5–193·0 kN m^{-2} for 5–8 h.
Gruyère	40·0–64·0 × 8·0–13·0	35·0–40·0	Press and turn cheese for 2–3 days at 60·0–70·0 kN m^{-2}.
Kingston	—	0·45–2·2	Press at 15·0 kN m^{-2} for 4 h minimum.
Lancashire	—	18·0–22·5	First day apply no pressure; second day press at 24·9 kN m^{-2} increasing to 99·6 kN m^{-2}; third day scald at 60°C for 30 s and press at 124·6 kN m^{-2}; fourth day bandage and re-press at 49·8 kN m^{-2} for 12 h.
Leicester	45·0 × 11·0	18·0–22·5	First day press at 24·9 kN m^{-2} increasing to 49·8 kN m^{-2} in 2 h and turn and re-press at 74·7 kN m^{-2} overnight; second day turn and press at 99·6–149·4 kN m^{-2} for 24 h.
Parmesan	35·0–45·0 × 17·0–22·0	30·0	Press at 12 kN m^{-2} for 1 h, turn and re-press for 12–24 h at same pressure.
Wensleydale (White)	—	1·5–5·0	First day apply no pressure: second day press at 24·9 kN m^{-2} for 2 h, bandage and re-press at same pressure for 3 h.

Adapted from Scott (1986).

TABLE XXIII

Cheese	*Size* ($L \times W \times D$ *cm*)	*Weight* (*kg*)	*Pressure* ($kN\ m^{-2}$)
Emmental	66 × 40 × 20	38–45	2845–4315
Cheddar	36 × 28 × 18	18	300– 500
Cheshire	36 × 28 × 18	18	200– 300
Dunlop	36 × 28 × 18	18	150– 300

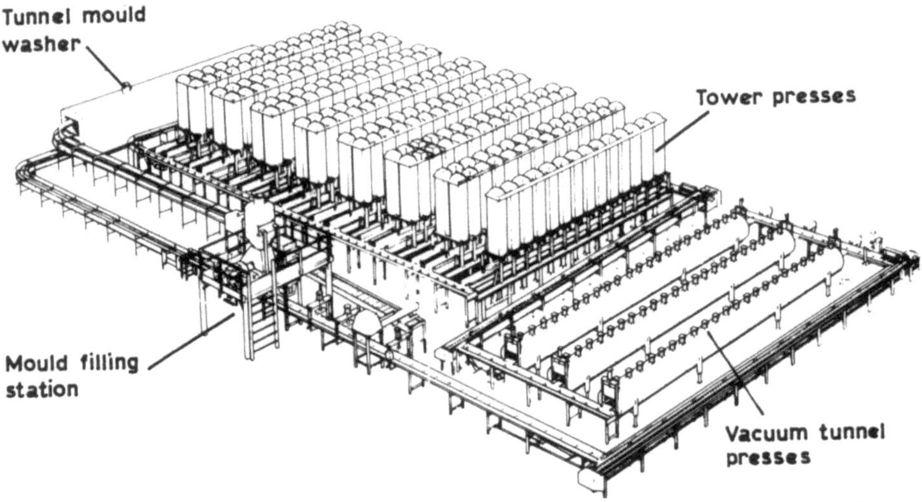

Fig. 25. A Cheddar cheese pressing system. (Reproduced by courtesy of Tebel BV, Leeuwarden, The Netherlands.)

lidded and distributed via a conveyor belt to each of the tube/tunnel presses. When each row of presses is filled, pressing commences from the top using compressed air (Anon. 1985*c*). The pressed cheese is emptied from the bottom end of the press and is transferred to the brining section; the empty moulds are cleaned and sanitized ready to be filled with pre-pressed curd (Hansen, 1986*d*, 1988*d*). According to the supplier, some features of the Gadan tubular cheese press system-type TP-2000 (Fig. 26(B)) are as follows.

— The press consists of a number of tubes and is completely enclosed to ensure a high standard of hygiene and minimum pollution of the environment.

(A)

(B)

Fig. 26. Presses used during the manufacture of semi-hard cheeses. (A) Four conveyor presses in cheese factory in Europe where a variety of Dutch cheese is pressed in Kadova plastic moulds. (Reproduced by courtesy of Tebel BV, Leeuwarden, The Netherlands.) (B) Illustration of tubular cheese press-type TP-2000. (Reproduced by courtesy of Gadan A/S, Them, Denmark.)

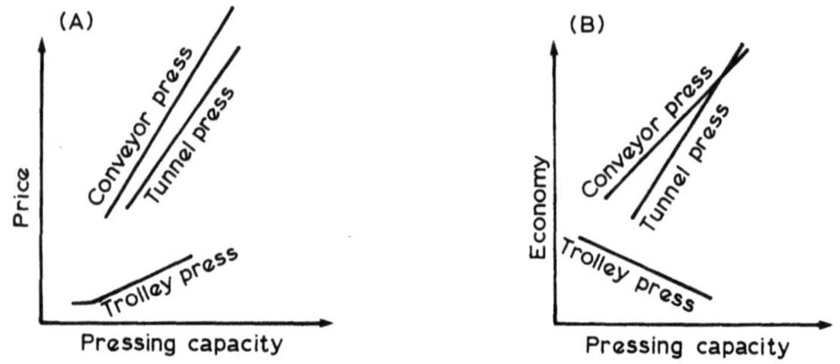

Fig. 27. Comparison of efficiencies and economics of different presses manufactured by Tebel. In (B) the following aspects are taken into account: depreciation, space, labour time, number of moulds and maintenance. (Reproduced by courtesy of Tebel BV, Leeuwarden, The Netherlands.)

- — All handling of the moulds and lids is controlled automatically by the computer of the pressing plant.
- — The press is capable of handling stainless-steel or plastic moulds, and depending on the size and shape, single or multi-moulds can be used.
- — The pressing tubes and the enclosed sections are equipped with a CIP system, and all CIP liquids and rinsing water are collected to be pumped away.
- — The expressed whey is collected in a buffer tank, and is later pumped to a storage tank.
- — The pressing system is also equipped with a mould-washing machine, where the moulds and lids are cleaned between each pressing cycle.
- — As an option, the Gadan press type TP-2000 can be supplied with a vacuum system enabling the cheese, e.g. Cheddar, to be pressed under vacuum, thus improving the texture of the cheese.

For further information regarding the application of conveyor and/or tunnel presses see Hansen (1985*c*, 1986*b*, 1987*a*, 1989*a*), Damerow (1988), Holmstrom (1991) and Larsen and Lynggaard (1991).

Rotary presses

Hermann Walder. This rotating cheese press consists of guided rails for receiving the moulds, and a clamp to hold each mould under the pressing cylinder. The whole unit is enclosed within a stainless-steel frame which is cylindrical in shape and 3 m in diameter (Hansen, 1979*e*).

Pre-pressing commences when each horizontal section of the press is filled with cheese moulds, and the press rotates one-twelfth of the circumference of the rotor before full pressure is applied. As a result, the cheese moulds rotate during the pressing period. The process can be made continuous by installing two presses in parallel, and furthermore, as the filled press can rotate, it is possible to empty the pressed moulds and re-fill with moulds to be pressed simultaneously.

Tebel – MKT tunnel press. This type of press (Fig. 28(A)) has been developed by Tebel-MKT BV, Jarvenpaa in Finland for pressing 'block'-shaped Emmental cheese (84 kg). The pre-pressed cheese is moulded and transferred by conveyor to the tunnel presses. Each press holds 12 × 84 kg cheese moulds, and inside the press, the lids are positioned automatically. Pressing commences as each press is filled, and the duration of pressing is 20 h. During the pressing period, the moulds are rotated slowly in order to ensure a uniform moisture content in the cheese. The rotary action of the press simulates the traditional method of turning the Emmental cheese several times a day during pressing.

APV cheese press. The cheese moulds are constructed on both sides of a supporting structure which is rotated by a motor up to 180° (see Fig. 28(B)), and this type of press has been developed by APV-Rosista AG in Switzerland. It is suitable for the production of Gruyère or Sbrinz cheeses.

The presses mentioned above are all designed for CIP cleaning.

Vacuum presses

The development of the vacuum press was primarily aimed at reducing the time required to press a cheese, but it also helps to cool the cheese during the pressing operation. Examples of such presses are as follows:

Tebel tunnel press. This press is shown in Fig. 25 for pressing Cheddar cheese, and the first stage of pressing takes place in a tunnel under vacuum, and the second stage of pressing is carried out in a Tower Press. In certain instances, the cheese can be packaged directly after the vacuum pressing stage, but alternatively, the cheese can be retained in another pressing system until the packaging shift commences work the following day.

Fig. 28. Presses used during the manufacture of Swiss cheeses. (A) Tebel-MKT presses at Yhteisjuustola Emmental Cheese Factory in Finland, and notice the Tebel-MKT-S cheese vats in the background. (Reproduced by courtesy of Tebel-MKT BV, Jarvenpaa, Finland.) (B) APV-Rosista cheese presses which are used during the production of Gruyère and Sbrinz cheeses. (Reproduced by courtesy of APV-Rosista AG, Worb, Switzerland.)

Miscellaneous presses. For example, the rotary and cassette presses, which were used for pressing the curd during the manufacture of Edam and Cheddar Cheese, respectively, are not produced any more; for further information see Hansen (1982*a*,*b*).

Block former. This machine, which was known as the Wincanton Bolck Former, has been designed as a continuous system for the fusion of the salted, curd chips under vacuum, so avoiding the need for cheese moulds and presses. The Block Former resembles a rectangular tower around 6–7 m high, and recently a round tower has been developed for the production of 'wheel'- or 'cylindrical'-shaped cheeses. By mid-1991, the total number of Block Formers installed all over the world amounted to 385. The salted curd is air-blown to the tower, and depending on the condition of the curd being fed to the tower, the filling time is around 20 min; the residence time of the curd in the tower is approximately 35 min. In theory, the weight of the curd mass in the tower, and the vacuum condition, is able to produce a block of cheese every 1·5 min once the tower has been filled with salted curd. The sequence of block discharge is as follows:

(a) the elevator platform, which is enclosed at the bottom of the tower, is raised beneath the guillotine;
(b) the guillotine is withdrawm and the column of curd (i.e. supported by the platform) is lowered to the height of a standard block of cheese; then the guillotine moves forward to sever a block from the column of curd;
(c) the door opens and an ejector discharges the block of cheese into a bag-loader; and
(d) the door closes and the cycle is repeated; the overall sequence of operation has been reported by Wegner (1979) and Hansen (1982*c*).

Wincanton cheese tower. The Cheese Tower can be classified as a 'blockforming' cheese system under vacuum (Fig. 29). It consists of the following sections: (i) base, (ii) tower, (iii) interceptor, (iv) exhauster, and (v) control panel. The tower inner is constructed of heavy-gauge stainless-steel and is a rectangular section, similar in size (i.e. width and length) to a rindless block of Cheddar cheese. The inside of the tower is specially tapered to achieve a highly efficient whey-draining system. The outer shell of the tower is cylindrical in shape to provide greater

(A)

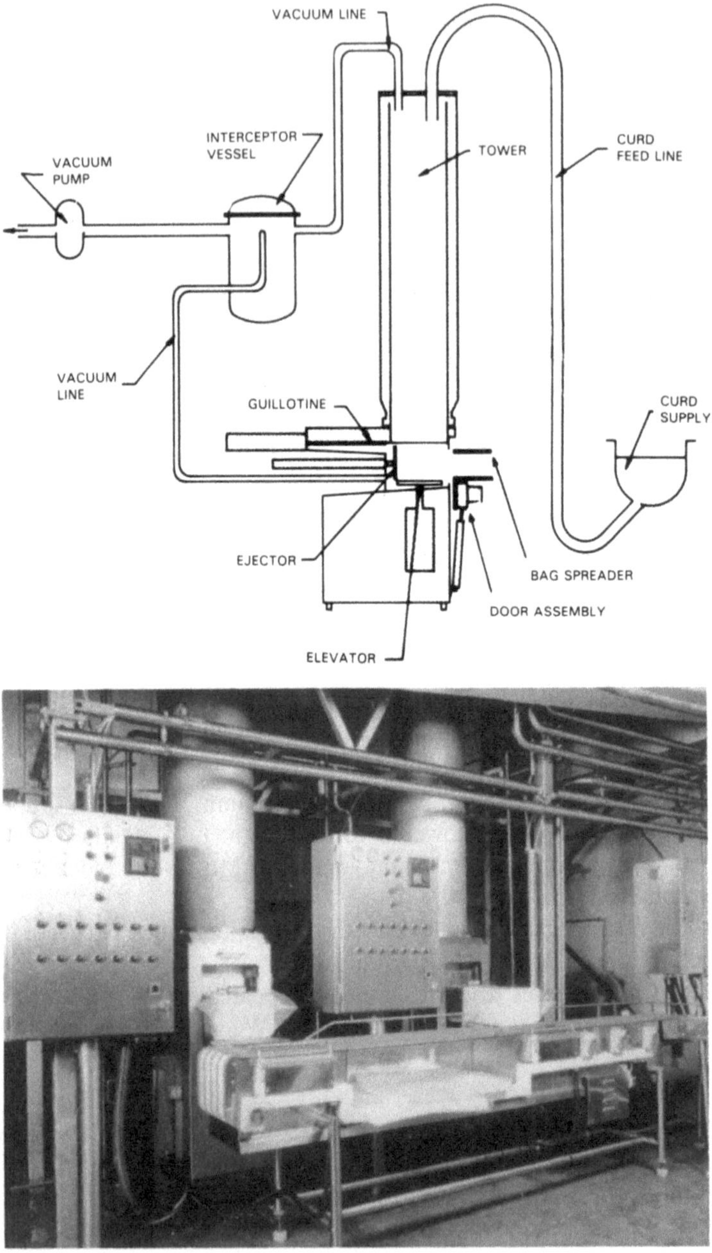

(B)

Fig. 29. (A) Schematic diagram of the Wincanton Cheese Tower. (B) On-site illustration of the Cheese Tower in a factory showing the ejection and bagging of cheese blocks. (Reproduced by courtesy of Wincanton Engineering Ltd, Dorset, UK.)

strength, and improve cleanability. The overall height of the tower is around 6.85 m. After priming the tower with salted curd (~20 blocks of cheese) which takes around 20 min, cheese blocks of adjusted weight and height will be guillotined and ejected into the bagging system at a rate of 40 blocks h^{-1}. The sequence of cheese discharge is as follows:

— the tower is filled with curd to the level probe, with the chamber platform in its raised position and the guillotine closed;
— the guillotine is withdrawn and the column of curd rests on the platform which is then lowered to a predetermined height into the chamber;
— the guillotine severs the 'fused' curd and the block is pressed by the platform under maximum pressure;
— the platform descends to its lowest level, the door opens and the ejection cylinder extends to push the cheese through the door into a plastic bag.
— all cylinders return to their previous positions to repeat the above cycles.

This machine has been recently developed and marketed by Wincanton Engineering Ltd, and the number of installations sold worldwide in 1990–91 amounted to 10.

Carousel Press®. This press has been developed by D. C. Norris & Co. (Engineering) Ltd in the UK for the pressing of salted curd chips in plastic moulds under vacuum. This type of press was first installed in the South Caernarvon Creameries. The press consists of 24 stacks of cylindrical, stainless-steel canisters which are supported by a box-section frame rotating on a circular track on nylon wheels. Filled moulds pass through a curd levelling unit before the lids are added. The moulds are stacked in each canister (i.e. to hold six or three moulds in alternating stacks) via a vertical feed conveyor. When each canister is fully loaded, the Carousel® moves slowly around to the next position, and the mechanically controlled door is closed (Pope, 1984). Each canister is fitted with special guide rails suitable for holding 'block' or 'wheel'-shaped moulds. Over each mould, a pneumatically operated cylinder presses the cheese while the vacuum is drawn to the required level. The pressing specifications for a 20 kg block of Cheddar cheese are as follows: vacuum 80–84 kN m^{-2}, air line pressure 275 kN m^{-2} and the duration of pressing is 1–2 h. After pressing the cheese, each canister is emptied, the moulds are transferred to a de-lidding and knock-out station, and the cheese is packaged; the moulds are washed ready to be refilled with curd.

Bulk presses

Salted curd can be pressed in bulk, and this development was aimed at eliminating the use of a large number of individual moulds. Examples of such pressing systems are as follows:

Large hoop or 'ton' press. This type of press was developed roughly three decades ago in New Zealand for Cheddar cheese pressing, and the specifications of the system have been reported by Crawford (1976) and Scott (1986). In brief, the press consists of a rectangular chamber fitted with a piston from the bottom end. The piston is hydraulically operated to both compress the curd, and extrude the pressed curd. The sequence of operations of the 'ton' press are as follows:

(a) fill the chamber with salted curd;
(b) close the door of the press;
(c) apply vacuum, e.g. 95 kN m^{-2} for 20 min;
(d) maintain vacuum and apply low pressure at 6200 kN m^{-2} for 10 min;
(e) release vacuum and press at 9300–10 340 kN m^{-2} overnight (during peak season, the pressing time is only 8 h);
(f) after pressing, the cheese is pushed to the height of a 'block'-shaped cheese and guillotined, and then further cut to give 4 × 18 kg cheeses at each extrusion.

This type of press incorporates a microprocessor control system for operation, fault warning and the CIP programme.

Block-system (290 kg) press. This is a Damrow pressing method, and the complete processing line is illustrated in Fig. 30. As can be observed, the cheese, e.g. Cheddar, is pressed under normal atmospheric conditions followed by pressing under vacuum in specially designed chambers. The sequence of operations of the Damrow block system, including the time required for each hoop pressing (in parentheses), is

— Station 1: assemble the hoop on a special cart fitted with four castors (3 min).

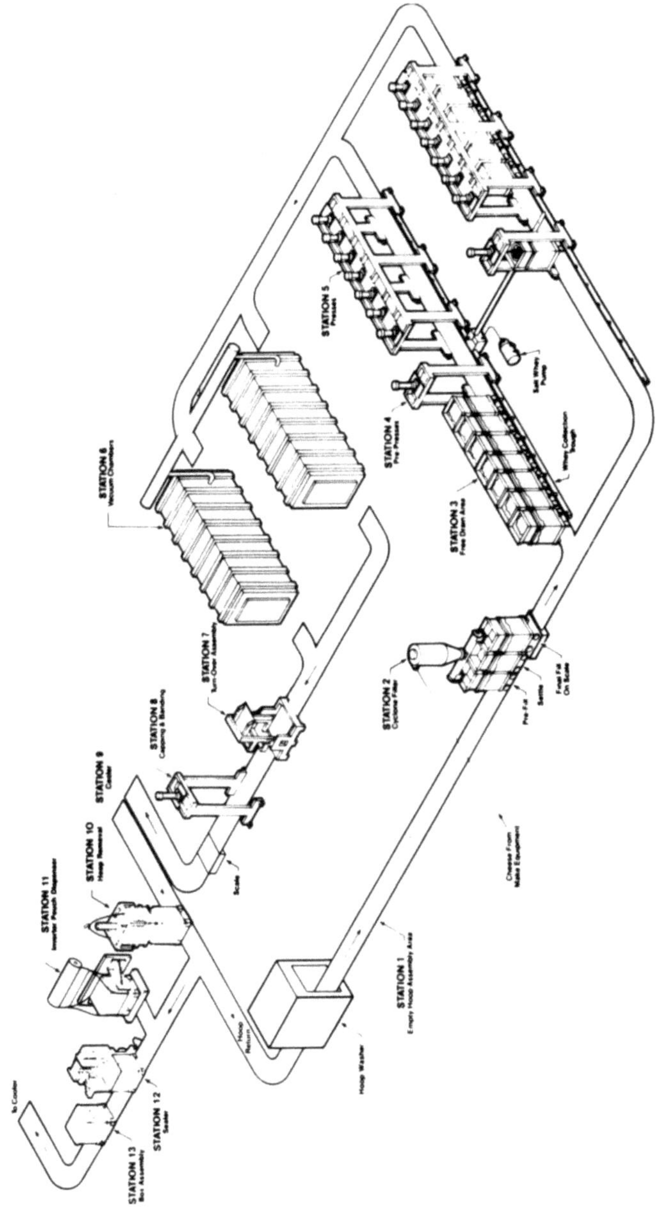

Fig. 30. The Damrow block-system pressing. (Reproduced by courtesy of Gadan A/S, Them, Denmark.)

- Station 2: pre-fill the hoop with salted curd using a cyclone filler, let the curd settle and finally fill the hoop to the desired weight on scales (2·5–4 min).
- Station 3: apply no pressure and allow whey to drain freely (20 min).
- Station 4: pre-press each hoop (3 min).
- Station 5: main pressing area (120 min).
- Station 6: vacuum chamber where the hoop is retained for a minimum period (30 min).
- Stations 7 and 8: turn the hoop over and carry out the necessary capping and banding (3 min).
- Station 9: before transferring the hoop to this section, weigh the cheese (1·5 min) and store cheese at 4–7°C (4–5 days).
- Station 10: remove hoop.
- Stations 11, 12 and 13: bulk packaging of cheese.

A slightly modified Damrow pressing line for Cheddar cheese has been reported by Hansen (1982*d*) in which vacuum pressing is not employed and, after storage in a cold room (3–4 days at 2–4°C), the cheese is cut into 18 kg blocks and packaged. Around 300–400 hoops are required for a cheese plant procesing 250 000 litres of milk per day into Cheddar cheese.

Large temperature differences between the sides and centre of a Cheddar cheese block (290 kg) can affect moisture transfer from high temperature to low temperature areas, and cause variability in salt content and pH within the block of cheese (Reinbold and Ernstrom, 1988). The same authors recommended that slow cooling could reduce this temperature differential, and that this could be accomplished by insulating the cheese block during cooling; hence minimising the faults mentioned above. Furthermore, a reduction in the moisture content of the salted curd before pressing, using whey suction probes, and further developments in pressing and vacuum treatments could also minimise differences in moisture.

Barrel-system (225 kg) press. This method of bulk pressing of Cheddar cheese is widely used in the US. The barrels are fitted with a polyliner, filled with cheese curd and placed under vacuum. Following the vacuum probing, the barrels are transferred to a tipping press (i.e. inverted to remove any free whey) and later removed to a cooler at 3·3°C for storage (Elliot, 1985).

Alvac (290 kg) system. This system is designed to take 16 × ~18 kg cheese blocks from the Block Former where they are assembled into layers of four using a special machine called the 'Collator'. The blocks of cheese are placed in a heavy-duty cardboard, steel or plastic box lined with a poly-film which is placed on a trolley fitted with wheels, and has a detachable base.

After stacking the box with cheese blocks, it is then moved to the next station where the open, top end of the plastic film is heat-sealed, capped and turned over. Subsequently, the box moves forward to Alvac vacuum chamber where the air is removed and the plastic film is sealed. After leaving the vacuum chamber, the box is capped and then transferred to the cold store (see also Hansen, 1987e).

Cheese pressed in individual moulds is transferred to a de-moudling machine or 'knock-out' station before any further handling of the cheese is carried out (this equipment removes the lid from the mould, inverts the mould and removes the cheese). In instances where mould liners are used, they must be removed (manually) before the cheese is packaged. However, 'traditional' cheeses are retained in their cloth bandages until the cheese is fully matured.

Cheeses (e.g. Gouda and Edam) pressed in plastic moulds are easy to remove from the mould compared with cheeses pressed at high pressure, i.e. Cheddar cheese in Lauda moulds. However, the pressed Cheddar cheese is easily removed from the mould using compressed air, and such a system has been described and schematically illustrated by Hansen (1978*b*).

Bulk packaging the cheese

After the pressing stage, the handling and packing of the cheese is dependent on its variety, and examples may include the following treatments:

Emmental cheese

A pressed, 'block'-shaped Emmental cheese (84 kg) is wrapped in cling-film and stacked on a specially designed pallet (see Fig. 31), which can be mechanically turned during the ripening/storage period (Hansen, 1984*a*). The smaller 'block'-shaped Emmental could be packaged mechanically in a Cryovac®-BK 1L bag. This type of packaging material is a laminate of different layers of plastics, and the sequence of operations using this system is as follows:

— The bag loader indexes one of the taped bags into position, opens it, and then automatically loads the block of cheese before

Fig. 31. Mechanical handling for turning large 'block'-shaped Emmental during the maturation period. (Reproduced by courtesy of Tebel-MKT BV, Jarvenpaa, Finland.)

depositing the bagged product onto the in-feed conveyor of the vacuum chamber.

— An operator straightens the neck of the bag to ensure that a perfect, wrinkle-free, heat-seal closure is obtained after air has been removed within the chamber of the machine.
— A second operator carries out visual inspection as the cheese passes through a check/weighing station and a weight label is applied.
— The cheese passes through a shrink tunnel and is placed in a plastic box or cardboard container.

The principle behind the Cryovac® packaging system (Registered trade mark of W. R. Grace & Co.) ensures that the cheese is matured under the following conditions:

(a) in an atmosphere free from oxygen, thus preventing surface mould growth;
(b) the shrinking of the packing material provides a tight wrapping of the cheese free from surface wrinkles;

(c) the packaging material prevents moisture loss, physical damage and contamination of the cheese; and
(d) the Cryovac®-BK 1L bag is impermeable to oxygen, but permeable to carbon dioxide; this latter property is necessary to prevent gassing during the secondary fermentation stage of Emmental cheese.

Cheddar and related cheese varieties
Different systems have been employed for the packing of Cheddar cheese and other British territorials. These may include the following:

Pukkafilm®. This type of packaging material consists of a waxed cellulose laminate. First, the cheese block is wrapped with the laminate; secondly, it is over-wrapped with waxed cellulose; and thirdly, the cheese is placed in a chamber for sealing by the application of heat and pressure (Hansen, 1975; Crawford, 1976).

Unibloc system®. The pressed cheese is wrapped with a plastic film, e.g. Saran® which is manufactured by the Dow Chemical Co. Ltd, and over-wrapped with a layer of paper prior to packaging within six wooden slats (British Patent 937441). The cheese is compressed within the slats by a specially designed machine, and the pressure is maintained by placing four metal straps around the cheese. In some instances, the wrapped cheese is placed within a thin cardboard box before final packaging. This box serves as a dispatch unit when the cheese leaves the factory, and the wooden slats are retained on the premises. The film wrapping of the cheese could be mechanised, but in many cases, the process is carried out manually. However, a comparative costing of different packaging equipment, including the Unibloc® System, for wrapping 81 kg blocks of cheese at a rate of 5–6 min^{-1} has been reported by Gray (1975). A new development in the Unibloc® system is the use of lightweight plastic tray/lids in conjunction with four, wooden, side slats. This development provides rigidity, and the telescopic effect maintains the shape of the block (i.e. overcomes the tapered edge), reduces labour requirements to a minimum, and the steel bands for strapping are not required. Incidentally, the cheese block is packaged, in a vacuum pouch or heat-shrink bag and not in a Saran® film.

Storpac®. The packaged cheese, e.g. in a vacuum pouch or heat-shrink bag, is wrapped in a thin cardboard box (optional), and is placed in a wooden box with a loose cover (British Patent 1433361). The latter piece is held onto the box using a plastic band for strapping. On dispatch, the cheese is removed from these boxes which are retained in the factory.

Heat-shrink bags. An example of such a bag is the Cryovac®-BB 4L bag, which consists of three main layers: polyolefine; a PVDC barrier layer against oxygen and moisture; a cross-linked polyolefine. The sequence of operations of packaging Cheddar cheese in this system is similar to the method described for packaging Emmental cheese. For 18 kg blocks, the bag is heat-sealed after the vacuum stage; however, a perfect seal can be achieved using a metal clip when packaging small Cheddar cheese truckles, or 'baby' Gouda and Edam. The clipping machine is equipped with a bag-trimming facility after the sealing stage; however, an alternative machine, which is widely used in the industry for Cheddar cheese truckles or 'baby' Gouda and Edam, is the Cryovac® VC10. Gas production in Cheddar cheese during the maturation period is considered a serious problem, and a quick remedy is to package the cheese in a carbon dioxide permeable material, e.g. Cryovac®-BK 1L bag.

Vacuum pouches. Different types of plastic film laminates can be used to package Cheddar cheese, and such pouches should provide a barrier against oxygen ingress and moisture loss. One such example is the Diolon® pouch which consists of 20 μm nylon (polyamide) and 60 μm polyethylene. The sequence of operations can be summarised as follows:

— place the 18 kg block of cheese in the pouch;
— transfer to vacuum chamber;
— remove the air and heat-seal; and
— package cheese in a shipping container, e.g. Unibloc®, plastic box or cardboard container.

It is evident that different types of packaging material are used in the cheese industry, and some relevant technical specifications of these laminates are shown in Table XXIV. Cheese packaging line(s) can be highly mechanised, and Fig. 32 illustrates the packaging of block Cheddar cheese (18 kg) in the UK. However, irrespective of how the cheese is packaged, e.g. in vacuum pouches or heat-shrink bags, the wrapping of the cheese in the shipping container (i.e. cardboard,

TABLE XXIV
Some properties of cheese packaging materials

Film	*Laminate*[a]	*Thickness* (*μm*)	*Moisture vapour transmission*	*Oxygen permeabiity*[b]	*Carbon doxide permeability*[b]
Saran®	PE, PVC, PP, polyester, foil and paper	25	3·1 g m^{-2}/24 h/38°C/ 90% RH	12–17 cm^3 m^{-2}/24 h/23°C	59–93 cm^3 m^{-2}/24 h/23°C
Diolon®	Nylon/PE	20/60	6·5	70 cm³ m⁻²/24 h/75% RH/25°C	NA[c]
		50/70	6·0	35	NA
		80/100	4·2	20	NA
Diomex®	Polyester/PE	12/50	5·0 g m^{-2}/24 h/ 38°C/90% RH	100 $cm^3$$m^{-2}$/24 h/0% RH/22°C	NA
Metallised film	Polyester/foil/low density ethylene butene copolymer	12/12/50	<1·0	<1·0 cm^3 m^{-2}/24 h/75% RH/25°C	NA
Pukkafilm®	Celulose/wax/ Pukkacote	NA	<0·5 g m^{-2}/24 h/25°C/ 75% RH	<20 cm^3 m^{-2}/24 h/23°C	NA
Novaflex II®	Nylon/polythene	30/50	2·3–2·6 g m^{-2}/24 h/25°C/ 75% RH	30–40 cm^3 m^{-2}/24 h/50% RH/23°C	NA
Cryovac®[d]—BB 4L	PO/PVDC/PO	60	Nom.[e] 20 g m^{-2}/24 h/ 38°C/100% RH	Nom. 35 cm^3 m^{-2}/24 h/ 50% RH/23°C	Nom. 150 cm^3 m^{-2}/24 h/ 50% RH/23°C
—BK 1L	As above		Nom. 15	Nom. 175	Nom. 850

[a]PE: polyethylene, PVC: polyvinyl chloride, PP: polypropylene, PO: polyolefine. PVDC: polyvinylidene choride polymer.
[b]Pressure applied during testing is $1{\cdot}01 \times 10^2$ kN m^{-2}. Note that the permeabilities are at different test conditions, and only if the expression is the same can the figures for each packaging material be compared.
[c]NA: not available.
[d]See text.
[e]Nom.: nominal.

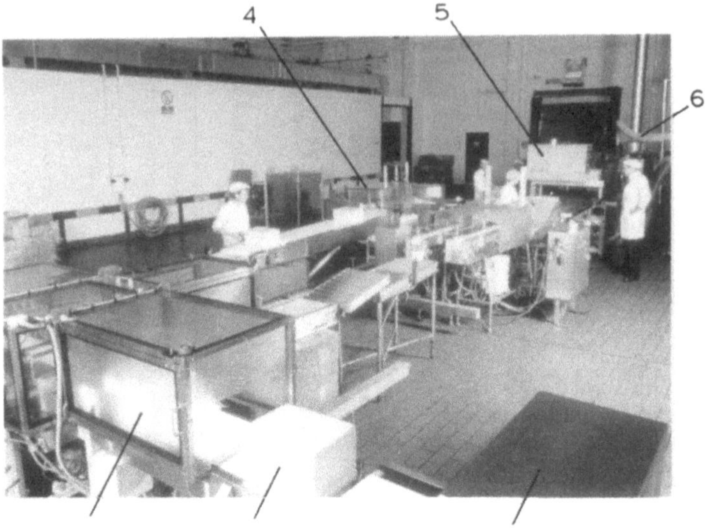

Fig. 32. Bulk packaging of Cheddar cheese (6 blocks min^{-1}) at the Lockerbie Creamery in Scotland. (1) Metal detector, (2) de-moulded block of cheese, (3) cheese cutter, i.e. 4 × 5 kg size, (4) bagging station, (5) vacuum and heat-sealing machine, (6) heat shrink tunnel, drying area, weighing, labelling and packaging in plastic container. (Reproduced by courtesy of Express Foods Group (International) Ltd, Middlesex, UK.)

plastic or wooden containers) and palletising is still carried out using semi-automatic machines relying heavily on manpower.

Over the past decade, a completely mechanised system has been developed by Odenberg Engineering Ltd in Ireland to handle the cheese being packaged. The system (Fig. 33) consists of: (i) Blockmaster, (ii) weighing and labelling, (iii) Coolmaster (optional), and (iv) palletiser. The Blockmaster (Fig. 33(A)) wraps the vacuum-packaged block of cheese, under compression, in a corrugated case and liner. This ensures that the shape and the finish of the cheese is maintained during the storage period, thus minimising the off-cuts during the portioning of the block of cheese and retail packaging. The Blockmaster is a fully automatic machine with its own independent control and on-board PLC controller. The sequence of operations is as follows:

— The sealed block of cheese is delivered to the In-feed Station by means of two air-operated, side-pushers which eliminate the risk of damaging the vacuum bag; a liner is inserted for additional

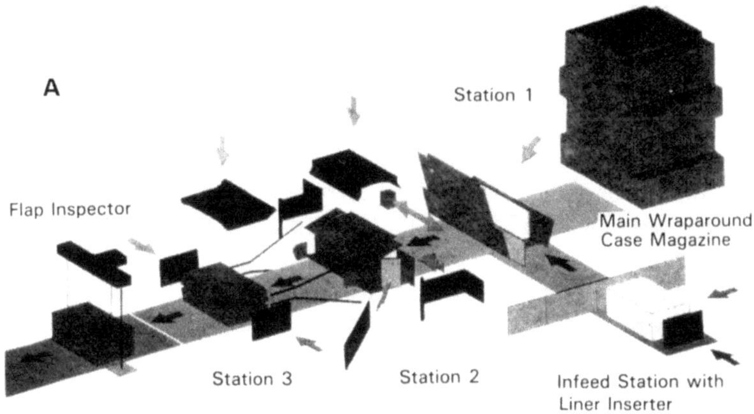

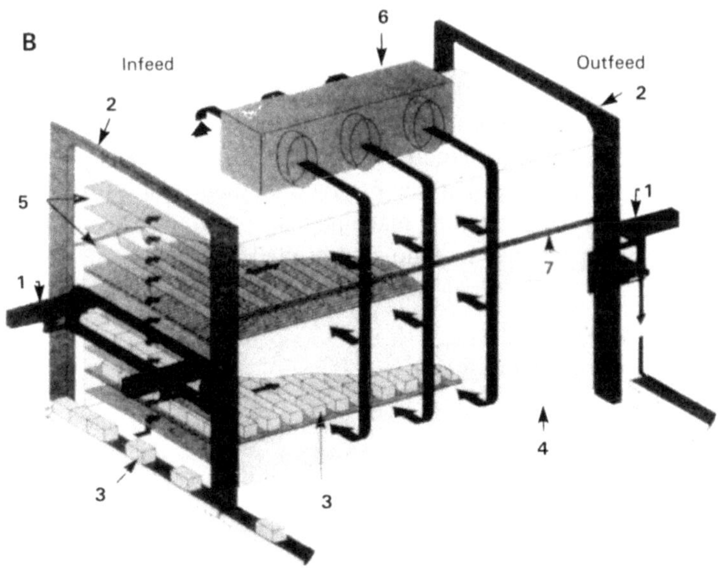

Fig. 33. The Odenberg cheese block packing and cooling system. (A) The Blockmaster. (B) The Coolmaster. (1) In-feed and out-feed traveling lifts, (2) in-feed and out-feed lift support, (3) product, (4) cold storage framework, (5) storage layers and roller mechanism, (6) cooling plant, (7) roller support, large and small arrows illustrate movement of cold air and product respectively) (Reproduced by courtesy of Van Den Bergh & Partners Ltd, Windsor, UK.)

stacking strength. Any deformed cheese blocks are re-squared by means of the compression plates fitted in this station.

- A blank case is withdrawn from Station 1 by suction and, after initial folding, the cheese block and liner are loaded onto the case blank and indexed forward.
- The minor flaps of the case and the end-flap of the liner are folded in Station 2, and hot melted glue is applied to the top seam. The top and bottom surfaces of the corrugated case are compressed, the top seam is formed, and the pack moves foreward to the next station.
- The major flaps of the box are folded and glued in Station 3, followed by the application of pressure to all sides of the cheese case. When the adhesive is set, the wrapped cheese is ready to exit the machine; however, a checking device is fitted to the machine which ensures that no cheese leaves the packer with open flaps.

This machine, which is built with safety features and a fault alarm system, has stainless-steel/aluminium contact parts and open construction for ease of access; it is capable of packaging eight blocks of cheese per minute. By 1990, the total number of Blockmasters in Ireland, UK, US, Australia and Spain amounted to 30.

For efficient management (i.e. production, quality and inventory control), a weighing and labelling machine is incorporated into this packaging line which provides the following printed information on each block of cheese:

- net weight;
- date of manufacture;
- vat number;
- code of factory; and
- bar code.

Incidentally, a metal detector is fitted before the weighing and labelling station in order to ensure the safety of the consumer.

The main purpose of the Coolmaster is to ensure rapid and even cooling of individual blocks of cheese before palletising and stacking in the maturation store. Palletised cheeses require a long time to cool in the store (Conochie, 1974; Conochie and Birtwistle, 1974), which can affect the quality of the matured cheese. Since 1980, work carried out at the New Zealand Dairy Research Institute (Fryer, 1982; Lawrence and Gilles,

1987) suggests that quick cooling of the blocks of Cheddar cheese helps to overcome the problems of uneven quality of matured cheese. Thus, the comparative rate of cooling of palletised cheese blocks (multi-layer and single stack) *vis-à-vis* with the Coolmaster is shown in Fig. 34. The sequence of operation of the Coolmaster (Fig. 33(B)) is as follows.

— The wrapped cheese blocks are conveyed automatically to the In-feed System from the weighing and labelling station, after being inverted to ensure a smooth top surface. When a full row has been accumulated, the In-feed System lift ascends, takes the row, lifts it to the appropriate layer and pushes the row of cheese into the Coolmaster. The pusher then retracts and the lift returns to the loading level to recommence the cycle. This procedure is repeated until each layer is filled.

— When the pusher pushes a row of cheese into a layer, the flight-bar mechanism advances the rows of cheese cases towards the Out-feed System. This mechanism maintains the correct spacing between the rows of cheese to ensure proper and even circulation of cool air. The flight-bar mechanism is activated independently by the In-feed and Out-feed lift systems. The layers are moved on rollers, with the cheese blocks being supported by four rollers at any one time.

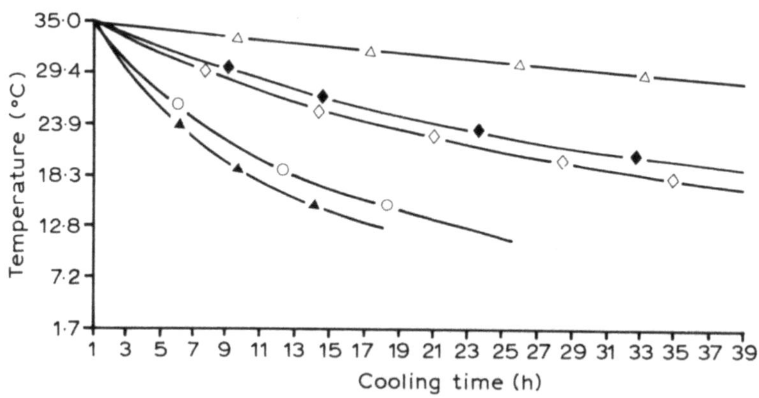

Fig. 34. Cooling curve for 18·5 kg block of Cheddar cheese wrapped in corrugated box. (△) Middle, (◆) bottom, and (◇) top layer of a pallet containing 50 blocks of cheese and cooled at 5·6°C; (○) single layer on a pallet cooled at −2·2°C, and (▲) cheese blocks cooled in the Accu-Cooler a 2·2°C. (Adapted from D. Kellett (pers. comm.).

— By activating the discharge conveyors, the Out-feed System lift starts to take a row of cheese blocks that have been in the Coolmaster for a pre-determined time, e.g. 18 h. The load is then transferred onto the Out-feed system lift which moves to the unloading level, where the load is then turned over and discharged to the outlet conveyor.

The Coolmaster is constructed of stainless-steel (i.e. contact parts) and has a heavy-duty galvanised framework. The rollers are Rislan® coated steel, and the operation and control of the Coolmaster is fully automated. The Out-feed lift unloading mechanism is completely independent of the In-feed system, and the cheese blocks are handled on a 'first-in', 'first-out' basis so that the identity of each vat is maintained. The capacity of the Coolmaster (e.g. 1000 to 9000 blocks of cheese) is dependent on the number of layers and the overall length, and it is supplied in two standard widths. The capacity figures are based on a cooling time of 18 h which corresponds to one day's production. In 1990, the total number of Coolmaster installations amounted to 15.

On completion of the cooling period, the cartoned cheeses are automatically discharged onto a conveyor belt, and then onto a sampling station before palletising. One block of cheese is removed on the basis of one sample per vat, i.e. drawn from the middle of each batch, and each pallet holds 50 blocks. At the entry to the palletiser, each carton is scanned by a laser to ensure that all the cheeses originate from the same vat. The palletising rate is ~ 10 blocks min^{-1} in layers of 10 using a pack pattern of 4–3–3. Finally, on exiting, the fully loaded pallet is wrapped automatically with a stretch film before dispatch to the cold-store. A pallet manifest is generated by the system giving the following information:

— individual block number;
— individual block weight;
— total weight of pallet;
— vat/batch number.

At the end of the day, a complete report and summary becomes available for total blocks of cheese produced and weight of cheese packed.

Gouda, Edam and related cheeses

After the brining stage, the cheese is plasticised twice to prevent mould growth during the ripening period, and this process is repeated several times if the cheese is to be stored for long periods. Prior to dispatch, the cheese is washed, dried and coated with paraffin wax and overwrapped with a red cellophane film (the latter packaging material is optional). The mechanical handling of the cheese in the store, and the waxing equipment are discussed below.

An alternative approach for the packaging of the 'loaf', 'block' or 'round' Dutch cheeses is to wrap the product in a heat-shrink bag, which is then either sealed by heat or with a metal clip.

Storage of the Cheese and Miscellaneous Handling

The quality of any cheese variety is dependent on many factors, such as the quality of the milk, the activity of the starter cultures, and the manufacturing stages; however, other criteria, for example, the bulk handling of the cheese in the store and the conditions provided during the maturation period, are important. The latter aspects can influence the biochemical changes that take place in the curd during the storage period, and can ultimately affect the quality of the cheese. The handling and storage conditions of cheese may vary from one variety to another, and some typical examples include the following:

Swiss cheeses

The Emmental and Gruyère-type cheeses undergo a secondary fermentation during the early stages of storage, and 'eye' formation takes place in the cheese due to the metabolic activity of the propionic acid bacteria. Thus, the 'wheel'-shaped Swiss cheeses are matured under controlled temperature and relative humidity (RH) conditions, and examples of the handling of the cheese in the store are as follows:

Gruyère

Store the cheese at 10°C for 3 weeks, and then at 15–20°C for 2–3 months in an atmosphere of 90–95% RH; complete storage at 12–15°C and 85% RH for 8–12 months (Scott, 1986). The cheese is turned reguarly and rubbed with a damp cloth (i.e. soaked in brine solution) to aid the growth of those bacteria that provide a red–brown smear coat.

Emmental

Store the cheese at 10–15·6°C for 10–14 days at 90% RH, then at 20–24°C for 3–6 weeks at 80–85% RH, followed by storage at 7·2°C or less for 6–12 months at 80–85% RH. The cheese is also turned and the surface is wiped with a cloth soaked in brine (Scott, 1986). Prior to dispatch, the 'wheel'-shaped cheeses are cleaned, stamped, waxed and packaged in a shipping container.

'Block'-shaped Swiss cheeses

Since the cheese is packaged in a barrier-type material, only the storage temperature is taken into account and humitidy control is not necessary. For example, the handling of 'block'-shaped Emmental (84 kg) in a factory in Finland (Fig. 31) is as follows: store the packed cheese at 8°C for 4 weeks, then at 23–24°C for 6 weeks and then 6°C for a few weeks. The cheese is only ripened for 3 months (Hansen, 1984*a*). According to Scott (1986), a 'block'-shaped Emmental (e.g. 18–45 kg) is handled as follows: dry the cheese after brining for 4 days at 7°C; package in barrier film material; store at 17–18°C for 10–15 days until 'eye' holes have formed (1·3–1·9 mm in diameter) and finally store at 8–12°C until the cheese is mature. Turning of small, 'block'-shaped Emmental in the store is not necessary.

As mentioned elsewhere, temperature control during the pressing, secondary fermentation and storage of large Swiss cheeses is critical, and an extensive study of the effect of temperature variation on the rate of development of D(−) and L(+) lactic acid has been reported by Gehriger (1979); in addition, an extensive review on Emmental-type cheeses, which also covers the published data, has been published by Kasprzyk *et al.* (1983).

British cheeses

The traditional types of British cheese are stored at different temperatures (e.g. 7·2–18·3°C) depending on the variety (Scott, 1986), and most likely at 85% RH to prevent dryness and crack formation in the rind. It is necessary for the cheese to be turned during the maturation period, and this process is carried out manually in small cheese factories or using rotating shelves. The latter process is illustrated by Davis (1965).

'Block'-shaped cheeses are normally matured at around 10°C, and plastic laminates (i.e. impermeable to oxygen and moisture) are used as the packaging material. Therefore, turning of the cheese in the store is not necessary. The stacking of cheese blocks in the maturation room is dependant on factors such as type of shipping or outer packaging container, volume of daily production and/or mechanised system employed. Some examples of systems for handling 'block'-shaped cheeses are as follows:

(a) Cheeses wrapped in cardboard boxes are normally stacked manually (2 × 18 kg blocks high) on wooden or metal shelves to avoid deformation in the shape of the block. The system can be mechanised using specially designed metal pallets, and one such type is constructed of angle-iron uprights with indentations/feet for stacking the pallets on top of each other. The pallet contains two folding shelves, so that the cheese blocks can be stacked on both the metal base and the shelves. In total, the pallet holds 60 blocks of cheese (12 cheeses in each layer), and the stacking of these pallets is carried out using a forklift truck. Cheddar cheese packed in cardboard boxes can be stacked mechanically or manually using a 'Boxpallet' (Hansen, 1982*e*), in which the cheese is distributed on four levels with 12 cheeses on each level.
(b) Cheeses wrapped in Unibloc® (see Fig. 35(A)) or Storpac® are stacked on wooden pallets, and each pallet holds 45 cheeses.
(c) Cheeses wrapped in plastic containers are stored in metal pallets similar to the type mentioned above, but without the folding shelves, and the capacity of each pallet is 60 cheeses (Fig. 35(B)).
(d) Cheeses wrapped in corrugated boxes and automatically palletised, for example the Odenberg system (50 per pallet), are stored on metal racks and stacking is achieved using a forklift truck (Fig. 35(C)). It is evident that different systems can be used for storing Cheddar (18 kg) and related cheese varieties, and some examples are illustrated in Fig. 35.
(e) Cheese, which is not wrapped in any type of outer/shipping container, can be stored in the OR-Maturing System developed by Olavi Rasanen OY in Mikkeli in Finland (see Fig. 36). All the different sections are made out of wood, no cartoning of the cheese is required and no metal straps are necessary when stacking. Each pallet holds six separate decks of 8 × 18 kg block cheese. The collar supports the framework and the construction of each deck and, in this system, every block of cheese is subjected to an individual pressure.

When the OR-Maturing System is employed in a cheese factory, the collar section is removed after 3–4 weeks for constructing new units.

Fig. 35. Different illustrations of modern cheese stores in the UK. (A) A view of cheese packaged using the Unibloc® system at Campbeltown Creamery. (Reproduced by courtesy of Campbeltown Creamery Ltd, Argyll, UK.) (B) The Express method 'Blue Boxes' for maturing cheese at Lockerbie Creamery. (Reproduced by courtesy of Express Foods Group (International) Ltd, Middlesex, UK.) (C) The storage of Cheddar cheese in cardboard boxes using the Odenberg system at Maelor Creamery. (Reproduced by courtesy of Dairy Crest Ingredients, Thames Ditton, Surrey, UK.)

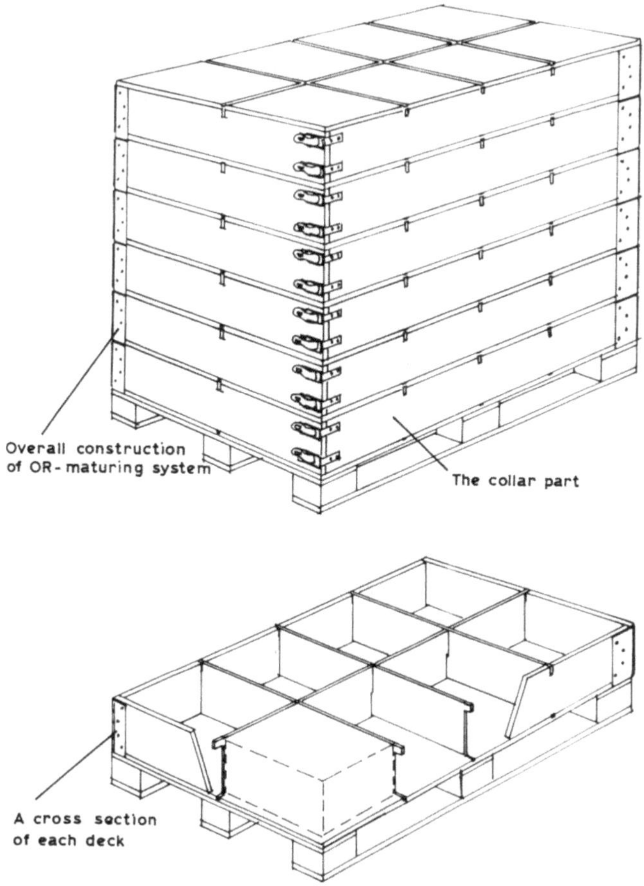

Fig. 36. The OR-maturing system. (The construction of each unit is strong enough for six pallets to be stacked on top of each other.) (Reproduced by courtesy of Olavi Räsänen, Mikkeli, Finland.)

Prior to dispatch, the cheeses are wrapped in thin cardboard boxes, and the wooden parts are retained in the creamery.

Dutch cheeses

Gouda, Edam and other related cheese varieties are stored for at least 28 days after manufacture at 12–13°C and 85–90% RH (J. H. Dijkstra, pers. comm.), and then at a temperature of 10°C and 85–90% RH until the cheese is mature (Scott, 1986). The cheese is turned many times during the maturation period, and this process is highly mechanised in large cheese factories. At present there are three ways of handling Dutch cheese varieties in The Netherlands, and they are known as (i) the compact system, (ii) the loose shelves system, and (iii) the box system. Detailed descriptions of these systems have been reported by Elten (1979, 1981), Weenik (1975) and Hansen (1981*d*,*e*,*f*, 1982*f*, 1985*a*, 1987*f*, 1988*e*,*f*). The control of air temperature and circulation is important, and some technical aspects have been reported by Bouman (1979), Viswat (1979, 1988) and Anon. (1985*d*). Improvements in air circulation under the blocks of Gouda and Edam cheeses have been achieved using a newly designed, plastic shelves known as the 'Fakir', which is manufactured by Arend BV in The Netherlands (Anon. 1988*b*).

The brined cheeses are plasticised a few times to prevent mould growth, and to prevent soiling of the shelves in the store; the cheese is plasticised from one side only, so that when it is turned, the dried section of the cheese is in contact with shelf. At the end of the maturation period, the cheese is washed, dried, coated with paraffin wax (red or yellow) at 90–120°C and, possibly, over-wrapped with a cellophane material before it is dispatched to market. Mechanical handling of Dutch cheeses has been reported by Anon. (1979, 1982*b*) and Hansen (1980*b*, 1983*b*). In addition, Dutch cheeses could be packed in an impermeable material, for example, by employing the Cryovac® packaging system.

Illustrations of systems for wrapping 'ball'- and 'wheel'-shaped cheese are shown in Fig. 37. The Alpma K-65 wraps Edam cheeses up to 13 cm in diameter, whereas a different model, the K-62, can package cheese balls up to 20 cm in diameter. In both machines, a piston pushes the cheese plus the cellophane through a 'brush die-box' where the packaging material is placed tightly around the upper half of the product. At the lower end of the cheese, the packaging material is pleated by a lamella-closure followed by heat-sealing; the fitting of a labelling device for self-adhesive labels is also possible. The capacity of the K-65 and K-62 is 1800 and 1200 packages h^{-1}, respectively, and while the former model is fed via a conveyor-chain, the K-62 is served by 'in-feed' arms (see Fig. 37).

The Alpma RO-62/S and RO-61/S (Fig. 37) machines have been specially developed for wrapping Gouda cheeses of up to 5 and 15 kg in weight, respectively. The capacity of these models is up to 1300 and 1100 cheeses h^{-1}, respectively, and they can also handle 'rectangular'-shaped blocks of cheese.

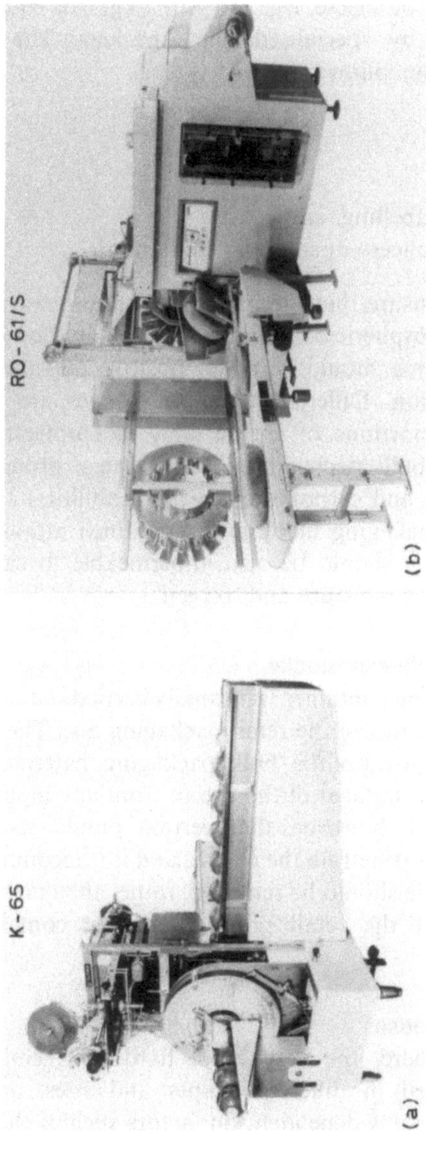

Fig. 37. Automatic packaging machines for wrapping 'ball' and 'wheel'-shaped cheeses. (Reproduced by courtesy of Alpma Hain & Co. KG, Rott am Inn, Germany.)

Retail Packaging of Cheese

Consumer packaging of cheese, e.g. 250 g to 1 kg, can be carried out in cheese factories, or by specialised dairy packers. The sequence of operations involves the following:

— unwrapping;
— portioning;
— packaging;
— weighing and labelling; and
— packaging the cheese in shipping containers.

It is necessary to ensure that the handling of the cheese is carried out under controlled atmospheric conditions, i.e. air should be filtered, and if possible, the whole area should be under positive pressure to minimise airborne contamination. Different types of plastics are used for the packaging of retail portions of cheese, and the protective properties associated with the bulk packaging of cheese (e.g. protection against moisture loss, oxygen and carbon dioxide permeabilities) are also enforced. However, the packaging material for modified atmosphere cheese packs, i.e. gas-flushing, should be also impermeable to carbon dioxide and nitrogen as well as moisture and oxygen.

Unwrapping of the cheese blocks

Removal of the shipping container is normally carried out manually in an area isolated from the rest of the retail packaging line. The second stage of operation is the removal of the 'bulk' packaging material, and in some instances, cleaning the surfaces of the cheese from any mould growth.

It has been argued, however, that certain mould species produce mycotoxins which can penetrate the cheese, and it is recommended that a 'slice' off the whole side should be removed, rather than mere scraping, in order to ensure that the retail portions do not contain any toxic substance(s).

Portioning of the cheese

As mentioned elsewhere, the very hard, hard and semi-hard cheese varieties are produced in different shapes and sizes, and the retail portioning of the cheese is dependent on factors such as the following:

— original shape of cheese;
— consumer size and shape required; and

— minimising wastage (i.e. off-cuts) during the portioning of the cheese.

In general, cheese is portioned into different shapes, e.g. wedge, rectangular or square, and Fig. 38 illustrates how some cheese varieties are cut. Different cheese portioning machines are available on the market, and some examples are as follows.

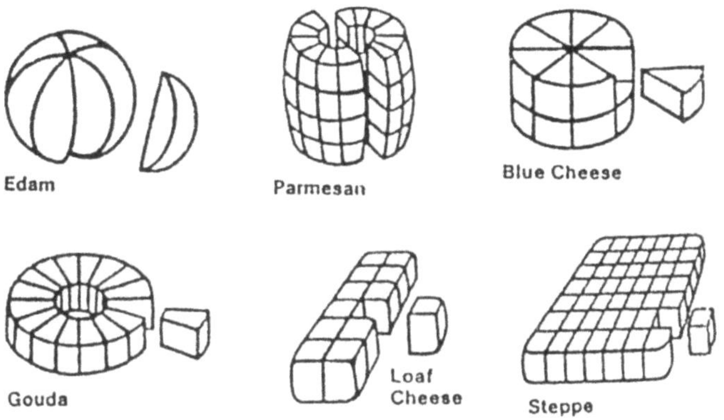

Fig. 38. The portioning of different cheeses. (Reproduced by courtesy of Alpma Hain & Co. KG, Rott am Inn, Germany.)

Alpma KSA 90

This cutting machine is designed for handling the 'block' and 'wheel'-shaped Emmental cheese (80 kg) (see Fig. 39). The cheese is cut into bars which are then fed through a high-speed portioning machine type Alpma Cut 21 or 22. These latter machines have a capacity of 30–80 cuts min^{-1} depending on the variety of cheese. The mode of operation of the Cut 21 can be described as follows: (i) the bars of cheese are positioned side by side, or one after another, on the 'in-feed' belt, (ii) an electronic timer controls the forward or step interval of the belt which corresponds to the cutting length of the bar of cheese, and (iii) the cutting device sections the cheese, and the portions are discharged by another transfer belt. In principle, the Cut 22 is similar to the Cut 21 model, but it includes automatic control of the forward or step length, and a balance which provides the higher accuracy for cutting the bar of cheese into equal portions. The cutting length of the cheese bars of the Cut 21 and 22 ranges from 4 to 999 or 4 to 110 mm, respectively.

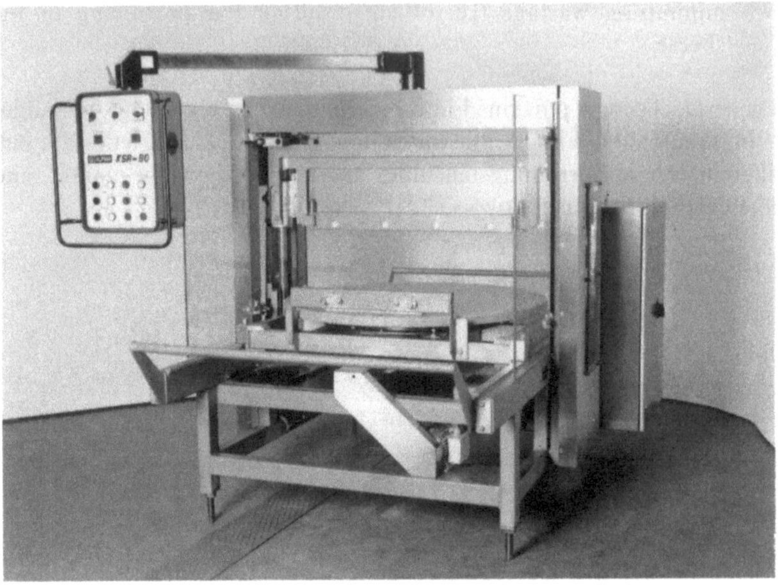

Fig. 39. An automatic machine type KSA 90 for cutting 'block' and 'wheel'-shaped Emmental up to the weight of 80 kg. (Reproduced by courtesy of Alpma Hain & Co. KG, Rott am Inn, Germany.)

Alpma HT II

This is a hydraulic cheese cutter suitable for portioning Edam, Gouda and all hard cheese types including Parmesan and Grana cheeses. The latter two varieties are normally cut using special wires, and thus the portions have smooth surfaces which are not broken.

Wright Pugson cheese cutting machines

Different models of cheese cutting machine are manufactured by Wright Pugson Ltd in Dorset (UK). These machines are entirely pneumatic in operation, constructed in stainless-steel with polypropylene cutting heads, and some examples of these portioning machines are:

— block cutter model C 21 using a two-stage cutting process for portioning 18 kg blocks of Cheddar cheese; the portion size ranges from 200 g to 5 kg, and the cutting rates are 3 and 4 blocks min^{-1}, respectively;
— the automatic cutter model C 23 is a high-speed cutter, and the rate of cutting is up to 4 blocks min^{-1}.

Codat 400 and 500 range machines

These machines were developed in the early 1980s and manufactured in the UK, but production of these machines was stopped a few years ago. However, the Codat 400 and 500 were designed to cut 18 kg blocks of cheese into retail size portions and place them automatically in form–fill–seal-type packaging machines. Hansen (1980*c*) reported the first installation of these machines in a cheese factory in the UK capable of handling 1680 × 18 kg blocks of Cheddar cheese per 7 h working day.

Equipment for cheese packaging

A multitude of high-speed, cheese portion packaging machines is available on the market. It would be impractical, of course, to discuss all the different types of machines in detail but, from a technical point of view, certain important specifications must not be overlooked, for example:

- capital cost;
- type of packaging material used;
- proposed method of packaging, e.g. over-wrapping, vacuum, vacuum heat-shrink, and/or gas-flushing;
- sanitary standards;
- versatility and reliability of the machine;
- power and labour requirements; and
- other specifications, such as availability of price labelling, date marking and safety measures.

In general, the cheese portion packaging machines can be divided into the following types: (i) over-wrapping, (ii) chamber, (iii) automatic form–fill–seal, (iv) skin-packaging, and (v) miscellaneous. The mode of packaging of types (ii), (iii) and (iv) is illustrated in Fig. 40. Some examples of portion-packaging machines are as follows:

Over-wrapping machines

These machines merely over-wrap the cheese portion with a plastics film, e.g. Saran® or polypropylene laminate, and the estimated shelf-life of the product is up to 3 weeks. The object of packaging the cheese by this method is to provide the customer with a 'fresh-look' appearance to the cheese.

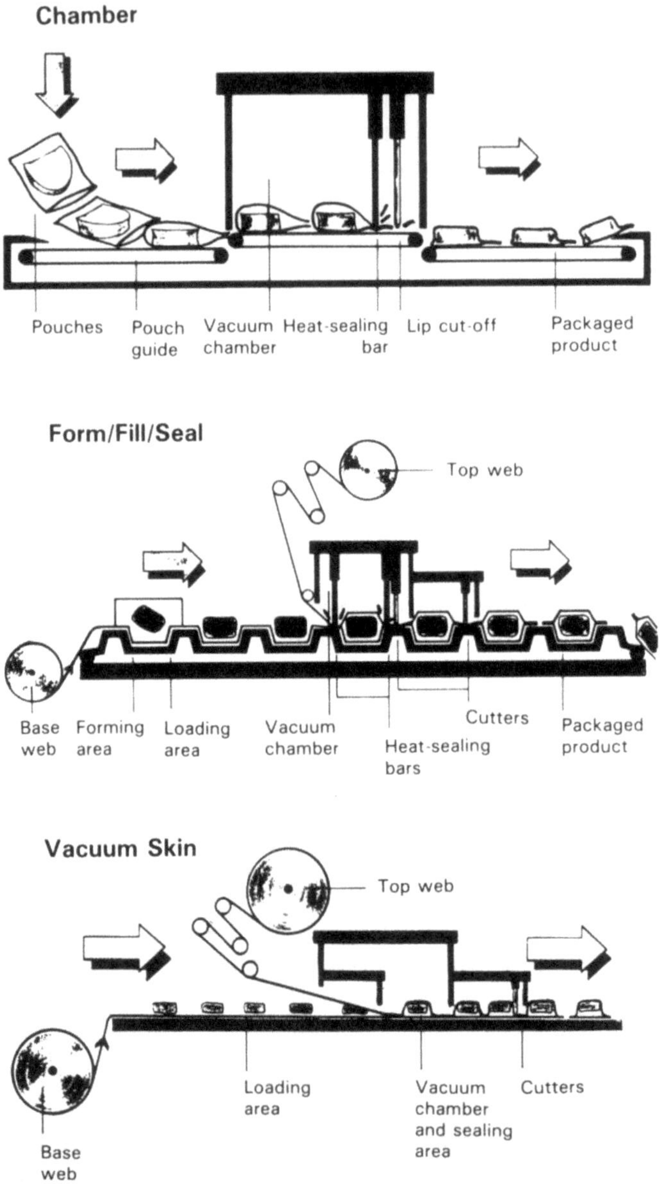

Fig. 40. Illustrations showing the different types of cheese portion packaging equipment. (Reproduced by courtesy of PLM Redfearn Flexpack Ltd, Lancashire, UK.)

A typical example is the Alpma U-64 series, which is manufactured by Alpma Hain & Co. Kg in Germany. The main characteristic of the machines is their ability to wrap a variety of cheese portions (i.e. rectangular, wedges, and/or 'baby ball'-shaped), and the sequence of wrapping operations is as follows:

- The cheese portion is placed in a specially designed pocket on the 'in-feed' chain which carries it to the exact position for wrapping.
- The packaging material is drawn from a reel to the correct length required, and the length is automatically adjusted to ensure accuracy of wrapping even if the product size varies by $\pm 5\%$.
- The die sets in the wrapping chamber are specifically designed to provide an exact cut and an attractive fold on the package.
- These machines can handle a wide range of packaging materials and, in some models (e.g. U-64/A, U-64/ASa and U-64/ASch), they are equipped with heat-sealing and heat-shrink devices to obtain a tight final pack.

Different accessories can be fitted to these machines, such as print mark control unit, labelling and coding devices, cartoning equipment and boxing unit. The capacity of the Alpma U-64 series is up to 3000 portions h^{-1} depending on the product (i.e. soft or hard), portion shape and the packaging material used.

Vacuum packaging machines

There are three different types of packaging machine which can be used for packing portions of cheese (see Fig. 40), and some examples are as follows.

(a) *Chamber machines*—This type of equipment uses mainly pre-made pouches (e.g. nylon/polyethylene laminate and polyester, or cellulose/polyethylene laminate or related plastic films). The pouch is filled with the product, loaded into the chamber which evacuates the air from the pack and then heat-seals it. As a result, the product is in an airtight container and the shelf-life of cheese packaged in this system (i.e. kept under refrigeration) is up to 3 months. In small operations, this system of packaging is carried out manually by placing the portion of cheese inside the pouch and transferring it to the vacuum chamber. When the lid of the chamber is closed, the vacuum is activated followed by heat-sealing. However, the fully automatic version, which is used by large throughput packers, is shown in Fig. 40.

(b) *Automatic thermoforming machines*—The majority of high-speed, vacuum packaging machines are known as form–fill–seal machines. One of the leading manufactures is the Multivac® Verpackungs maschinen in Germany, and examples of these machines are the M-855F, R-5200 and R-7000 which have a capacity in excess of 100 packages min^{-1}. Most of the form–fill–seal, machines have been adapted for gas-flushing (see later). The packaging material is delivered to the dairy in large reels of plastic sheet, and two reels are fed into the machine. The base reel, which is a thermoplastic material, e.g. Diolon® 50/70 μm or 20/60 μm (see Table XXIV) or up to 200 μm thick, is warmed, and vacuum or compressed air stretches the film down into cooled, multi-row moulds which transforms the sheet into the desired shape. The cheese portions are placed into the well and, as the cheese container moves forward to the vacuum and heat-sealing chamber, it is covered by a plastic sheet from the top reel, e.g. Diomex® 12/50 μm or 20/60 μm (see Table XXIV). The following stage consists of removing the air from the chamber, heat-sealing the thermoplastic materials to each other, and finally longitudinal and cross cutting units separate the cheese portions before transferring them to the weighing and labelling section. Illustrations of such machines are shown in Fig. 41.

Another type of a high-speed, automatic, thermoforming, vacuum packaging machine is the Fuji® Fw-3710/B which is manufactured by Fuji Packaging Machines Ltd, Nagoya in Japan. This machine is the horizontal-type, form–fill–seal method of packaging. This particular model comes in two different specifications, i.e. the B140W and B114W, where the operating speeds can be up to 80 and 100 packs min^{-1}, respectively. Some of the features of the Fuji FW-3710/B series are as follows:

— It has a memory in the computer for 30 products, and this controls the drive and slave motors to match the product speed, film condition and seal temperature; thus, rapid change-over is achieved by product selection within the computer with minor manual changes.
— The construction of the machine is stainless-steel which can be washed down.
— The type of motor drive system requires no maintenance or lubrication.

(A) (B) (C) (D)

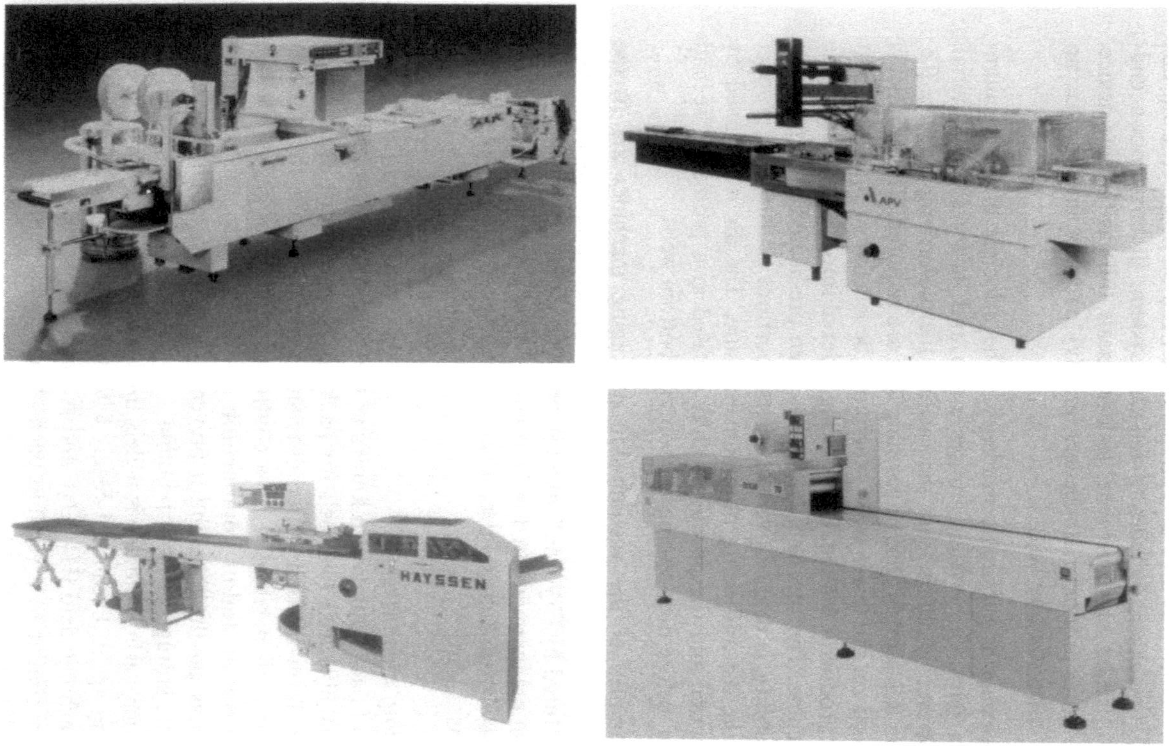

Fig. 41. Illustrations of machines for packaging retail portions of cheese. (A) Multivac® R-7000. (Reproduced by courtesy of Multivac UK Ltd, Swindon, UK.) (B) APV RF 452 Flowpak®. (Reproduced by courtesy of APV Packaging Machinery Ltd, Leeds, UK.) (C) Hayssen® RT 218 (Reproduced by courtesy of Hayssen Europa SRL, Zingonia, Italy and PackAnalysis, Evsham, UK.) (D) Dixie Union® Pak 70. (Reproduced by courtesy of Dixie Union (UK) Ltd, Milton Keynes, UK.)

— Other features or options, which can be obtained from this machine, include: (i) fail-safe system, (i) box-motion end-sealer which provides leak-free seals, (iii) gas-flushing device, (iv) product auto-feeders, and (v) 'packless' function, i.e. no product no bag.

(c) *Vacuum-skin packaging machines*—The principle of the vacuum-skin packaging system utilises a top and base web of material. However, it differs from the vacuum thermoforming machine (i.e. form–fill–seal) in that no forming dies are required. The product is placed on the lower web, which may be left flat or formed to suit the product if desired, before being indexed into the chamber of the machine. The top web, which has been pre-heated to improve its flexibility, is draped over the cheese portions utilising the product as the die and using minimum pressure. The air is removed by vacuum, and the product is fully encased in a protective 'second skin' before heat-sealing the top and bottom webs.

Some examples of the vacuum-skin packaging machines are (i) the Multivac® CD-6000 using the DARFRESH® packaging material which is manufactured by Cryovac®, and (ii) the Dixie Union® Pak 50E, 70 and 100S which are manufactured in Germany (Fig. 41). The latter equipment supplier also manufactures deep-drawn vacuum and shrink-bag packaging systems.

Gas flushing machines

This type of packaging is also known as the 'modified atmospheric pack', and the machine uses two reels for packing the cheese just as described above, i.e. form–fill–seal. After evacuating the air from the vacuum chamber, it is then replaced by a controlled mixture of gas(es) (carbon dioxide, nitrogen and/or combination of these). Thus, this system is referred to as gas-flushing, and it prevents the spoilage of cheese during intermediate storage prior to retailing.

In practice, carbon dioxide is widely used, compared with nitrogen as a gas-flushing agent, because it reacts with the moisture in the cheese to produce carbonic acid (H_2CO_3), and so gives the product a neat, 'snug' appearance similar to a vacuum pack. Also, the carbonic acid acts as a mould inhibitor.

Some examples of gas-flushing/cheese packaging equipment are the APV RF 452 Flowpak® (previously known as Rose Forgrove), and the Hayssen® RT 218.

The APV RF 452 Flowpak® (Fig. 41) has a capacity of 20–120 packs min^{-1} depending on the product, feeding arrangement and wrapping material. The barrier film (e.g. polyester/polyethylene or nylon/polyethylene with or without the addition of a sandwich coating of polyvinyldichloride laminate) is fed into the machine from a reel, and a pillow-type pack with fin-sealed longitudinal and end seams is produced. As the wrapping material is formed into a tube in the folding box, a portion of cheese is delivered and, before fin-sealing the edges, the oxygen is replaced by gases, (e.g. carbon dioxide or a carbon dioxide/nitrogen mixture); the residual oxygen is estimated at less than 1%. This type of packaging is referred to as the 'flexible' system; however, the thermoforming model uses a rigid or semi-rigid base material which is heated to form the tray. The product is loaded into the tray, then moves forward into a chamber where the air is evacuated and the desired gas or gases introduced. Simultaneously, a top lidding material is fed over the tray and sealed to the lip edge. This machine can also handle preformed trays, and the thermoforming and pillow wrapping types can be fitted with gas analysis equipment to monitor the residual oxygen level, or the concentration of the flushed gases (see also Guise, 1983).

The Hayssen® RT 218 horizontal form–fill–seal machine is manufactured in Italy. Depending on the overall configuration of the packaging line, speeds in excess of 80 and 100 packs min^{-1} can be achieved on 454g and 227g cheese portions, respectively. At present, a large number of these machines are operating throughout the UK, principally packaging random weight cheese portions, and 90% of the cheese wedges packed in this country use these machines (P. Bennett, pers. comm.). This type of gas-flushing packaging machine (Fig. 41) has a unique method of making the back-seal using Accutrak and a rotating die-wheel. This method of sealing the pack ensures a longer time for sealing when compared with other conventional methods of sealing and, as a result, lower sealing temperatures can be used if compatible with the packaging material. Thus, by using a suitable plastic film and providing the correct setting of the machine, the oxygen levels and rate of leakers is maintained below 1%. In addition, the packaging function of the Hayssen® machine is microprocessor-controlled which ensures a placement of the product, when transferred from the 'In-feed' conveyor, at a very high accuracy, i.e. $\sim$5 mm tolerance.

One of the recent developments within the Hayssen® machine is the packaging of cheese portions in a peel pack (e.g. similar to a tobacco pouch) with the initial seals still being hermetic, or a 'zip'-pack (i.e. a side-opening pack complete with a tear strip and a reclosable bead feature). Incidentally, such types of packaging can be achieved on Hayssen® RT 218 machines by purchasing the appropriate spare parts.

Heat-shrink packaging machines

In principle, these machines are similar to the type used for the bulk packaging of cheese. In Scotland, Cheddar and Dunlop cheeses produced on different islands (i.e. in 'square' block or 'wheel'-shapes of 200–450 g) are packaged in heat-shrink bags, e.g. Cryovac®, and normally the closure is a metal clip. However, this method of closure is being replaced by heat-sealing of the bag, after the application of vacuum and prior to heat-shrinking.

Miscellaneous equipment

Recent innovations in the retail packaging of cheese have been primarily geared towards protecting and benefiting the consumer, and the changes include the following: first, improvement in the 'ease' of opening the cheese pack; secondly, the introduction of 'peel-off' packs supplied with reclosable lids; however, the material used in some of the 'peel-off' type packs tends to change colour, i.e. become opaque, when opened; and thirdly, the 'zip'-type closures (Fig. 42). Incidentally, the reclosable packs are normally made from a semi-rigid plastic material.

The 'zip'-type closure consists of two parts: (i) the tear-off strip, and (ii) the 'zip' section (see Fig. 42). During the packaging of cheese portions, the 'zip' is wedged between the two webs in the vacuum chamber of the thermoforming machine (see Fig. 40) and heat-sealed with packaging material after removal of the air. The sequence of opening by the consumer is as follows:

- tear-off the strip;
- pull the 'zip' apart;
- remove the portion of cheese; and
- close the 'zip' and place in the refrigerator.

The cheese packs with reclosable lids or the 'zip'-type closure ensure minimal dehydration of the cheese in the domestic refrigerator, and hence protection of the product. Furthermore, the 'zip'-packs and some types of 'peel-off' packs provide the consumer with 'tamper-proof' cheese packs.

Most of cheese portion packaging machines are equipped with a weighing and labelling unit. The latter machine provides the customer with the price per kg, and the actual price of the cheese portion, the

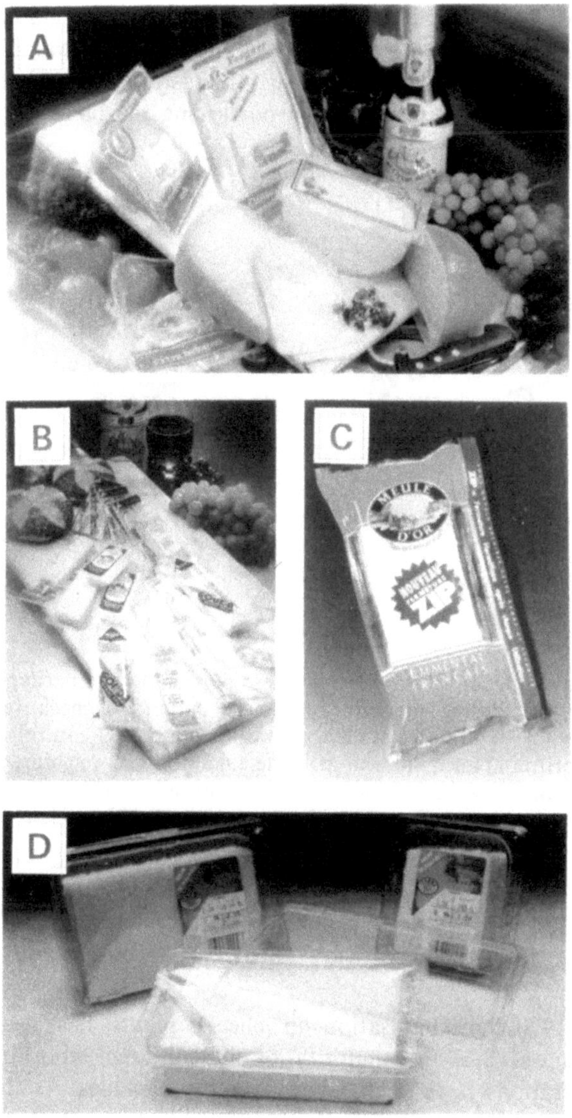

Fig. 42. Different types of packaged cheese portions. (A) and (B) Form–fill–seal pack under vacuum. (C) 'Zip'-type closure. (D) Peal-off wrap with press-on lid. (Reproduced by courtesy of Multivac® UK Ltd, Swindon, UK.)

expiry date, the type of cheese packaged and place of origin (e.g. English, Scottish, Irish, New Zealand or Canadian Cheddar). The final handling of the retail cheese pack involves placing it into shipping containers ready for dispatch to the retail shops, and Fig. 43 illustrates a highly mechanised, retail cheese packaging line from portioning the block to cartoning the packaged product.

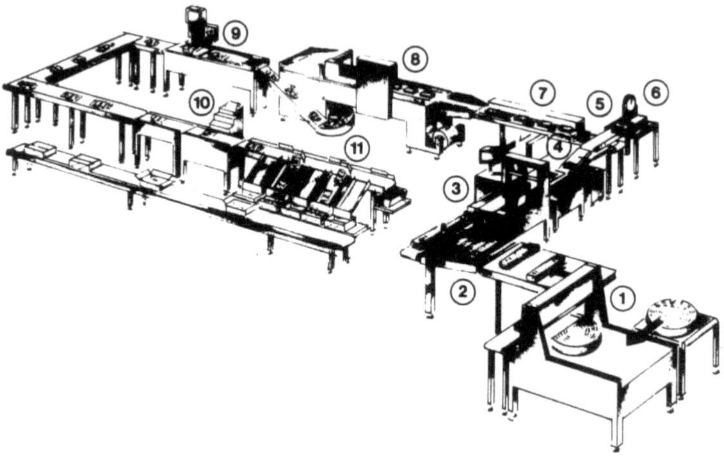

Fig. 43. Flow diagram of the Alpma cheese portioning and packaging line. (1) Block cutter, (2) feeding magazine, (3) piece cutter, (4) check-weigher with rejection flap, (5) 90° transfer system, (6) taring station, (7) control and in-feed system, (8) horizontal form–fill–seal machine, (9) automatic weighing, pricing and labelling system, (10) weighing station, and (11) installation for weight classification with boxing station. (Reproduced by courtesy of Alpma Hain & Co. KG, Rott am Inn, Germany.)

MECHANISATION OF CHEESE PRODUCTION AND PLANT DESIGN

Since the 1970s, mechanisation in cheese plants has been extensively introduced and the majority of difficult and laborious cheese-making operations have been taken over by machines. As the scale of cheese production increases or becomes more centralised, the use of mechanisation to handle the milk, curd, whey, pressing and packaging of the cheese becomes inevitable. A wide range of equipment is available for the manufacture of different cheese varieties, but the final choice of any mechanised cheese system is governed by the following:

— scale of production;
— type of cheese(s) produced;
— degree of mechanisation;
— degree of automation, e.g. 'push button' or fully integrated system;
— labour saving; and
— capital cost.

In the past, cheesemaking equipment that had been in contact with the product had to be dismantled and cleaned by hand at least once a day. However, in the mid-1950s, cleaning-in-place (CIP) was introduced by dairies and the equipment had no longer to be dismantled for cleaning. The CIP programmes were designed so that processing machines could be cleaned by circulating detergent solutions.

The beginning or birth of automation in cheese factories could be attributed to this development. Centralisation of cheesemaking has led to a rapid expansion of production capacity and, as a result, the number of operations required or needed to be executed has increased substantially. Therefore, the production line contains more valves and motors that have to be operated at a specific time during cheesemaking, and the timing of these operations becomes critical, especially in large capacity factories. For example, opening a valve too soon or late can involve product loss, and every malfunction in the process or wrong decision made by the operator, can have serious economic consequences.

The degree of automation in a cheese plant is mainly governed by the scale of operation, i.e. semi-automatic control for small- and medium-size factories and fully automatic control for large dairies. The control tasks of an automation system are divided into the following categories: (i) on/off or digital control, (ii) analogue control, (iii) monitoring, and (iv) reporting.

In view of the different types of equipment which can be used for the manufacture of cheese, the plant may be divided into sections (see Table XXV) in order to achieve a high degree of mechanisation and control.

Numerous mechanised and automated cheese systems are available on the market, and these can either be 'custom built' or obtained as a modular unit (see Figs 44–50). Whatever system is chosen, trained

TABLE XXV
Possible different sections in a large cheese factory

Processing area	*Function/duties*
Department 1 Milk handling	(a) Milk reception, cooling and storage. (b) Filtration, clarification, standardisation, bactofugation and/or MF/Bactocatch. (c) Heat treatment of milk.
Department 2 Cheese production I	(a) Production of the coagulum. (b) De-wheying and curd handling. (c) Milling, salting and mould filling.
Department 3 Cheese production II	(a) Pre-pressing (b) Brining } of the curd. (c) Pressing
Department 4 Cheese production III	(a) Bulk packaging of the cheese. (b) Storage and stock control. (c) Washing and waxing of certain cheese varieties.
Department 5 Whey handling	(a) Recovery of curd fines. (b) Fat separation. (c) Miscellaneous treatment(s) of the whey depending on its utilisation.
Department 6 Starter production	(a) Preparation of the starter culture milk. (b) Control of starter culture production. (c) Quality control of starter cultures.
Department 7 CIP station(s)	(a) Cleaning of equipment. (b) Cleaning programmes could be de-centralised.
Department 8 Quality control	(a) Monitoring the chemical and microbiological quality of the incoming raw materials and the finished cheese products. (b) Providing analytical services to other departments in the cheese factory.
Department 9 Effluent treatment	(a) Refer to Tamime and Robinson (1985) for further detail.
Department 10 Retail packaging	(a) Unwrapping of cheese blocks followed by cutting. (b) Packaging of portions and cartoning. (c) Waxing of certain cheese varieties. (d) Stock control.
Department 11 Provision of energy	(a) Hot water and steam. (b) Boiler maintenance. (c) Handling of fuel (oil and/or coal).

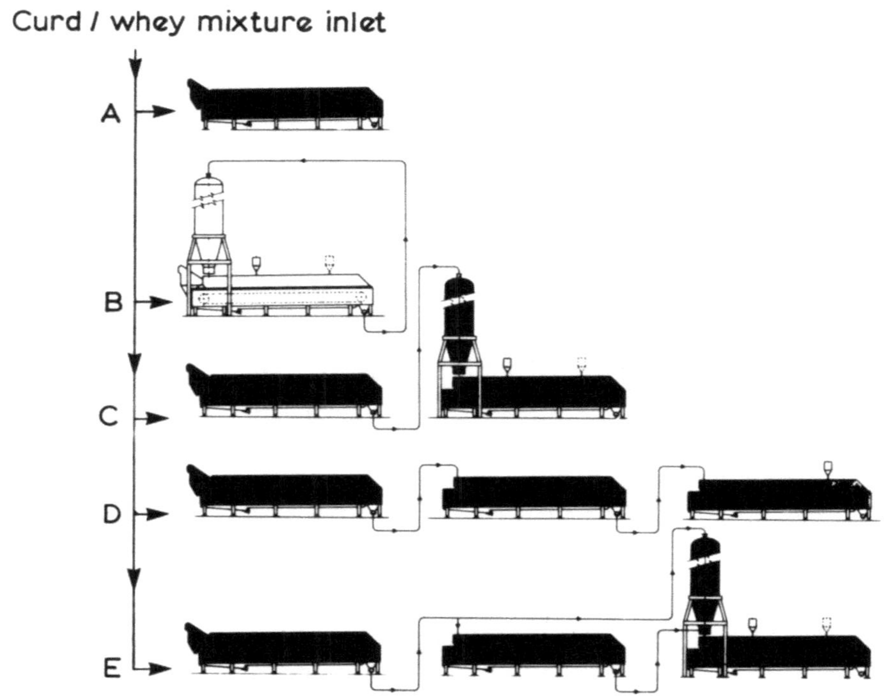

Fig. 44. Cheddarmaster installations for the production of different cheese varieties. (A) Feta or soft cheeses, (B) Cheddar or Mozzarella, (C) Cheddar or unwashed Colby, (D) stirred Cheddar or British territorials, (E) Cheddar or stirred British territorials. (Reproduced by courtesy of Van den Bergh and Partners Ltd, Windsor, UK.)

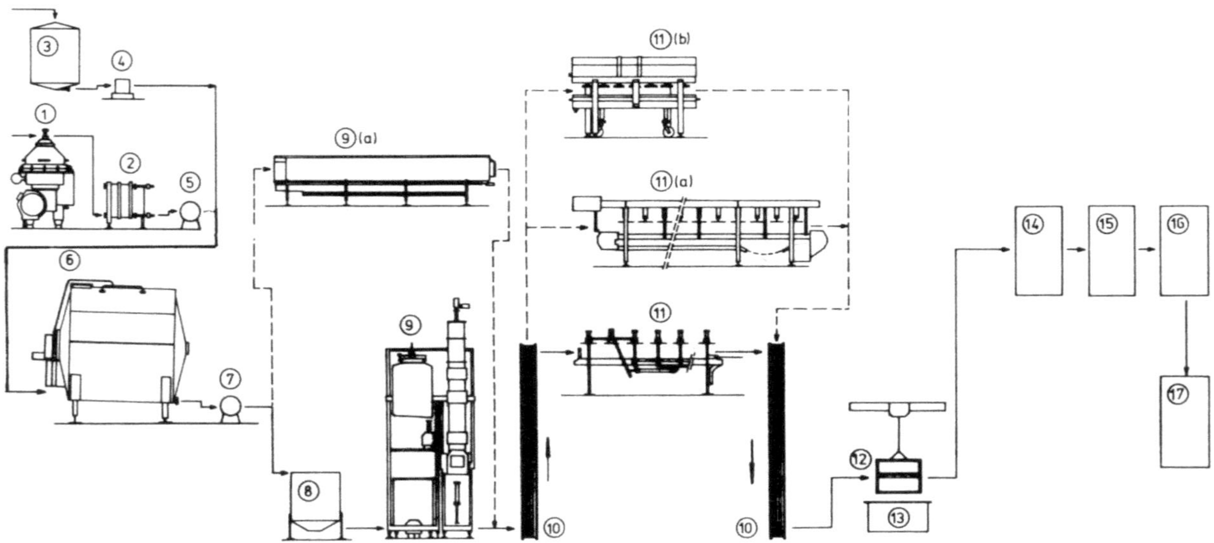

Fig. 45. Alfa-Laval production lines for Swiss, Gouda and Edam cheeses. (1) and (2) Equipment for preliminary treatment of milk, (3) bulk starter tank, (4) starter proportioning pump, (5) centrifugal pump, (6) cheese vat OST IV, (7) special pump for curd–whey mixture, (8) buffer tank, (9) Casiomatic, (9)(a) DBS strainer vat, (10) conveyor, (11) conveyor press, (11)(a) tunnel press, (11)(b) trolley press, (12) container, (13) basin for brine, (14) green store, (15) surface treatment, (16) ripening store, and (17) slicing and packaging. (Reproduced by courtesy of Alfa-Laval Cheddar Systems Ltd, Somerset, UK.)

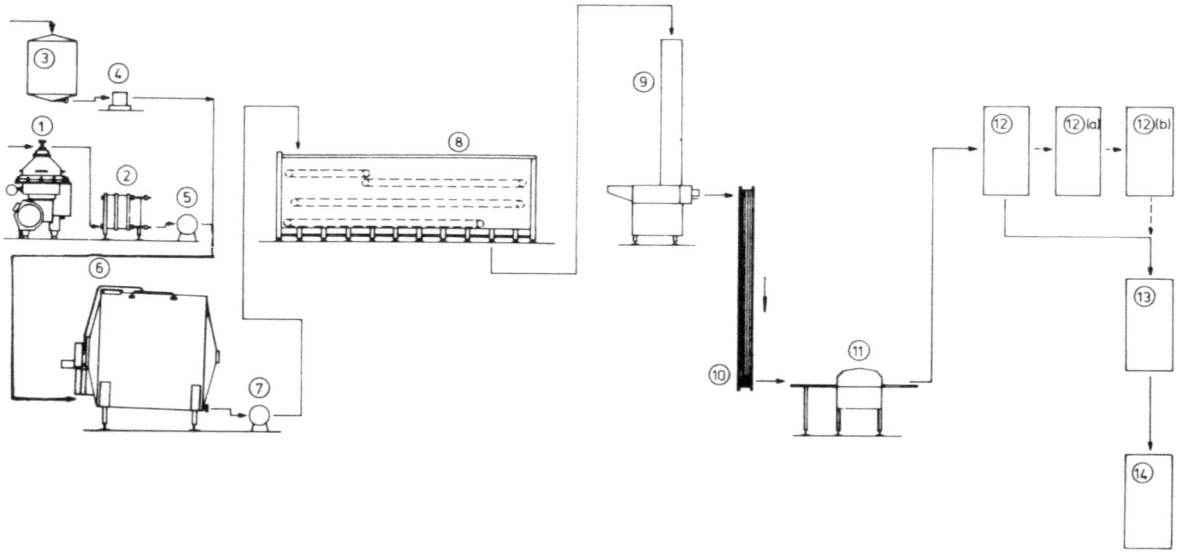

Fig. 46. Alfa-Laval Cheddar cheese production line. (1)–(7) See Fig. 45, (8) Alf-O-Matic, (9) Block Former, (10) conveyor, (11) vacuum packaging, (12) packaging–wrap around carton, (12)(a) packaging–shoe box carton, (12)(b) packaging–Unibloc®, (13) cheese store, and (14) slicing and packaging. (Reproduced by courtesy of Alfa-Laval Cheddar Systems Ltd, Somerset, UK.)

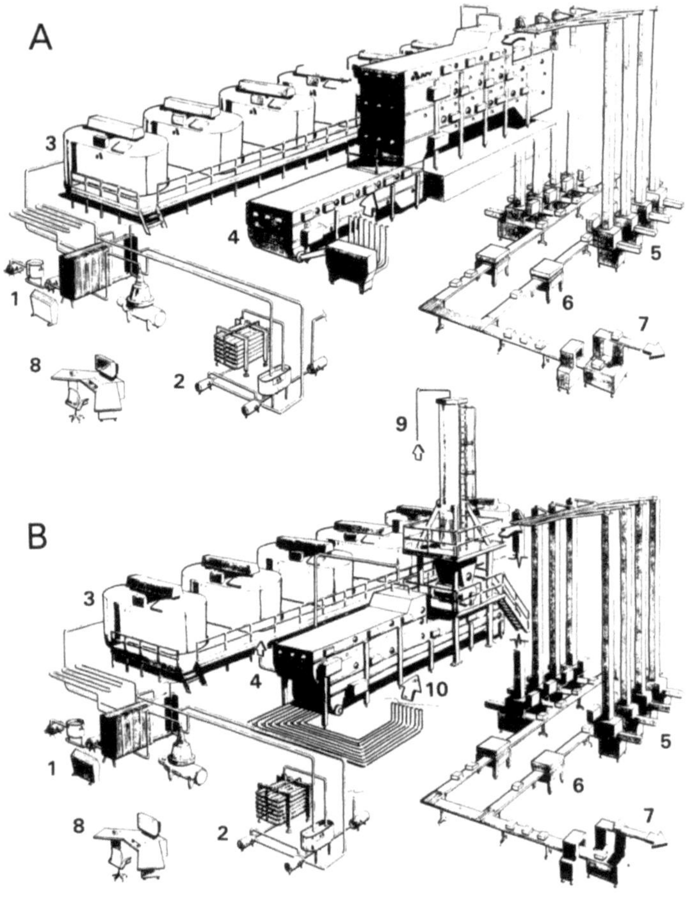

Fig. 47. Different process lines for the production of Cheddar and related cheeses. (A) (1) Pasteurisation and fat standardisation, (2) protein standardisation with UF milk, (3) cheesemaking vats, (4) draining conveyor, (5) block-formers, (6) vacuum packaging, (7) cheese block packing, (8) process control panel. (**B**) (1)–(3) and (5)–(8) are similar items mentioned in (A), (4) draining conveyor, (9) cheddaring tower and (10) curd from salting and mellowing conveyor. (Reproduced by courtesy of APV-Pasilac AS, Silkeborg, Denmark.)

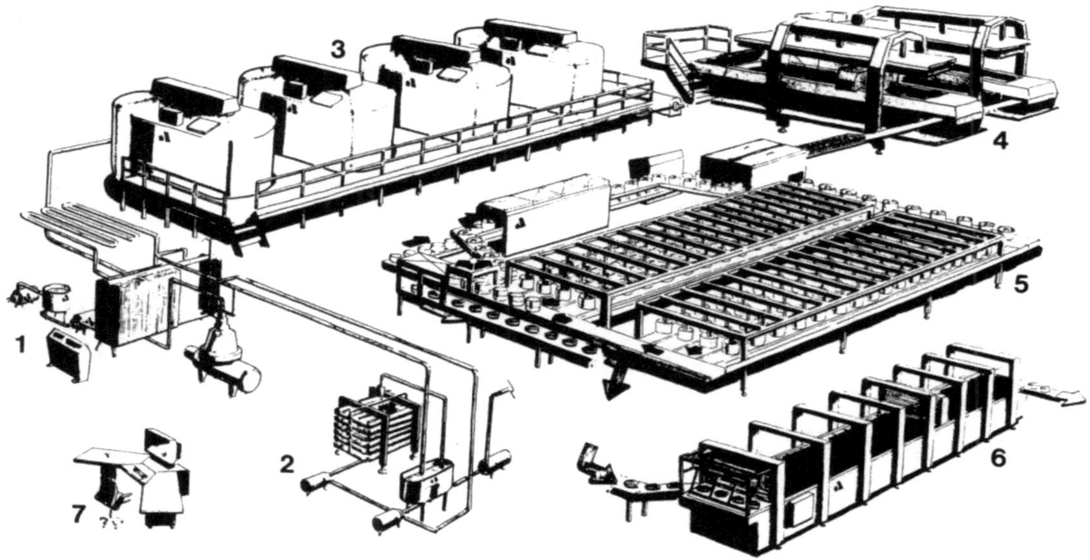

Fig. 48. APV-Passilac production line for the manufacture of Gouda or Tilsit cheese. (1) Pasteurisation and fat standardisation, (2) protein standardisation, (3) cheesemaking vats, (4) batch pre-presses (note: for further illustrations see Fig. 50), (5) after press, (6) brining plant, (7) main process control panel. (Reproduced by courtesy of APV-Pasilac AS, Silkeborg, Denmark.)

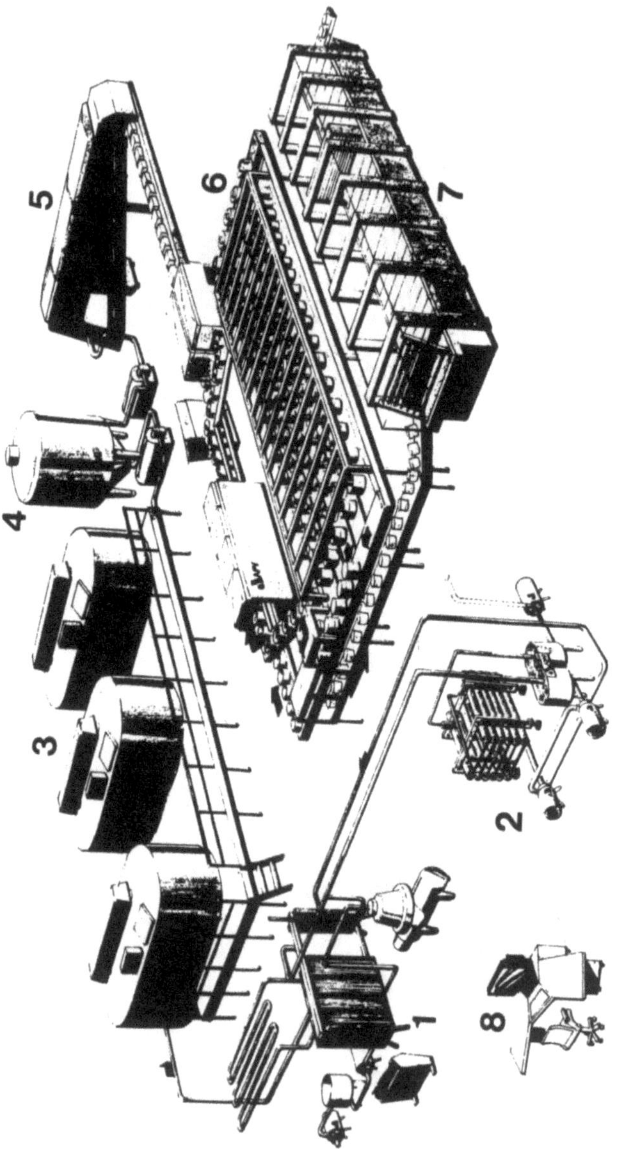

Fig. 49. An illustration of Gouda cheese processing line. (1) Pasteurisation and fat standardisation, (2) protein standardisation, (3) cheesemaking vats, (4) curd buffer tank, (5) continuous pre-press, (6) after press, (7) brining plant, (8) main process control panel. (Reproduced by courtesy of APV-Pasilac AS, Silkeborg, Denmark.)

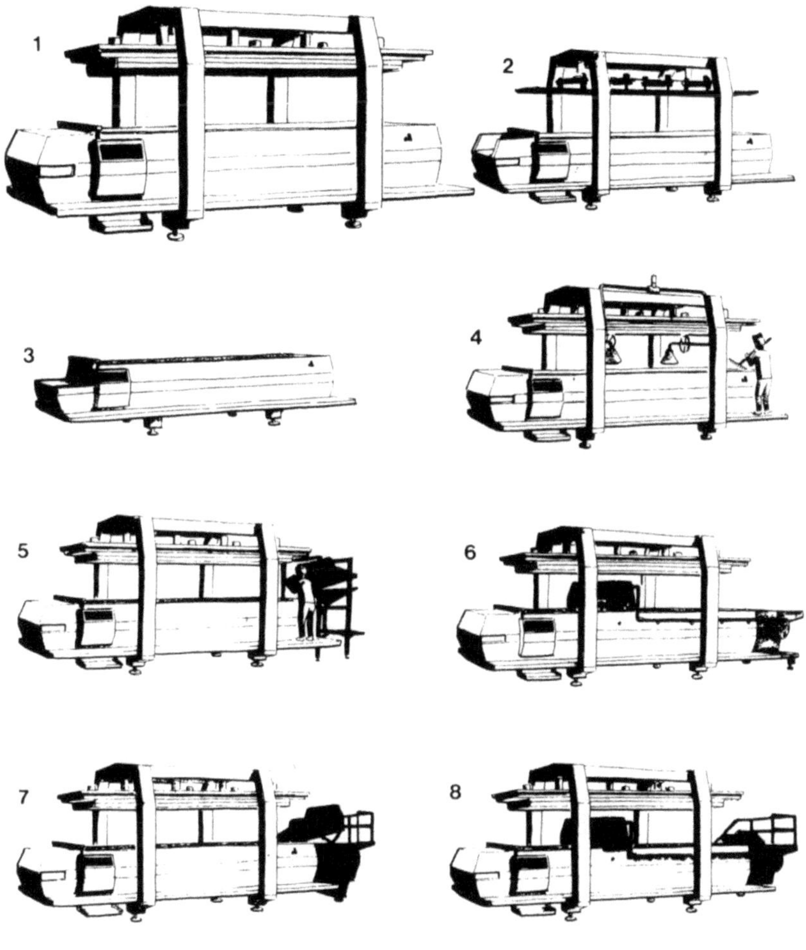

Fig. 50. Illustrations of batch pre-presses and curd distributors for the manufacture of Gouda and Tilsit cheeses. (1) Batch pre-press type OPD, (2) OPD pre-press without a lid, (3) strainer vat type OPD-S, (4) and (5) manual curd distribution; (6)–(8) automatic curd distribution. (Reproduced by courtesy of APV-Pasilac AS, Silkeborg, Denmark.)

personnel are required to operate and look after the plant (Anon., 1987*e*), and the real attractions of automation could be summarised as follows:

(a) improved efficiency of the operator under more congenial conditions;
(b) improved quality and productivity; and
(c) better utilisation of floor processing area.

The design and layout of a cheese factory plays a major role in achieving the ultimate objectives mentioned above and, in an ideal situation, the construction of a cheese factory and plant installation are carried out simultaneously. But the permutations available, particularly within existing buildings, mean that each cheese plant design has to be considered in its own right. In theory, and in practice, the design of a cheese production line should assess the following factors:

(a) the throughput of the milk processing plant should take into account the need for a continuous flow of milk to the cheese vats;
(b) the capacities of, and duration of emptying of, the cheese vats should be synchronised for the continuous flow of cheese curd to the curd handling equipment;
(c) the number of working shifts required during a normal processing day;
(d) the time required to wash the cheese vat before refilling in order to avoid the diversion of milk;
(e) the decentralisation of the CIP station is recommended, so that different equipment could be cleaned when required rather than cleaning all the plant at one time;
(f) the provision of an ample supply of hot water and steam in order to avoid an unnecessary shut down of the plant;
(g) the systems of cheese pressing and brining could influence the degree of automation adopted, and for the former system, whether or not cheese moulds are required;
(h) the area available for storage is also critical in order to ensure adequate space available, especially if long maturing cheeses are produced, and the degree of automation employed to handle the cheese in the store; and
(i) the design of any cheese plant should accommodate some flexibility for future increase in production without incurring high capital investments or tremendous changes in the existing equipment.

As mentioned elsewhere, one of the advantages of cheese mechanisation is increased productivity and a reduction in labour costs. The latter aspect has already greatly improved, and according to Wilbrink (1982), the labour requirements to produce 100 kg of Edam cheese (in 2 kg size) and the same amount of Gouda cheese (in 10 kg size) in 1965 were 149 and 94 man minutes respectively; however, in 1981 these figures were reduced to 42 and 12 man minutes, respectively. Table XXVI illustrates the improved productivity and reduction of manpower obtained using the Gadan A/S mechanised pre-pressing and pressing systems during the manufacture of a wide range of semihand cheeses. It is evident that a cheese plant is a highly complexed structure, and Fig. 51 illustrates the layout of a factory for the production of Emmental and Edam cheeses in Finland.

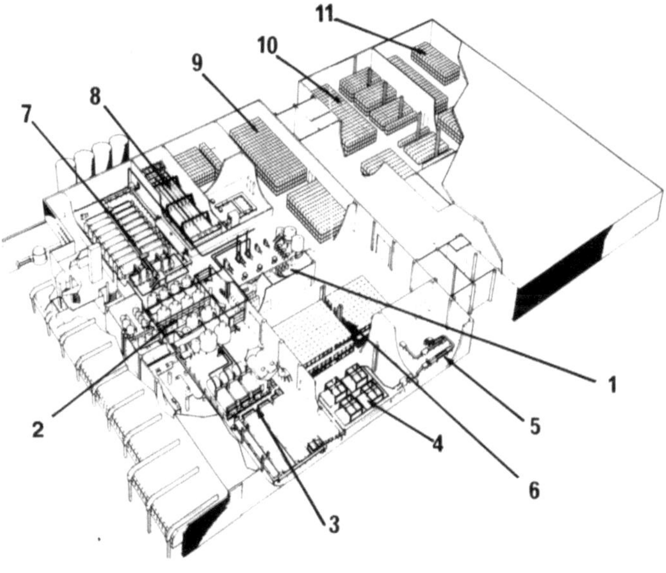

Fig. 51. Layout of the Yhteisjuustola cheese factory in Lapinlahti, Finland. (1) Milk treatment, (2) MKT-S cheese vats. Edam section, — (3) de-wheying, mould filling and pressing equipment, (4) brining tanks, (5) and (6) packaging and maturation rooms. Emmental section, — (7) Tebel-MKT tunnel presses, (8) brining tank, (9)–(11) maturation rooms at 8°C, 23–26°C and 4–6°C, respectively, (12) cheese wrapping and packaging section). (Reproduced by courtesy of Tebel-MKT BV, Jarvenpaa, Finland.)

TABLE XXVI

Examples of labour force required when using the Gadan mechanised cheesemaking equipment[a]

Type of cheese	Production ($t\ day^{-1}$)	Size of batch (litres)	Capacity of plant ($litres\ h^{-1}$)	Size of cheese (kg)	Space required (m^2)	Number of operators	Equipment used
Danbo, Samso, Maribo, Elbo, Herregardsost, Trope Fynbo	12	15 000	22 500	5–14	345	1–2	1 × 15 000 litre pre-pressing vat with automatic pressing plate and cutting unit. Three 'final presses' with automatic filling, emptying and cleaning the moulds.
Gouda, Jarlsberg	27	15 000	37 500	13	525	1–2	2 × 15 000 litre pre-pressing vats with curd distributor. Three automatic 'final presses' for mould handling. One waiting section for acid development in the cheese.
'Ball' Edam	3·6	6000	4850	1	50	1	1 × 6000 litre automatic pre-pressing vat including cutting, filling and emptying units. One press equipped with 20 yokes where press hose secure correct pressure for 600 cheeses.
Tilsiter, Gouda, Edam	16	16 000	16 000	3–15	150	1	1 × 16 000 litre automatic pre-pressing vat with curd distributor. One automatic press and change of moulds.
Polar, Edam	NR[b]	16 000	13 600	13·5	420	1–2	4 × 16 000 litre cheese tanks. 2 × 16 000 litre automatic pre-pressing vat. Two pressing lines in two tiers. De-moulding is carried out automatically by means of vacuum.

[a]Data was compiled from Gadan A/S technical information sheets number 1-0019-1, 1-0021-1, 2-0006-1, 2-0010-1 and 3-0004-1.
[b]NR: not reported.

Application of Ultrafiltration for Cheesemaking

Ultrafiltration (UF) is a membrane separation process which is carried out under pressure using a semi-permeable membrane made out of a high-polymer substance or cellulose acetate. The milk constituents (i.e. water, lactose, mineral salts, organic acids and other low-molecular-weight material) that pass through the membrane are referred to as the permeate, and the materials (e.g. fat and proteins) that do not pass through the membrane are known as the retentate. Thus, in theory, the application of UF in the cheese industry resembles the traditional concept of cheesemaking, in which the milk fat and proteins are concentrated by means of whey separation.

Over the past two decades, a wide range of data has been published on the application of UF in the dairy industry, including the handling of milk, whey and cheese (Green *et al.*, 1984; Glover, 1985, 1986; Green, 1985, 1987; Cheryan, 1986; Jensen *et al.*, 1987; Lawrence, 1987; Lelievre and Lawrence, 1988; Renner and Abed El-Salam, 1991). The application of UF in cheesemaking is classified into the following categories based on the concentration factor of the fat and proteins in the retentate: (i) *low* concentration (up to two-fold) so that conventional cheesemaking techniques can be employed using traditional or existing mechanised equipment, (ii) *medium* or pre-concentration of the milk (three- to six-fold) and, as a result, the subsequent cheesemaking process has to be modified; however, some whey drainage will take place, but different equipment is normally used, and (iii) *high* concentration, i.e. six- to eight-fold, where the dry matter content in the retentate is the same as in the cheese; this concentrated milk is known as 'pre-cheese', and whey drainage does not occur; the cheesemaking equipment is entirely of new design (Anon., 1987*e*).

Some of the potential advantages of UF technology for the manufacture of cheese are as follows:

- increased cheese yield due to improved recoveries of protein and/or fat, and the retention of whey proteins which otherwise will be lost;
- reduction in the cost of ingredients used per tonne of cheese, i.e. lower addition rates of coagulant, starter cultures, colouring matter or calcium chloride are required;
- reduction in some of the manufacturing costs, such as energy, equipment and labour;
- improvements in the quality and compositional control of certain cheese varieties (e.g. soft/fresh and semi-soft), but the changes may have a different significance in hard and semi-hard cheeses; and
- increased output of cheese per day.

At present, UF has found its greatest application in the manufacture of cheeses, such as Quarg, Cottage, Ricotta, Feta, Mozzarella, Camembert, Brie or Queso Blanco (see Chapter 3 of this volume; Coton, 1986; Tamime and Kirkegaard, 1991).

Some of the problems, which may be encountered in UF cheesemaking, are texture, flavour, moisture retention and rate of acid development. Nevertheless, by understanding UF cheese technology, such problems have been avoided by, for example, subjecting the retentate to a high heat treatment (i.e. $>80°C$), homogenisation of the concentrated milk and, in some instances, the application of diafiltration to adjust the lactose, salt or the mineral content. As a result, fresh/soft and semi-soft cheeses have been produced successfully in large cheese factories in different countries. However, as UF cheesemaking is gradually being accepted as a technology for the manufacture of hard and semi-hard cheeses, great progress has taken place over the past few years, and some selected data using ultrafiltration for the manufacture of Esrom, Danbo, Havarti and Gouda have been reported by Jacobson (1985), Qvist *et al.* (1985, 1987), Qvist (1987), Skovhauge (1987), Spangler *et al.* (1989, 1990) and Borgstrom (1990). An illustration of the manufacture of these cheeses, including Cheddar, using UF technology is provided in Fig. 52.

An important aspect, which has to be considered when using UF technology for the manufacture of hard and semi-hard cheeses, is the economic cost and profitability of a cheese factory. Barbano *et al.* (1989) carried out a comparative study on the economics of Cheddar cheese manufacture (traditional method versus UF) using the 'Economic Engineering Approach' of different size cheese factories. They concluded that, by using UF retentate at six-fold concentration, the process was most profitable when compared with low UF concentration (i.e. < two-fold) or the traditional process. A mechanised and fully automated process for the manufacture of Cheddar cheese from UF retentate (five-fold concentration) was developed by CSIRO and APV Bell Bryant Pty in Australia in 1985. The process is known as the APV-SiroCurd (Garrett, 1987*a*,*b*; Jameson, 1987, Anon., 1987*f*, 1988*c*).

A process outline of the APV-SiroCurd system is shown in Fig. 53, and such a plant offers continuous and automated cheesemaking for the production of Cheddar, Colby and washed curd cheeses. The sequence of operations of the APV-SiroCurd can be described as follows:

(a) *Ultrafiltration*—The standardised milk is pasteurised at 72°C for 15 s and concentrated to five-fold by UF in a multi-stage plant in

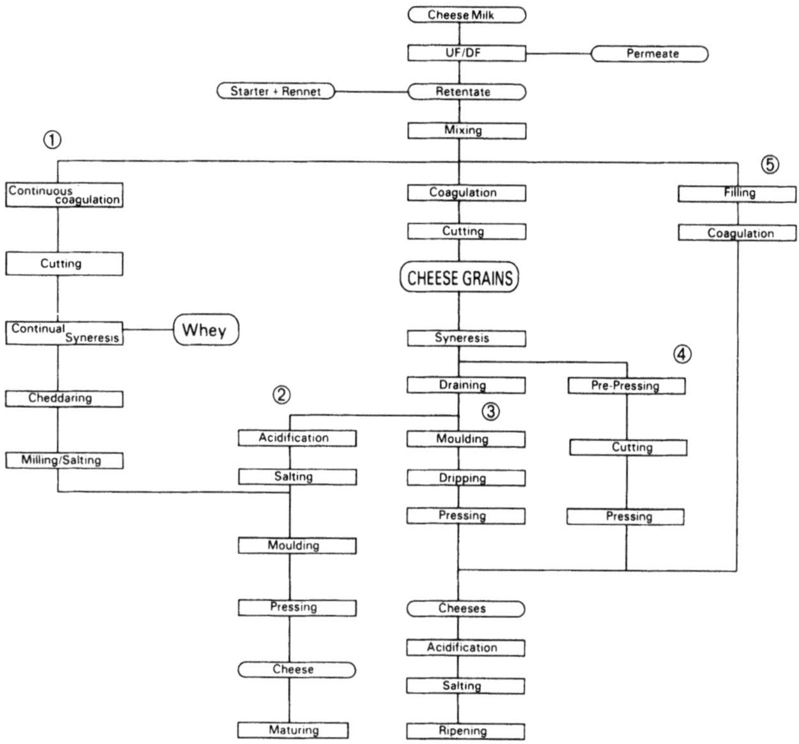

Fig. 52. Process diagram for ultrafiltrated matured cheeses. (1) Cheddar cheese using the APV-SiroCurd process, (2) barrel Cheddar cheese, (3) Havarti, Esrom, Montery Jack, Feta or Danablue cheese, (4) Danbo or Gouda cheese, and (5) Danbo, Saint Paulin, Camembert, Feta, Domiati or Akkawi cheese. (Adapted from Osterggard (1986) and Phillips (1989).)

order to minimise the shear damage. At the same time, diafiltration is carried out in conjunction with UF in order to standardise the salt balance. The retentate (i.e. ~40% total solids) is stored for a short time in a buffer tank(s) so guaranteeing a constant supply of milk to the processing plant regardless of any fluctuations that may occur during UF.

(b) *Fermentation*—A portion of the retentate, ~10%, is sterilised at high temperature, cooled, inoculated with starter culture and fermented overnight (i.e. ~18 h). Due to the fact that the retentate is a highly buffered starter growth medium, a starter population of $1\text{–}2 \times 10^9$ cfu g^{-1} can be easily attained with minimum acid-induced cell damage (Jameson, 1987). The remaining 90% of the retentate is blended with fermented retentate made the previous day at a ratio of 90:10. This latter blend, which is known as a ripened retentate, is then fed directly into the coagulator.

(c) *Coagulation*—The continuous coagulator (see Fig. 53) is made out of six large diameter tubes. The retentate blend (i.e. 90:10 mixture) is mixed with the coagulant solution and/or colouring agent before being fed into the bottom of the coagulator, sequentially filling the coagulating tubes. The residence time in the tubes is 16 min, and this is accurately controlled by the automation system. By the time the retentate blend reaches the outlet of the coagulator, the milk has already coagulated due to the fact that gel formation in UF retentate occurs when only a fraction of the κ-casein has been hydrolysed by the acid proteases. However, aggregation of casein micelles in milk occurs after ~80% of the κ-casein has been hydrolysed (see Tamime and Kirkegaard, 1991). Thus, the coagulation/aggregation of the retentate is much more rapid than with conventional cheesemaking, and the volume of the curd made from this UF retentate *vis-à-vis* milk is 50 and 10%, respectively.

The coagulum emerging from the coagulator tubes is very firm, and it is cut into 'curd strips' as it passes through a stationary set of vertical and horizontal knives. A rotating knife cuts the curd strips into 10 mm cubes. Timing of the cutting of the coagulum is critical, because (i) cutting too early causes shattering of the curd and

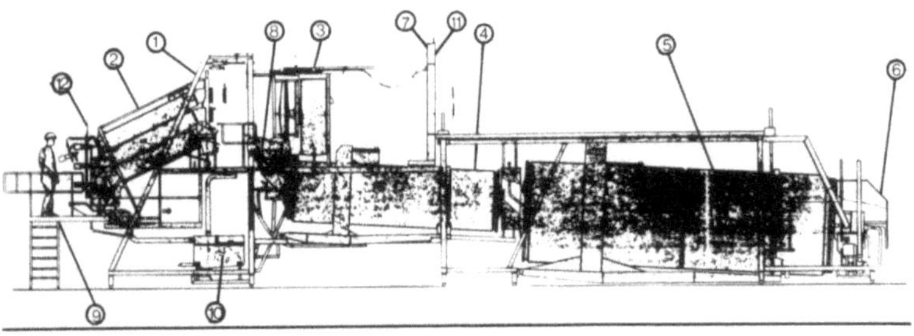

Fig. 53. Schematic illustration of the APV-SiroCurd continuous coagulator and syneresis unit. (1) Coagulator and drum lifting frames assembly, (2) coagulator tubes assembly, (3) rotary cutter working area, (4) small syneresis drum assembly, (5) large syneresis drum assembly, (6) curd and whey transfer assembly, (7) hydraulics layout, (8) cutting grid handling and cleaning-in-place (CIP), (9) stairway and access platforms arrangement, (10) CIP collection, (11) electrical layout and (12) pneumatic layout. (Reproduced by courtesy of APV Crepaco Inc., Wisconsin, USA.)

excessive losses in the whey, and (ii) cutting too late can retard synersis (Jameson, 1987). The curd mass is continuously forced out of the coagulator tubes by the incoming blend of retentate for the next fill cycle.

(d) *Syneresis*—The curd cubes are delivered to the inlet of the first of two, slowly rotating drums which are known as syneresis drums. These drums are slightly inclined, fitted with internal vanes and have end-openings. As the curd moves forward through these drums, the curd is continuously agitated and steadily heated which simulates conventional cheesemaking. The residence time is ~40 min, and by regulating the velocity of the drum rotation as well as retention time and temperature, the cheese maker can control both the rates of acid development and retention of moisture in the end-product in the same manner as for curd in a conventional cheese vat.

(e) *Curd handling*—The curd leaving the syneresis drums is delivered to conventional, mechanised cheddaring, milling, salting and pressing equipment, for example, the Cheddarmaster, Alf-O-Matic, Damrow DMC, Scherping CCDM or the Wincanton DCC (refer to earlier text in this chapter).

At present the total number of the APV-SiroCurd installations amount to two, and Fig. 54 shows one such on-site unit in a cheese factory in the US (see also Honer, 1989).

The concentration of milk by UF for the manufacture of cheese is normally carried out in factories. However, since milk production and handling on the farm in most milk-producing countries is highly mechanised, French scientists (Maubois and Mahaut, 1974) in the mid-1970s explored the possibility of introducing UF technology to dairy farms. The warm milk from the milking parlour is ultrafiltered to give a two-fold concentration, thermised and finally cooled. Thus, because the volume of milk is halved, the concentrated milk will be collected once every 2 days without requiring larger storage facilities on the farm. This approach becomes especially attractive in remote places, because the transportation costs of milk will be reduced and the permeate can be fed to the cows. Such milk will be suitable for manufacturing purposes, such as fermented milks, cheese, high-protein powder, butter or chocolate crumb. Zall and Chen (1981) evaluated the system in the US, followed by cheese trials under factory conditions in California (Zall, 1987*a*,*b*). The technical and economic studies of the on-farm UF of milk have been reported by Slack *et al.* (1982*a*,*b*) who concluded that this process is economically viable for herd sizes $\geqslant$100 cows.

Fig. 54. On-site illustrations of the APV-SiroCurd equipment at the Land O'Lakes-Perham cheese factory. (A) View of the buffer tank, platform and fermentation vessel, (B) coagulator tubes. (C) syneresis drums 1 and 2, (D) outlet end of the syneresis drum 2 and associated dischage hopper. (Reproduced by courtesy of APV Crepaco Inc., Wisconsin, USA.)

Between 1987 and 1989, a collaborative research programme was carried out between SAC-Auchincruive, the Scottish Milk Marketing Board (SMMB) and Alfa-Laval in Sweden to evaluate on-farm UF for the production of Scottish Cheddar cheese. Preliminary studies using either a 500 litre open vat or 3000 litre Tebel Ost-IV vat for the production of Cheddar cheese were good, and only slight modification of the conventional cheesemaking process (i.e. cutting the coagulum into slightly larger cubes and reducing the cooking temperature by 4°C) was required to produce a quality Cheddar cheese with around 35–37% moisture (A. Y. Tamime, unpublished data). However, in 1990 when two UF plants were instaled on commercial farms where the combined milk was sufficient to run a semi-industrial scale investigation at least once every month, the research programme was abandoned due to (i) changes in the strategic marketing policy of the SMMB, and (ii) availability of government funding for 'near market' research projects. Nevertheless, the advantages and disadvantages to the farmer, haulier and the manufacturer of using such developed technology can be summarised as follows:

I FARMER
 (A) Advantages
  1. Reduction in cooling cost;
  2. improvement in microbiological quality of milk;
  3. milk collection on second or third day rather than daily;
  4. lower cost of concentrate because the permeate is fed to the cows (i.e. energy); and
  5. premium component pricing of milk from enhanced value to the manufacturer.
 (B) Disadvantages
  1. High initial cost of UF equipment;
  2. higher operational cost e.g. spare parts, special detergent and requirement of high-quality water (the latter aspect may not pose a problem in Scotland in terms of calcium and magnesium hardness); and
  3. statutory regulations.

II HAULIER
 (A) Advantages
  1. Lower transport cost; and
  2. possible reduction in the number of tankers required.
 (B) Disadvantages
  N/A

III MANUFACTURER
 (A) Advantages
  1. Better quality of incoming milk;
  2. lower cost of processing:
   (a) lower energy requirement (electricity, oil, steam etc);
   (b) reduced manpower;

(c) shorter daily production period;
(note: (a), (b) and (c) are achieved because the same volume of cheese could be produced by using UF milk equivalent in volume to 50% of the original milk);
(d) reduction in whey disposal, handling and effluent treatment; and
(e) reduction in the amount of starter and coagulent required;

3. by using two-fold concentrated UF milk, no specialised cheesemaking equipment is required; and
4. possible increase in cheese yield.

(B) Disadvantages

1. Adoption of new technology in cheesemaking.

It is most likely that on-farm UF of milk will become an accepted and widely practiced technology in the future, depending mainly on the following aspects: first, the economies mentioned above will be realised if milk production on the farm and processing of milk in factories are within one corporate body; and secondly, relaxation in statutory regulations so that UF milk becomes a permitted raw ingredient for the manufacture of cheese.

FUTURE DEVELOPMENTS AND CONCLUSION

The process of cheesemaking (i.e. very hard, hard and semi-hard varieties) is an effective method of preserving the nutritional components of milk for a long period of time. It is likely, however, that the future developments in this field, based on current scientific research work, will aim to fulfil some of the following requirements.

(a) *Milk as a raw material*—Seasonal and other variations in the chemical composition of milk, which ultimately affects the quality of milk, can be overcome by standardisation of the casein and fat components to a desired, fixed ratio. The concentration of milk by UF is orientated towards better utilisation of the cheese milk (i.e. improved productivity, reduced cost and increased yield), but more research is required to optimise the cheesemaking process.
(b) *Starter cultures*—Selection and identification of cheese starter culture strains is likely to increase, in order to overcome certain problems during manufacture (e.g. attack by bacteriophages), and to produce cheese of similar flavour characteristics within desired and well-defined parameters. An understanding of the physiology

of these cultures may help to improve their function during cheesemaking, including the application of genetic manipulation. Over the past decade, much research has been conducted on lactic acid bacteria in this field in laboratories all over the world, and some selected information has been reported by Gasson and Davies (1984), Gasson (1990), Kok (1990), McKay and Baldwin (1990), McIntyre and Harlander (1990) and Vos (1990). It is possible to suggest that genetically manipulated cheese starter cultures will be widely used in the industry by the turn of the century. The direct acidification of milk by organic acids (e.g. lactic, acetic or phosphoric) is likely to be only applicable to non-ripened cheese varieties, (i.e. fresh and semi-soft cheeses).

(c) *Milk coagulants*—Since the enzyme rennet (i.e. a mixture of chymosin and pepsin) is widely used in the cheese industry and is in short supply worldwide, the major development in this area will be the search for an alternative and acceptable replacement, or to manipulate genetically certain microorganisms to produce chymosin more economically (Beppu, 1990; Teuber, 1990). At present (1991), the current prices per litre of different coagulants in the UK are as follows: calf rennet (£10·00), microbial coagulant (£6·50) and recombinant chymosin (£5·70).

(d) *Mechanisation of cheesemaking*—Mechanisation will continue with a view to improving productivity and reducing labour costs. A high degree of automation, including the incorporation of microprocessor control, has been achieved in the cheese industry over the past two decades—but there are still areas in cheesemaking where further process controll could be an advantage.
For example:

- an in-tank instrument for the measurement of milk gel rigidity before cutting of the coagulum;
- an in-tank pH recorder of the acid development during cheesemaking; and
- an instrument for measuring the ripening indices during the maturation of cheese in store.

(e) *Accelerated ripening of cheese*—Different methods have been used to accelerate the ripening process of Cheddar and other cheeses, and comprehensive reports have been compiled by Law (1976, 1984, 1990), El Soda (1986), Conner (1988), Fox (1988–89), and El Soda and Pandian (1991). What is highlighted from current research is the need for a better understanding of the biochemistry of the maturation of cheese, including the mechanisms of flavour development, in order to overcome obstacles, e.g. undesirable flavours, discolouration, and body and texture defects. However, enzymes such as proteases and lipases extracted from certain bacterial and fungal species, including dairy starter culures, have been successfully used to accelerate the ripening of cheese. It is safe to assume that the main industrial application of these enzymes is to produce cheese which will be utilised for the manufacture of processed cheese.

(f) *Packaging materials and equipment*—Packaging technology of cheese has developed greatly over the past few years, and the future will see an increase in the rate of packaging (i.e. bulk and/or retail portion) and improved specifications of the packaging materials at a reduced cost. Future developments in the field of cheese packaging may include the following: first, the use of plastics which are environmentally friendly; secondly, greater use of modified atmosphere packs, and third, wider application of tamper-evident packs to protect the consumer.

(g) *Flavouring of cheese*—Natural cheeses flavoured with herbs have been produced for some time, and a recent development is the addition of alcoholic beverages intended to increase cheese consumption. Tamime (1984) concluded that the addition of concentrated wine flavour to Cheddar cheese did not affect the quality, and such cheese has a potential market.

(h) *Low-fat cheese*—The world demand for cheese is expected to expand in the future at a rate in excess of growth in population (Sliter, 1990*a,b*); however, some of the potential expansion is going to be in the low-fat cheese market, for example, a Cheddar-type cheese (Banks *et al.*, 1989). Low-fat Cheddar cheese production in Scotland has increased six-fold since 1985 when the production figure was 240 t; in 1990–91, it had increased to 1440 t (J. Russell, pers. comm.), and the future trend is for increased production.

(i) *New method(s) of making cheese*—The manufacture of cheese in a shorter period helps to contain the rising cost of production and helps to increase the throughput of existing plants. A typical example is the 'Short Method' of Cheddar cheese manufacture which was developed in Australia. Czulak *et al.* (1954), Hammond (1979, 1982), Hammond and Freeman (1983), Radford (1984) and Hutkins *et al.* (1986) have reported that the manufacturing times (i.e. from cutting → de-wheying and from de-wheying → milling) based on the conventional method to produce Cheddar cheese are 140 and 90 min; the 'Short Method' takes 120 and 30 min, respectively. Thus, the manufacturing time is reduced by 80 min; however, the starter cultures consisted of mesophilic lactic acid bacteria and an active strain of *Str. salivarius* subsp. *thermophillus* (TS3 or SD1). It is possible to suggest that the 'dry stirring' method for the manufacture of Cheddar cheese in which the matting of the curd, the cheddaring stage and the milling of the curd are omitted should be considered also as a 'short method' of production. For example, the processing time (i.e. from renneting → mould filling) requires up to 4 h as compared with the traditional method of 5 h (A. Y. Tamime, unpublished data).

(i) *UF and cheesemaking*—Further research in UF technology for cheesemaking is still required for the manufacture of very hard, hard and semi-hard cheeses of excellent quality. When UF technologies are fully developed, then almost all cheese varieties (e.g. from Cottage to Emmental and Cheddar cheeses) can be made. It can be argued, however, that the first stage of development in this field will be achieved if the on-farm UF of milk becomes widespread, especially in areas where the milk is primarily utilised for cheesemaking.

ACKNOWLEDGEMENTS

The author acknowledges and is grateful for the enthusiastic response and assistance of all the companies in providing extensive data required, and to Mr E. C. McCall (SAC-Auchincruive) for the preparation of the figures.

REFERENCES

Al-Darwash, A. K. (1983). Changes in the characteristics and properties of milk from production to consumption—cheese manufacture and quality. PhD thesis, University of Glasgow, Glasgow, UK.

Al-Obaidi, G. Y. (1980). A study of the use of coagulants in cheddar cheesemaking. PhD thesis, University of Glasgow, Glasgow, UK.

Anon. (1959). In: *Cheesemaking*. Bulletin No. 43, HMSO, London, UK.

Anon. (1976). *North European Dairy J.*, **42**, 248.

Anon. (1979). *North European Dairy J.*, **45**, 264.

Anon. (1980). *North European Dairy J.*, **46**, 36.

Anon. (1982*a*). In: *Yearbook of Agricultural Statistics—1978/1981*, Brussels, Belgium.
Anon. (1982*b*). *North European Dairy J.*, **48**, 103.
Anon. (1985*a*). *North European Dairy J.*, **51**, 136.
Anon. (1985*b*). *Food Technol. NZ*, **20** (10), 23.
Anon. (1985c). *North European Dairy J.*, **51**, 262.
Anon. (1985*d*). *North European Dairy J.*, **51**, 205.
Anon. (1986). *North European Dairy J.*, **52**, 290.
Anon. (1987*a*). *North European Dairy J.*, **53**, 305.
Anon. (1987*b*). *North European Dairy J.*, **53**, 109.
Anon. (1987*c*). *North European Dairy J.*, **53**, 84.
Anon. (1987*d*). *North European Dairy J.*, **53**, 52.
Anon. (1987*e*). In: *Dairy Handbook*. Alfa-Laval A/B, Lund, Sweden, p. 319.
Anon. (1987*f*). *Rural Res.*, **136**, 9.
Anon. (1988*a*). In: *The World Market for Dairy Products—1988*, International Dairy Agreement, 9th Annual Report, General Agreement on Tariffs and Trade (GATT), Geneva, Switzerland, p. 29, 57.
Anon. (1988*b*). *North European Food Dairy J.*, **54**, 107.
Anon. (1988*c*). *Caseus*, **December**, 11.
Anon. (1989). *North European Food Dairy J.*, **55**, 205.
Banks, J. M. (1990). *J. Soc. Dairy Technol.*, **43**, 35.
Banks, J. M. and Tamime, A. Y. (1987). *J. Soc. Dairy Technol.*, **40**, 64.
Banks, J. M., Muir, D. D. and Tamime, A. Y. (1984*a*). *J. Soc. Dairy Technol.*, **37**, 83.
Banks, J. M., Muir, D. D. and Tamime, A. Y. (1984*b*). *J. Soc. Dairy Technol.*, **37**, 88.
Banks, J. M., Griffiths, M. W., Phillips, J. D. and Muir, D. D. (1986). *Dairy Indust. Inter.*, **51**(7) 31.
Banks, J. M., Griffiths, M. W., Phillips, J. D. and Muir, D. D. (1988). *Food Microbiol.*, **5**, 9.
Banks, J. M., Brechany, E. and Christie, W. W. (1989). *J. Soc. Dairy Technol.*, **42**, 6.
Barbano, D. M., Hurst, S. and Aplin, R. D. (1989). Paper presented at the *1st Symposium on Advances in Dairy Product Technology—Application of Membrane Separation Processes*, San Luis Obispo, California, USA.
Beppu, T. (1990). In: *Proc. XXIII International Dairy Congress*, Vol. 3, Mutual Press, Montreal, p. 1604.
Berg, G. van den (1990). In: *Proceedings of the XXIII International Dairy Congress* (Vol. 3), Mutual Press, Montreal, Canada, p. 1864.
Berg, G. van den, Daamen, C. B. G. and Stadhouders, J. (1989). *North European Food Dairy J.*, **55**, 63.
Bines, V. E., Young, P. and Law, B. A. (1989). *J. Dairy Res.*, **56**, 657.
Bogh-Sorensen, T. (1989). *Scand. Dairy Indust.*, **3**(1), 48.
Bojgaard, S. E. (1988). *North European Food Dairy J.*, **54**, 126.
Borcherds, K. B. (1983). In: *Miles Jaarlikse Haasmakers Symposium*. Miles Lab. (Pty) Ltd, Cape Town, Republic of South Africa, p. 45.
Borcherds, K. B. (1985). In: *SERAVA Annual Cheesemakers Symposium*. Miles Lab. (Pty) Ltd, Cape Town, Republic of South Africa p. 28.
Borgstrom, U. (1990) *Scand. Dairy Inform.*, **4**(3), 42.
Bouman, S. (1979). *North European Dairy J.*, **45**, 4.

Bramley, A. J. and McKinnon, C. H. (1990) In: *Dairy Microbiology* (Vol. 1, 2nd edn), ed. R. K. Robinson. Elsevier Applied Science Publishers Ltd, London, UK, p. 163.
Brochu, E., Dumas, R., Julien, J. P., Nadeau, J. P. and Riel, R. (1985). *Dairy Science and Technology—Principles and Applications.* La Fondation de Technologie Laitiere du Quebec Inc., Canada.
Brockwell, I. P. (1981). In: *Proc. 2nd Biennal Marschall International Cheese Conference*, Wisconsin, USA, p. 208.
Chamba, J. F. (1989). In: *Food Ingredients Europe—Conference Proceedings—Paris*, Expoconsult Publishers, Maarssen, p. 104.
Chang, S. F., Ayres, J. W. and Sandine, W. E. (1985). *J. Dairy Sci.*, **68**, 2846.
Cheeseman, G. C. (1981). In: *Enzymes and Food Processing*, ed. G. G. Birch, N. Blakebrough and K. J. Parker. Elsevier Applied Science Publishers Ltd, London, UK, p. 195.
Cheryan, M. (1986). In: *Ultrafiltration Handbook*, Technomic Publishing Co. Inc., PA, USA, p. 73, 171, 235.
Christiansen, F. A. (1991). *Scand. Dairy Inform. J.*, **5**(2), 8.
Conner, T. (1988). *Cultured Dairy Prod. J.*, **23**(4) 21.
Conochie, J. (1974). *Aus. J. Dairy Technol.*, **29**, 141.
Conochie, J. and Birtwistle, R. (1974). *Aus. J. Dairy Technol.*, **29**, 184.
Coton, G. (1986). *Dairy Indust. Inter.*, **51**(8), 29.
Crawford, R. J. M. (1960). A study of factors affecting the activity of lactic acid producing cultures in cheesemaking. PhD thesis, University of Glasgow, Glasgow, UK.
Crawford, R. J. M. (1976). *J. Soc. Dairy Technol.*, **29**, 71.
Cross, H. R. and Overby, A. J. (1988). In: *Meat Science, Milk Science and Technology*. Elsevier Science Publishers B. V., Amsterdam, The Netherlands.
Czulak, J. (1981). In: *Proc. 2nd Biennial Marschall International Cheese Conference*, Wisconsin, USA, p. 307.
Czulak, J., Hammond, L. A. and Meharry, H. J. (1954). *Aus. Dairy Rev.* **22**, 18.
Dalgleish, D. G. (1982*a*). In: *Food Proteins*, ed. P. F. Fox and J. J. Condon. Elsevier Applied Science Publishers Ltd, London, UK, p. 155.
Dalgleish, D. G. (1982*b*). In: *Developments in Dairy Chemistry—1 Proteins*, ed. P. F. Fox. Elsevier Applied Science Publishers Ltd, London, UK, pp. 157–87.
Dalgleish, D. G. (1987). In: *Cheese: Chemistry, Physics and Microbiology* (Vol. 1), ed. P. F. Fox. Elsevier Applied Science Publishers Ltd, London, UK, p. 63.
Dalgleish, D. G. (1990). In: *Proc. XXIII International Dairy Congress* (Vol. 2). Mutual Press, Montreal, Canada, p. 1513.
Damerow, G. (1988). *North European Food Dairy J.*, **54**, 267.
Davies, F. L. and Law, B. A. (1984). In: *Advances in the Microbiology and Biochemistry of Cheese and Fermented Milks.* Elsevier Applied Science Publishers Ltd, London, UK.
Davis, J. G. (1965). In: *Cheese* (Vol. I). J. & A. Churchill Ltd, London, UK.
Davis, J. G. (1976). In: *Cheese* (Vol. III). Churchill Livingstone, London, UK.
Davis, J. G. (1981). *Ins. Food Sci. Technol. Proc.*, **14**, 131.
Dijkhuizen, G. (1981). In: *Proc. 2nd Biennial Marschall International Cheese Conference*, Wisconsin, USA, p. 274.
Eck, A. (1984). *Dairy Sci. Abstr.*, **46**, 242.

Eck, A. (1987). In: *Cheesemaking—Science and Technology* (2nd edn), C. D. Thomson (translator). Technique et Documentation, Lavoisier, Paris, France.
Elliot, R. (1985). *Dairy Field*, **168**(3), 24.
El Soda, M. (1986). *J. Food Protect.*, **49**, 395.
El Soda, M. and Pandian, S. (1991). *J. Dairy Sci*, **74**, 2317.
Elten, G. J. van (1979). *North European Dairy J.*, **5**, 98.
Elten, G. J. van (1981). In: *Proc. 2nd Biennial Marschall International Cheese Conference*, Wisconsin, USA, p. 93.
FAO (1972). In: *Production Yearbook* (Vol. 26). FAO, Rome, Italy, pp. 208–14.
FAO (1982*a*). In: *1981 FAO Production Yearbook* (Vol. 35). FAO, Rome, Italy, p. 229.
FAO (1982*b*). In: *1981 FAO Trade Yearbook* (Vol. 35). FAO, Rome, Italy, p. 99.
FAO (1988). In: *1987 FAO Production Yearbook* (Vol. 37). FAO, Rome, Italy, p. 272.
FAO (1990*a*). In: *1989 FAO Production Yearbook* (Vol. 43). FAO, Rome, Italy, p. 271.
FAO (1990*b*). In: *1988 FAO Trade Yearbook* (Vol 42). FAO, Rome, Italy, p. 107.
FAO/WHO (1972). In: *Recommended International Standards for Cheeses and Government Acceptances*, CAC/C1-C25, FAO, Rome, Italy.
FAO/WHO (1978). In: *Joint Committee of Government Experts on the Code of Principles Concerning Milk and Milk Products*, CX 5/70–19th Session, FAO, Rome, Italy.
FAO/WHO (1984). In: *Code of Principles Concerning Milk and Milk Products, International Standards for Milk Products and International Individual Standards for Cheeses*, Codex Alimentarius (Vol. XVI). 1st edn, FAO, Rome, Italy.
Fox, P. F. (1981). In: *Proteinases and their Inhibitors, Structure, Function and Applied Science*, ed. V. Turk and L. J. Vitale. Pergamon Press, Oxford, UK, p. 245.
Fox, P. F. (1982*a*). In: *Use of Enzymes in Food Technology*, ed. P. Dupuy. Technique et Documentation Lavoisier, Paris, France, p. 135.
Fox, P. F. (1982*b*). In: *Proc. XXI International Dairy Congress* (Vol. 2). Mir Publishers, Moscow, CIS, p. 256.
Fox, P. F. (1984). In: *Developments in Food Proteins—3*, ed. B. J. F. Hudson. Elsevier Applied Science Publishers Ltd, London, UK, p. 69.
Fox, P. F. (1986). In: *Milk—the Vital Force—Proc. of the XXII International Dairy Congress*, D. Reidel Publishing Co., Dordrecht, The Netherlands, p. 61.
Fox, P. F. (1987). In: *Cheese: Chemistry, Physics and Microbiology* (Vols 1 and 2). Elsevier Applied Science Publishers Ltd, London, UK.
Fox, P. F. (1988–89). *Food Biotechnol.*, **2**(2), 133.
Fox, P. F. (1989). *J. Dairy Sci.*, **72**, 1379.
Fox, P. F. and Morrissey, P. A. (1981). In: *Enzymes and Food Processing*, ed. G. G. Birch, N. Blakebrough and K. J. Parker. Applied Science Publishers Ltd, London, UK, p. 213.
Fox, P. F. and Mulvihill, D. M. (1983). In: *Proc. IDF Symposium—Physicochemical Aspects of Dehydrated Protein—Rich Milk Products*. International Dairy Federation, Brussels, Belgium, 188–259.

Fryer, T. F. (1982). *Proc. XXI International Dairy Congress* (Vol. 2). Mir Publishers, Moscow, CIS, p. 485.
Galesloot, Th. E. (1958). *Neth. Milk Dairy J.*, **12**, 130.
Garrett, N. L. T. (1987*a*). *J. Soc. Dairy Technol.*, **40**, 68.
Garrett, N. L. T. (1987*b*). *North European Food Dairy J.*, **53**, 135.
Garvie, E. I. (1984). In: *Advances in the Microbiology and Biochemistry of Cheese and Fermented Milk*, ed. F. L. Davies and B. A. Law. Elsevier Applied Science Publishers Ltd, London, UK, p. 35.
Gasson, M. J. (1990). *FEMS Microbiol. Rev.*, **87**, 43.
Gasson, M. J. and Davies, F. L. (1984). In: *Advances in the Microbiology and Biochemistry of Cheese and Fermented Milk*, ed. F. L. Davies and B. A. Law. Elsevier Applied Science Publishers Ltd, London, UK, p. 99.
Gehriger, G. (1979). In: *Proc. 1st Biennial Marschall International Cheese Conference*, Wisconsin, USA, p. 311.
Gilmour, A. and Rowe, M. T. (1990). In: *Dairy Microbiology* (Vol. 1, 2nd edn), ed. R. K. Robinson. Elsevier Applied Science Publishers Ltd, London, UK, p. 37.
Ginkel, W. van, Vries, E. de and Daamen, C. B. G. (1987). In: *Test of a Curd-Making Tank, Type 3113–15000, Code No. 3113–1048, with a Capacity of 15000 Litres, Manufactured by Pasilac Silkeborg*. NIZO Report No. 125, Ede, The Netherlands.
Glover, F. A. (1985). In: *Ultrafiltration and Reverse Osmosis for the Dairy Industry*. Technical Bulletin 5, NIRD, Reading, UK, p. 73.
Glover, F. A. (1986). In: *Modern Dairy Technology* – Vol. 1: *Advances in Milk Processing*, ed. R. K. Robinson. Elsevier Applied Science Publishers Ltd, London, UK, p. 235.
Grant, G. R. (1987). In: *SERAVAC Annual Cheesemaker Symposium*, Miles Lab. (Pty) Ltd, Cape Town, Republic of South Africa, p. 70.
Gray, B. E. (1975). *J. Soc. Dairy Technol.*, **28**, 11.
Green, M. L. (1985). *J. Dairy Res.*, **52**, 555.
Green, M. L. (1986). *J. Dairy Res.*, **53**, 329.
Green, M. L. (1987). *J. Dairy Res.*, **54**, 303.
Green, M. L., Scott, K. J., Anderson, M., Griffin, M. C. A. and Glover, F. A. (1984). *J. Dairy Res.*, **51**, 267.
Griffiths, M. W., Banks, J. M., McIntyre, L. and Limond, A. (1991). *J. Soc. Dairy Technol.*, **44**, 24.
Gripon, J. C., (1989). In: *Food Ingredients Europe—Conference Proceedings – Paris*, Expoconsult Publishers, Maarssen, p. 96.
Guinee, T. P. and Fox, P. F. (1987). In: *Cheese: Chemistry, Physics and Microbiology* (Vol. 1), ed. P. F. Fox. Elsevier Applied Science Publishers Ltd, London, UK, p. 251.
Guise, B. (1983). *Food Process.*, **52**(4), 29.
Hammond, L. A. (1979). In: *Proc. 1st Biennial Marschall International Cheese Conference*, Wisconsin, USA, p. 495.
Hammond, L. A. (1982). *Aus. J. Dairy Technol.*, **37**, 71.
Hammond, L. A. and Freeman, N. H. (1983). UK Patent Application, GB 2 111 368A.
Hansen, R. (1975). *North European Dairy J.*, **41**, 301.
Hansen, R. (1976). *North European Dairy J.*, **42**, 127.

Hansen, R. (1977*a*). *North European Dairy J.*, **43**, 88.
Hansen, R. (1977*b*). *North European Dairy J.*, **43**, 93.
Hansen, R. (1978*a*). *North European Dairy J.*, **44**, 138.
Hansen, R. (1978*b*). *North European Dairy J.*, **44**, 195.
Hansen, R. (1979*a*). *North European Dairy J.*, **45**, 66.
Hansen, R. (1979*b*). *North European Dairy J.*, **45**, 241.
Hansen, R. (1979*c*). *North European Dairy J.*, **45**, 86.
Hansen, R. (1979*d*). *North European Dairy J.*, **45**, 274.
Hansen, R. (1979*e*). *North European Dairy J.*, **45**, 153.
Hansen, R. (1980*a*). *North European Dairy J.*, **46**, 126.
Hansen, R. (1980*b*). *North European Dairy J.*, **46**, 167.
Hansen, R. (1980*c*). *North European Dairy J.*, **46**, 152.
Hansen, R. (1981*a*). *North European Dairy J.*, **47**, 191.
Hansen, R. (1981*b*). *North European Dairy J.*, **47**, 208.
Hansen, R. (1981*c*). *North European Dairy J.*, **47**, 199.
Hansen, R. (1981*d*). *North European Dairy J.*, **47**, 22.
Hansen, R. (1981*e*). *North European Dairy J.*, **47**, 62.
Hansen, R. (1981*f*). *North European Dairy J.*, **47**, 106.
Hansen, R. (1982*a*). *North European Dairy J.*, **48**, 7.
Hansen, R. (1982*b*). *North European Dairy J.*, **48**, 223.
Hansen, R. (1982*c*). *North European Dairy J.*, **48**, 137.
Hansen, R. (1982*d*). *North European Dairy J.*, **48**, 70.
Hansen, R. (1982*e*). *North European Dairy J.*, **48**, 317.
Hansen, R. (1982*f*). *North European Dairy J.*, **48**, 19.
Hansen, R. (1983*a*). *North European Dairy J.*, **49**, 218.
Hansen, R. (1983*b*). *North European Dairy J.*, **49**, 253.
Hansen, R. (1984*a*). *North European Dairy J.*, **50**, 138.
Hansen, R. (1984*b*). *North European Dairy J.*, **50**, 112.
Hansen, R. (1984*c*). *North European Dairy J.*, **50**, 202.
Hansen, R. (1985*a*). *North European Dairy J.*, **51**, 103.
Hansen, R. (1985*b*). *North European Dairy J.*, **51**, 98.
Hansen, R. (1985*c*). *North European Dairy J.*, **51**, 118.
Hansen, R. (1985*d*). *North European Dairy J.*, **51**, 275.
Hansen, R. (1985*e*). *North European Dairy J.*, **51**, 170.
Hansen, R. (1985*f*). *North European Dairy J.*, **51**, 255.
Hansen, R. (1986*a*). *North European Dairy J.*, **52**, 138.
Hansen, R. (1986*b*). *North European Dairy J.*, **52**, 253.
Hansen, R. (1986*c*). *North European Dairy J.*, **52**, 262.
Hansen, R. (1986*d*). *North European Dairy J.*, **52**, 171.
Hansen, R. (1987*a*). *North European Dairy J.*, **53**, 34.
Hansen, R. (1987*b*). *North European Dairy J.*, **53**, 13.
Hansen, R. (1987*c*). *North European Food Dairy J.*, **53**, 126.
Hansen, R. (1987*d*). *North European Food Dairy J.*, **53**, 130.
Hansen, R. (1987*e*). *North European Food Dairy J.*, **53**, 252.
Hansen, R. (1987*f*). *North European Food Dairy J.*, **53**, 68.
Hansen, R. (1988*a*). *North European Food Dairy J.*, **54**, 197.
Hansen, R. (1988*b*). *North European Food Dairy J.*, **54**, 161.
Hansen, R. (1988*c*). *North European Food Dairy J.*, **54**, 50.

Hansen, R. (1988*d*). *North European Food Dairy J.*, **54**, 208.
Hansen, R. (1988*e*). *North European Food Dairy J.*, **54**, 117.
Hansen, R. (1988*f*). *North European Food Dairy J.*, **54**, 136.
Hansen, R. (1989*a*). *North European Food Dairy J.*, **55**, 195.
Hansen, R. (1989*b*). *North European Food Dairy J.*, **55**, 132.
Harding, F. and Royal, L. (1974). *Dairy Indust. Inter.*, **39**, 372.
Hellstrom, F. (1986). *North European Dairy J.*, **52**, 277.
Hicks, C. L., O'Leary, J. and Bucy, J. (1988). *J. Dairy Sci.*, **71**, 1127.
Holland, B., Unwin, I. D. and Buss, D. H. (1989). In: *Milk Products and Eggs* (4th Supplement to McCance and Widdowson's *The Composition of Foods* (4th edn)), The Royal Society of Chemistry and MAFF, London, UK, p. 56.
Holmstrom, P. (1991). *Scand. Dairy Inform.*, **5**(2), 34.
Honer, C. (1989). *Dairy Foods*, **90**(12), 54.
Hutkins, R., Halambeck, S. M. and Morris, H. A. (1986). *J. Dairy Sci.*, **69**, 1.
IDF (1979). In: *Market conditions and Trends for Cheese*, Doc. 110, International Dairy Federation, Brussels, Belgium.
IDF (1980). In: *Factors Influencing the Bacteriological Quality of Raw Milk*, Doc. 120, International Dairy Federation, Brussels, Belgium.
IDF (1981). In: *IDF Catalogue of Cheeses*, Doc. 141, International Dairy Federation, Brussels, Belgium.
IDF (1982*a*). In: *Consumption Statistics for Milk and Milk Products*, Doc. 144, International Dairy Federation, Brussels, Belgium.
IDF (1982*b*). In: *The World Market for Cheese*, Doc. 146, International Dairy Federation, Brussels, Belgium.
IDF (1985). In: *Milk-Clotting Enzymes*, Doc. 194, International Dairy Federation, Brussels, Belgium, p. 2.
IDF (1987*a*). In: *Consumption Statistics for Milk and Milk Products*, Doc. 213 International Dairy Federation, Brussels, Belgium.
IDF (1987b). In: *The Use of Lysozyme in the Prevention of Late Blowing in Cheese*, Doc. 216, International Dairy Federation, Brussels, Belgium, p. 2.
IDF (1990*a*). In: *Production of Chymosim by Micro-Organisms and its Use for Cheesemaking – Detection and Prevention of Anaerobic Spore formers and Cheese Quality*, Doc. 251, International Dairy Federation, Brussels, p. 3 and 17.
IDF (1990*b*). In: *Use of Enzymes in Cheesemaking*, Doc. 247, International Dairy Federation, Brussels, Belgium, p. 24.
IDF (1991*a*). In: *The World Dairy Situation*, Doc. 266, International Dairy Federation, Brussels, p. 3 & 50.
IDF (1991). In: *Consumption Statistics for Milk and Milk Products*, Doc. 254, International Dairy Federation, Brussels, Belgium.
Jacobson, M. K. (1985). *North European Dairy J.*, **51**, 38.
Jameson, G. W. (1987). *Food Technol.—Australia*, **39**(12), 560.
Jensen, L. A., Johnson, M. E. and Olson, N. F. (1987). *Cultured Dairy Prod. J.*, **22**(2), 6.
Johnston, K. A., Dunlop, F. P. and Lawson, M. F. (1991). *J. Dairy Res.*, **58**, 345.
Kamaly, K. M. and Marth, E. H. (1989). *J. Dairy Sci.*, **72**, 1945.
Kasprzyk, P., Michel, J. F., Seuvre, A. M. and Mathlouthi, M. (1983). In: *Maturation des Fromages a Pate Pressee Cuite de Type Emmental*, No. 331,

Actualites Scientifiques et Techniques dans les Industries Agro-Alimentaires, Department Biologique Appliquee, IUT, Dijon, France.
Khalid, N. M. and Marth, E. H. (1990). *J. Dairy Sci.*, **73**, 2669.
King, D. W. (1966). *VII Internat. Dairy Congress*, **D**, 723.
King, J. S. and Mabbitt, L. A. (1982). *J. Dairy Res.*, **49**, 439.
Kiuru, V. (1976). *North European Dairy J.*, **42**, 44.
Koch, N., Prokopek, D. and Krusch, U. (1986). *North European Dairy J.*, **52**, 273.
Kok, J. (1990). *FEMS Microbiol. Rev.*, **87**, 15.
Kosikowski, F. (1982). In: *Cheese and Fermented Milk Foods* (2nd edn). F. V. Kosikowski and Associates, New York, USA.
LaGrange, W. S. and Goering, D. H. (1991). *Dairy, Food Environ. Sanitation*, **11**, 136.
Lampert, L. M. (1975). In: *Modern Dairy Products* (3rd edn). Chemical Publishing Company Inc., New York, USA.
Larsen, N. E. and Lynggaard, T. (1991). *Scand. Dairy Inform.*, **5**(1), 46.
Law, B. A. (1976). In: *Special Address at Annual Session (1977–78) on Milk Production, Enterobacteriaceae, Enzymatic Methods, Cheese Manufacture, Rennet and Substitues, and Marketing Aspects of Dairy Products*, Doc. 108, International Dairy Federation, Brussels, Belgium, p. 40.
Law, B. A. (1984). In: *Advances in the Microbiology and Biochemistry of Cheese and Fermented Milk*, ed. F. L. Davies and B. A. Law. Elsevier Applied Science Publishers Ltd, London, UK, p. 187.
Law, B. A. (1987). In: *Cheese—Chemistry, Physics and Microbiology* (Vol. 1), ed. P. F. Fox. Elsevier Applied Science Publishers Ltd, London, UK, pp. 365–92.
Law, B. A. (1990). In: *Proc. XXIII International Dairy Congress* (Vol 2). Mutual Press, Montreal, Canada, p. 1616.
Law, B. A., Sharpe, M. E. and Chapman, H. (1976). *J. Dairy Res.*, **43**, 459.
Lawrence, R. C. (1987). In: *The Use of Ultrafiltration Technology in Cheesmaking*, Annual Sessions in Helsinki (Finland), B-Doc. 136, International Dairy Federation, Brussels, Belgium, p. 2.
Lawrence, R. C. and Gilles, J. (1987). In: *Cheese—Chemistry, Physics and Microbiology* (Vol. 1), ed. P. F. Fox. Elsevier Applied Science Publishers Ltd, London, UK, p. 1.
Lee, B. J. (1981). In: *Proc. 2nd Biennial Marschall International Cheese Conference*, Wisconsin, USA, p. 328.
Lefier, D., Duboz, G. and Grappin, R. (1990). *French Patent Application*, FR 2 648 318 Al.
Lelievre, J. and Gilles, J. (1982). *NZ J. Dairy Sci. Technol.*, **17**, 69.
Lelievre, J. and Lawrence, R. C. (1988). *J. Dairy Res.*, **55**, 465.
Lidberg, E. and Bredahl, B. (1991). *Scand. Dairy Inform.*, **5**(2), 20.
Lindsay, R. C., Hargett, S. M. and Bush, C. S. (1982). *J. Dairy Sci.*, **65**, 360.
Luquet, F. M. (1985). In: *Les Produits Laitiers* (Vol. 23). Technique et Documentation, Lavoisier, Paris, France.
Malmberg, R. and Holm, S. (1988). *North European Food Dairy J.*, **54**, 30.
Manap, M. Y. (1988). Effects of low temperature storage and thermisation on the quality of raw and heat treated milk. PhD thesis, University of Glasgow, Glasgow, UK.
Marth, E. H. (1953). *Milk Products J.*, **44**(10), 31.

Maubois, J. L. and Mahaut, M. (1974). *Revue Laitiere Francaise*, **322**, 479.
McIntyre, D. A. and Harlander, S. K. (1990). In *Proc. XXIII International Dairy Congress* (Vol. 2). Mutual Press Montreal, Canada, p. 1578.
McKay, L. L. and Baldwin, K. A. (1990). *FEMS Microbiol. Rev.*, **87**, 3.
McMahon, D. J. and Brown, R. J. (1984). *J. Dairy Sci.*, **67**, 919.
Meersohn, M. (1989). *North European Food Dairy J.*, **55**, 108.
Moore, W. E. C. and Holdeman, L. V. (1974). In: *Bergeys Manual of Determinative Bacteriology* (8th edn), ed. R. E. Buchanan and N. E. Gibbons. The Williams & Wilkins Co., Baltimore, MD, USA, p. 633.
Morris, H. A. and Anderson, K. (1991). *Cultured Dairy Prod. J.*, **26**(2), 13.
Muir, D. D. (1990). In: *Dairy Microbiology* (Vol. 1, 2nd edn), ed. R. K. Robinson. Elsevier Applied Science Publishers Ltd, London, UK, p. 209.
Muir, D. D., Kelly, M. E., Phillips, J. D. and Wilson, A. G. (1978). *J. Soc. Dairy Technol.*, **31**, 137.
Nielsen, P. G. (1990*a*). *European Patent Application*, EP 0 398 836 AI.
Nielsen, W. K. (1990*b*). *Scand. Dairy Inform.*, **4**(4), 50.
Nieuwoudt, J. A. (1987). In: *SERAVAC Annual Cheesemakers Symposium*, Miles Lab. (Pty) Ltd, Cape Town, Republic of South Africa, p. 90.
Osteras, M. (1989). In: *Special Address at the IDF Annual Sessions—Budapest 1988*, Doc. 244, International Dairy Federation, Brussels, Belgium, p. 30.
Osteras, M. (1990). In: *Dairying in the Changing World—Proc. XXIII International Dairy Congress* (Vol. 3). Mutual Press, Ottawa, Canada, pp. 2124–32.
Ostergaard, B. (1986). In: *Concentration and Drying of Foods*, ed. D. MacCarthy. Elsevier Applied Science Publishers Ltd, London, UK, p. 133.
Ostergaard, B. (1990). *Scand. Dairy Inform.*, **5**(2), 16.
Ostergaard, B. (1991). *Scand. Dairy Inform.*, **5**(2), 16.
Pappas, C. P. (1988). *Br. Food J.*, **90**(4), 163.
Park, W. J. (1979). In: *Proc. 1st Biennial Marschall International Cheese Conference*, Wisconsin, USA, p. 235.
Pearse, M. J. and Mackinlay, A. G. (1989). *J. Dairy Sci.*, **72**, 1401.
Pederson, C. S. (1979). In *Microbiology of Food Fermentation's* (2nd edn). AVI Publishing Co., CT, USA.
Peterson, S. D. and Marshall, R. T. (1990). *J. Dairy Sci.*, **73**, 1395.
Phillips, D. J. (1989). In: *Processing Engineering in the Food Industry—Developments and Opportunities*, ed. R. W. Field and J. A. Howell. Elsevier Applied Science Publishers Ltd, London, UK, p. 249.
Pope, V. J. (1984). *Diary Indust. Inter.*, **49**(6), 31.
Prokopek, D., Meisel, H., Krusch, U., Teuber, M. and Schlimme, E. (1990*a*). *Kieler Milchwirtschaftliche Forschungsberichte*, **42**(4), 597.
Prokopek, D., Brath, C. A., Klobes, H., Meisel, H., Schlimme, E. and Verse, M. de (1990*b*). *Kieler Milchwirtschaftliche Forschungsberichte*, **42**(4), 565.
Qvist, K. B. (1987). In: *Objective and Sensory Assessment of Texture of Danbo Cheese Made from Milk Concentrated 2-Fold Using Ultrafiltration.* Report No. 272, The Danish Research Institute for Dairy Industry, Hillerod, Denmark.
Qvist, K. B., Thomsen, D. and Jensen, G. K. (1985). In: *Manufacture of Semi-Hard Cheese with Round Holes from Milk Pre-Concentrated Using Ultrafiltration*, Report No. 266. The Danish Research Institute for Dairy Industry, Hillerod, Denmark.

Qvist, K. B., Thomsen, D., Forsingdal, K. and Hyldig, G. (1987). *Scand. Dairy Indust.*, **1**(3), 156.
Radford D. R. (1984). In: *Proceedings of a Dairy Culture Review Conference*, Australian Society of Dairy Technology—Technical Publication No. 27, Glenleg North, South Australia, p. 33.
Rashed, M. A., Mehanna, N. M. and Mehanna, A. S. (1986). *J. Soc. Dairy Technol.*, **39**, 62.
Raun, L. (1990). In: *Dairying in the Changing World—Proc. XXIII International Dairy Congress* (Vol. 3). Mutual Press, Ottawa, Canada, p. 2295.
Reddy, K. A. and Marth, E. H. (1991). *J. Food Protect.*, **54**, 138.
Reimerdes, E. H. (1990). In: *Enzymes in Industry—Production and Applications*, ed. W. Gerhartz. VCH Verlagsgesellschaft mbH, Weinheim, Germany, p. 119.
Reinbold, R. S. and Ernstrom, C. A. (1988). *J. Dairy Sci.*, **71**, 1499.
Renner, E. (1983). In: *Milk and Dairy Products in Human Nutrition*. Volkswirtschaftlicher Verlag GmbH, Munich, Germany, p. 359.
Renner, E. (1986). In: *Milk—The Vital Force—Proc. XXIII International Dairy Congress*. D. Reidel Publishing Co., Dordrecht, The Netherlands, p. 179.
Renner, E. (1988). In: *Meat Science, Milk Science and Technology*, ed. H. R. Cross and A. J. Overby. Elsevier Science Publishers BV, Amsterdam, The Netherlands, p. 393.
Renner, E. and Abed El-Salam, M. H. (1991). In: *Application of Ultrafiltration in the Dairy Industry*. Elsevier Applied Science Publishers Ltd, London, UK, p. 153.
Robertson, N. H. (1987). In: *SERAVAC Annual Cheesemakers Symposium*, Miles Lab. (Pty) Ltd, Cape Town, Republic of South Africa, p. 48.
Robinson, R. K. (1990). In: *Dairy Microbiology* (Vols 1 and 2, 2nd edn). Elsevier Applied Science Publishers Ltd, London, UK.
Robinson, R. K. and Tamime, A. Y. (1991). In: *Feta and Related Cheeses*, Ellis Horwood Ltd, London, UK.
Schelhaas, H. (1982). In: *Proc. XXI International Dairy Congress* (Vol. 2). Mir Publishers, Moscow, CIS, p. 290.
Scherz, H. and Kloos, G. (1981). In: *Food Composition and Nutrition Tables 1981/82* (2nd edn). Wissenschaftliche Verlagsgesellschaft GmbH, Stuttgart, Germany, p. 62.
Schroeder, C. L., Bodyfelt, F. W., Wyatt, C. J. and McDaniel, M. R. (1988). *J. Dairy Sci.*, **71**, 2010.
Scott, R. (1986). In: *Cheesemaking Practice* (2nd edn). Elsevier Applied Science Publishers Ltd, London, UK.
Scott, K. J. (1989). In: *Micronutrients in Milk and Milk-Based Food Products*, ed. E. Renner. Elsevier Applied Science Publishers Ltd, London, UK, p. 71.
Skovhauge, E. (1987). *North European Dairy J.*, **53**, 61.
Skovhauge, E. (1989). *Scand. Dairy Indust.*, **3**(1), 44.
Slack, A. W., Amundson, C. H. and Hill Jr, C. G. (1982*a*). *Proc. Biochem.*, **17**(5), 23.
Slack, A. W., Amundson, C. H., Hill Jr, C. G. and Jorgensen, N. A. (1982*b*). *Proc. Biochem.*, **17**(4), 6.
Sliter, J. (1989). In: *The World Dairy Situation—1988*, Doc. 243, International Dairy Federation, Brussels, Belgium.
Sliter, J. (1990*a*). In: *The World Dairy Situation—1989*, Doc. 249, International Dairy Federation, Brussels, Belgium.

Sliter, J. W. (1990*b*). In: *Dairying in the Changing World—Proc. XXIII International Dairy Congress* (Vol. 3). Mutual Press, Ottawa, Canada, p. 2087.

Spangler, P. L., El Soda, M., Johnson, M. E., Olson, N. F., Amundson, C. H. and Hill Jr, C. G. (1989). *Milchwissenschaft*, **44**, 199.

Spangler, P. L., Jensen, L. A., Amundson, C. H., Olson, N. F. and Hill Jr, C. G. (1990). *J. Dairy Sci.*, **73**, 1420.

Speckmann, E. W. (1979). In: *Proc. 1st Biennial Marschall International Cheese Conference*, Wisconsin, USA, p. 339.

Tamime, A. Y. (1984). *Dairy Indust. Inter.*, **49**(7), 30.

Tamime, A. Y. and Robinson, R. K. (1985). In: *Yoghurt: Science and Technology*. Pergamon Press Ltd, Oxford, UK.

Tamime, A. Y. (1990). In: *Dairy Microbiology* (Vol. 2, 2nd edn), ed. R. K. Robinson. Elsevier Applied Science Publishers Ltd, London, UK, p. 131.

Tamime, A. Y. and Kirkegaard, J. (1991). In: *Feta and Related Cheeses*, ed. R. K. Robinson and A. Y. Tamime. Ellis Horwood Ltd, London, UK, p. 70.

Tamime, A. Y., Dalgleish, D. G. and Banks, W. (1991). In: *Feta and Related Cheeses*, ed. R. K. Robinson and A. Y. Tamime. Ellis Horwood Ltd, London, UK, p. 11.

Taylor, S. L. (1985). In: *Histamine Poisoning Associated with Fish, Cheese and Other Foods*, WHO Publication VPH/FOS/85.1, Geneva, Switzerland.

Teuber, M. (1990). In: *Production of Chymosin by Micro-organisms and its Use for Cheesemaking/Detection and Prevention of Anaerobic Spore Formers and Cheese Quality*, Doc. 251, International Dairy Federation, Brussels, Belgium, p. 3.

.The Cheese (Scotland) Regulations (1966). No. 98-S.8; Amendment (1967) No. 93-S.8; Regulation (1970) No. 108-S.4; Amendment (1974) No. 1337-S.115; Amendment (1984) No. 847-S.84, Statutory Instruments, HMSO, Edinburgh, UK.

Tofte-Jespersen, N. J. and Dinesen, V. (1979). *J. Soc. Dairy Technol.*, **32**, 194.

Tschager, E. (1988). *North European Food Dairy J.*, **54**, 249.

Tunick, M. H. (1987). *J. Dairy Sci.*, **70**, 2429.

USDA (1978). In: *Cheese Varieties and Descriptions* (revised edn). Handbook No. 54, US Department of Agriculture, Washington, DC, USA.

Venema, G., Kok, J., Vossen, J. M. van der and Gruchte, M. van de (1990). *European Patent Application*, EP 0 380 823 AI.

Viswat, E. (1979). *North European Dairy J.*, **45**, 13.

Viswat, E. (1988). *North European Food Dairy J.*, **54**, 256.

Vos, W. M. de (1990). In: *Proc. XXIII International Dairy Congress* (Vol. 2). Mutual Press, Montreal, Canada, p. 1596.

Vries, E. de (1979). In: *Test of a Curd-Making Tank, Type OST III with Capacity of 10 000 Litres*, NIZO Report No. 110, Ede, The Netherlands.

Vries, E. de and Ginkel, W. van (1980). In: *Test of a Curd-Making Tank, Type Damrow Double 0 with Capacity of 16 000 Litres*, NIZO Report No. 113, Ede, The Netherlands.

Vries, E. de and Ginkel, W. van (1983). In: *Beproeving van een Gesloten Wrogebereider met een Inhoud van 16 000 Litres*, NIZO Report No. 118, Ede, The Netherlands.

Vries, E. de and Ginkel, W. van (1984). In: *Test of a Curd-Making Tank, Type OST IV with Capacity of 10 000 Litres*, NIZO Report No. 120, Ede, The Netherlands.

Walstra, P. and Jenness, R. (1984). In: *Dairy Chemistry and Physics*, Wiley-Interscience, New York, USA, p. 146.
Walstra, P., Dijik, H. J. M. van Geurts, T. J. (1987). In: *Cheese—Chemistry, Physics and Microbiology* (Vol. 1), ed. P. F. Fox. Elsevier Applied Science Publishers Ltd, London, UK, p. 135.
Weenik, A. J. H. (1975). *North European Dairy J.*, **41**, 408.
Wegner, F. (1979). In: *Proc. 1st Biennial Marschall International Cheese Conference*, Wisconsin, USA, p. 213.
Wilbrink, A. (1982). In: *Proc. XXI International Dairy Congress* (Vol. 2). Mir Publishers, Moscow, CIS, p. 170.
Wilbrink, A. (1985). *North European Dairy J.*, **51**, 142.
Zall, R. R. (1980). *Dairy Indust. Inter.*, **45**(2), 25.
Zall, R. R. (1987*a*). *Milchwissenschaft*, **42**, 3.
Zall, R. R. (1987*b*). *Milchwissenschaft*, **42**, 98.
Zall, R. R. and Chen, J. H. (1981). *J. Dairy Sci.*, **64**, 1540.

Chapter 3

Modern Cheesemaking: Soft Cheeses

M. B. Shaw

Dairy Crest Ltd, Development Division, Surbiton, Surrey, UK

CLASSIFICATION OF SOFT AND SEMI-SOFT CHEESE

Legislative Designations

The UK Cheese Regulations (1970) describe compositional standards for some 29 cheese varieties which are listed in a Schedule. These standards are expressed as minimum fat in the dry matter (FDM), and maximum moisture content in the cheese. All cheeses other than those in the Schedule, are categorised in the Regulations as either 'soft' or 'hard', depending on whether or not they are 'readily deformed by moderate pressure' (sic). Soft cheese must bear one of the descriptions given in Table I, depending on the fat and moisture content. Soft cheese may also be described as 'cream cheese' if it contains not less than 45% milk fat, or 'double cream cheese' if it contains not less than 65% milk fat.

TABLE I
Legal designations for soft cheese (UK Cheese Regulations)

Description	*Milk fat*	*Water*
Full-fat soft cheese	Not less than 20%	Not more than 60%
Medium-fat soft cheese	Less than 20% but not less than 10%	Not more than 70%
Low-fat soft cheese	Less than 10% but not less than 2%	Not more than 80%
Skim-milk soft cheese	Less than 2%	Not more than 80%

Legislation in other Member States of the EC also draws distinctions between hard and soft cheeses. In France, five categories are identified: pâtes fraîches (fresh or unripened soft cheese), pâtes molles (ripened soft cheese), pâtes pressées non-cuites (hard pressed cheese, unscalded), pâtes pressées cuites (hard pressed cheese, scalded) and pâtes persillées (blue veined cheese). Similarly, in Germany, legislation distinguishes between the following categories: hard, semi-hard, ripened soft, unripened soft and sour milk cheese.

Distinction is also drawn between 'soft' and 'hard' cheeses with regard to permitted ingredients. For example, in the UK Cheese Regulations, hard cheeses may contain 'common salt (sodium chloride), starter, coagulant, and various permitted miscellaneous additives such as calcium chloride and a range of permitted colouring matters including carotene and annatto'.

Soft cheeses may contain 'the ingredients mentioned above, together with flavourings, starches (whether modified or not), permitted emulsifiers and stabilisers, such as alginates, carrageenan, guar gum and locust bean gum, and the permitted miscellaneous additives lactic acid, citric acid, acetic acid, hydrochloric acid, orthophosphoric acid, and D-glucono-1,5-lactone'.

In addition, specific cheeses, such as Mozzarella, may contain titanium dioxide as a whitening agent; blue-veined cheeses, Feta and Provolone may contain chlorophyll and copper complexes of chlorophyll and chlorophyllins. Provolone and Pecorino-Romano may contain lipases from animal sources for the purposes of flavour production.

Harmonisation activities within the member states of the EC have led to consideration of a number of underlying principles involving a combination of minimum standards and denominations, possibly associated with Codes of Practice for example FAO/WHO Cheese Standards (1984). Products deviating from these minimum standards may be allowed to be marketed under more defined criteria as 'certain specialities'. For cheese, the favoured principles are as follows:

- a general definition for cheese;
- minimum compositional standards for different groups of cheese, e.g. dry matter or fat content;
- definition, denomination and manufacturing requirements of internationally acknowledged cheeses, e.g. as laid down in the FAO/WHO Cheese Standards (1984);
- protection of cheeses of special origin by means of an 'Appellation d'Origine' defined by the Community.

Functional Classification

While many cheese varieties will be encompassed by the terms 'hard' and 'soft' within the legal designations, quite often it will be found that the true nature of the product is not adequately described, particularly in the area of semi-hard and semi-soft cheese, as in the case of Edam and Saint Paulin for example. Thus, a number of classification systems have evolved where a cheese will be placed in a category based on consideration of such factors as consistency, fat content, moisture content, cooking (scalding) temperature and method of ripening (see Table V of Chapter 2).

Generally, the terms semi-soft and soft, when applied to consistency, are attributed to products having a moisture in fat-free cheese (MFFC) of 61–69%, and greater than 69%, respectively. Figures 1 and 2 illustrate the range of soft and semi-soft cheese covered in this chapter, sub-divided into groups according to the method of ripening.

FUNDAMENTALS OF SOFT AND SEMI-SOFT CHEESE MANUFACTURE

Process Principles

The manufacture of cheese is basically a means of preserving milk over the short to medium term, with the essential characteristics being the lowering of pH and water activity (a_w). Milk, with a water activity of 0·995 and a pH in the range 6·5–6·7 may be considered easily perishable, while cheeses will become more shelf-stable, that is less perishable, as the a_w and pH are progressively reduced. The relationship between pH and a_w for a number of cheeses is shown in Fig. 3.

During the initial stages of cheese manufacture, a_w is about 0·99 and this value will drop, depending on the processing regime applied for each particular cheese variety as described later in this chapter. Generally, the harder the cheese (lower MFFC percentage), the lower the a_w value and, conversely, the softer the cheese (higher MFFC percentage), the higher the a_w value. The a_w value is important when considering the ripening/maturation characteristics and subsequent shelf-life of cheese varieties. Generally, after salting and during maturation, the a_w values are lower than the optimum for the growth of starter bacteria, and hence the a_w value exerts a control over their metabolic activity and multiplication,

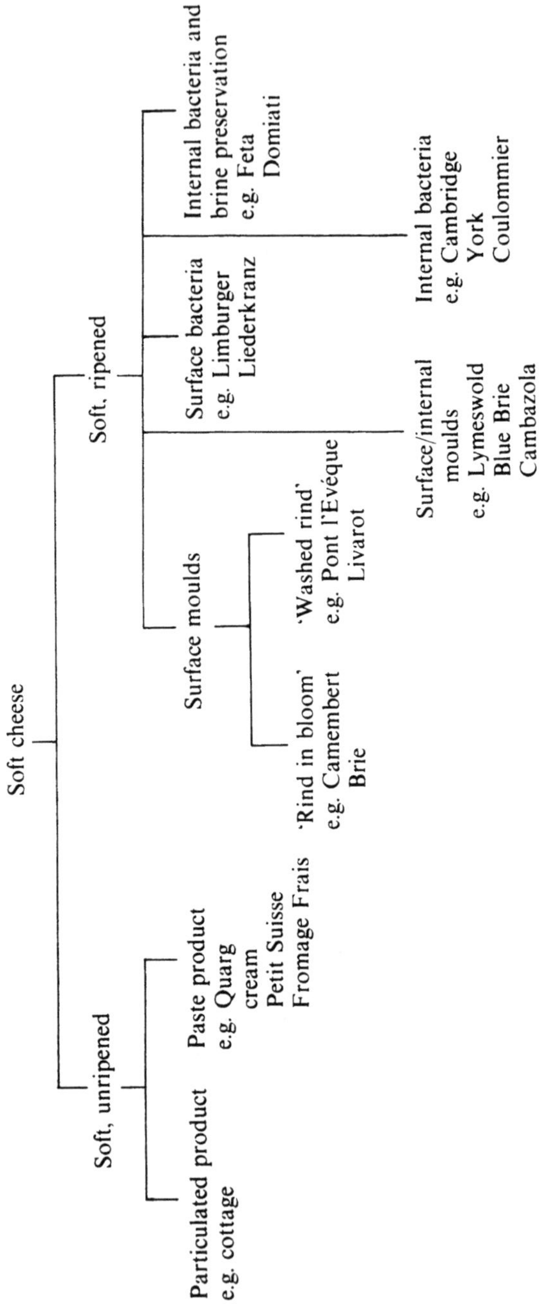

Fig. 1. Classification of soft cheese.

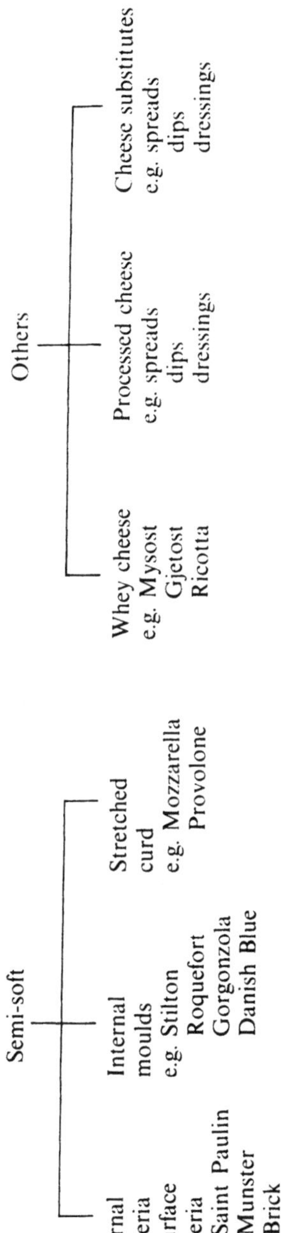

Fig. 2. Classification of semi-soft and other soft/semi-soft cheese.

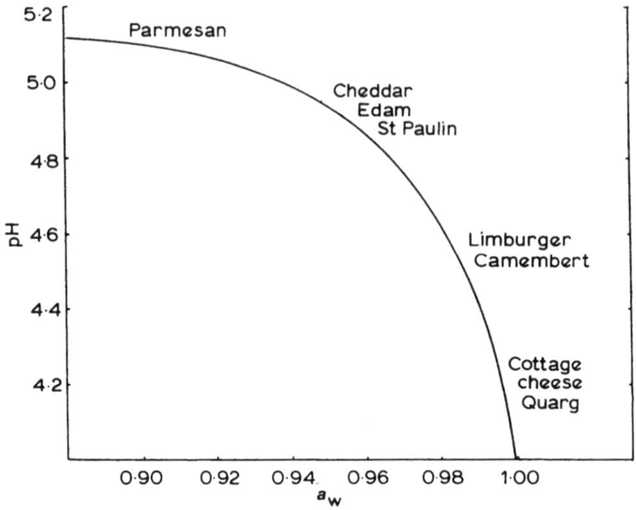

Fig. 3. Relationship between pH and water activity (a_w) of various cheeses.

and over the subsequent maturation rate and expected shelf-life of the cheese. Thus, cheese such as Parmesan, with a relatively low a_w value in the order of 0·90 will mature slowly, have a relatively high pH value, and may have a shelf-life in excess of 2 years.

At the other end of the spectrum, a soft cheese, such as Quarg or cottage cheese, will have a high a_w value in the order 0·99, and consequently the metabolic activity of the starter bacteria will be extremely high, resulting in a fairly low pH, and a comparatively short shelf-life in the order of 2 weeks. Intermediate products, such as Camembert, for example, which have an a_w value in the order of 0·98 and initial pH in the region of 4·6, will mature in a relatively short time (4–8 weeks) and have a maximum shelf-life in the order of 8–12 weeks.

Soft and semi-soft cheeses generally have a_w values in the range 0·96–0·99 and pH values in the range 4·3–5·0 with shelf-lives (excluding processed cheese) in the range 2–12 weeks at refrigeration temperatures. Thus, they are regarded as short shelf-life products, and require a complete 'cool chain' from the point of manufacture until consumption. On the other hand, processed soft and semi-soft cheeses are heat treated, resulting in 'commercial sterility', and hence product deterioration due to microbial activity is not evident; a shelf-life of up to 1 year is commonly achieved at ambient storage temperatures. In this case, packaging is of the upmost importance in order to avoid subsequent microbial contamination of the product.

The controlled transformation of milk into product is achieved by a number of stepwise procedures which are illustrated in Fig. 4. Obviously, the manufacture of any particular cheese variety might not involve every step shown, and mechanisation and new technologies, such as membrane processing, may have an influence on the unit processes carried out during manufacture.

The fundamental principle of soft/semi-soft cheese manufacture involves a reduction in pH and a_w brought about by a controlled lactic fermentation, accompanied by subsequent drainage of whey and salting of the curd. Milk coagulation is achieved by a combination of enzymic action by a coagulant protease (which may either be an animal rennet or an alternative microbial coagulant), and the production of lactic acid by lactic starter bacteria. The acidification profile during the various stages of manufacture, such as milk ripening, coagulation, curd draining and subsequent cheese maturation, is of prime importance.

The rate and quantity of lactic acid produced affects the degree of solubilisation of the calcium in the milk, and consequently, the rheological properties of the coagulum; it affects the extent of whey drainage, as well as the mineral composition and final texture of the cheese. In addition to the acid producing role, starter bacteria are also important during the maturation stage (if any), where they contribute to the overall flavour characteristics of the product. The degree of product acidification can be controlled by adjusting the amount of lactose available for fermentation. A number of methods are available such as curd washing, ultrafiltration and use of thermophilic starter cultures. The degree of specificity of proteolysis achieved by the coagulating enzymes is also of paramount importance in influencing the rheological properties of the coagulum, its subsequent whey drainage ability, the yield of product and the cheese maturation profile.

Where a ripening or maturation period is required for a particular soft/semi-soft cheese, this can be of variable duration dependent on the final flavour characteristic required. The time varies from zero for such cheeses as cottage, Quarg and York which are consumed fresh, to over 3 months for full-flavoured products, such as Stilton, Camembert and Saint Paulin. The degree of maturation and type of flavour developed will depend on a number of factors, which include the moisture and fat content of the cheese, the action of proteases and lipases derived from the coagulant, the species of starter bacteria, the surface or internal microflora of the cheese, and the ripening humidity, temperature and time.

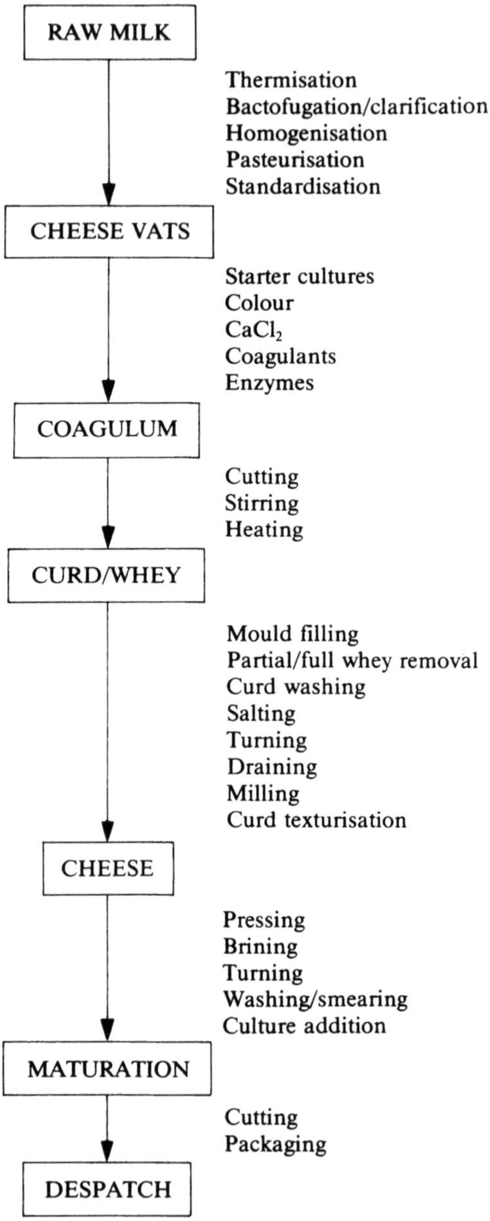

Fig. 4. Steps in soft/semi-soft cheese manufacture.

Food Safety

The outbreaks of disease brought about by the ingestion of food-poisoning organisms or their toxins brings sharply into focus the need for hygienic practices for the production and distribution of soft cheese. The linking of outbreaks of listeriosis, caused by *Listeria monocytogenes*, with soft cheeses made from raw and pasteurised milk in Switzerland (1983 and 1987) and California (1985) has served to increase awareness with respect to the need for hygienic manufacture. Post-processing contamination of the Swiss cheese was thought to have occurred, while contamination of the Mexican-style cheese in California was probably due to the presence of raw or under-pasteurised milk. The use of pasteurised milk in the production of cheese and strict attention to hygiene in the dairy premises are necessary to prevent contamination of the cheese with *Listeria* spp. Codes of Practice have been published in the UK to give guidance on the hygienic production of soft cheeses. A revision of the German Federal Cheese Ordinance made in 1986 requires that all cheese produced in that country be made from pasteurised milk. Industry support is growing for the adoption of the HACCP approach to food safety aimed at ensuring that all products are free from food-poisoning organisms and foreign matter. The HACCP concept is based on examining the product, all raw materials and processes used to make the product and asking: what can go wrong within the total system?, (see ICMSF, (1988) and Campden R. A. (1987)).

Cultures Used in Soft and Semi-soft Cheese Manufacture

Starter cultures

The selection, maintenance and use of starter cultures is perhaps the most important aspect of cheesemaking, particularly when considering the modern, mechanised processes where predictability and consistency are essential.

The reasons for using starters can be summarised as follows:

(a) it ensures consistent acidity development at a controllable rate;
(b) acid production aids rennet action and subsequent coagulum formation;
(c) acid aids expulsion of moisture (i.e. whey) from the curd;
(d) starters govern the flavour, body and texture of the cheese; and
(e) growth of undesirable bacteria in the curd is suppressed.

The organisms selected for use as cheese starters are shown in Table II, and the selection of particular cultures depends on the cheese variety being manufactured, for example:

(a) to produce a fresh, acid cheese, such as Quarg or cottage cheese, one would use a starter containing homofermentative, fast acid-producing strains of *Lactococcus lactis* subsp. *lactis* and *Lac. lactis* subsp. *cremoris*;
(b) to produce a cheese with some openness in texture, such as Stilton, one would use strains of *Lac. lactis* subsp. *lactis* and *Lac. lactis* subsp. *cremoris* together with *Lac. lactis* biovar. *diacetylactis* for gas and flavour production;
(c) to produce a sweet, washed curd cheese, such as Saint Paulin, one might use slower acid-producing strains of *Lac. lactis* subsp. *lactis* and *Lac. lactis* subsp. *cremoris*, together with a proportion of the flavour-producing *Leuconostoc* species;
(d) to produce a high-scald, stretched-curd cheese, such as Mozzarella, one would use *Str. salivarius* subsp. *thermophilus* and *Lac. delbrueckii* subsp. *bulgaricus* or possibly in addition, *Str. durans* or *Enterococcus faecalis* for flavour development.

TABLE II
Cheese starter organisms

Type	*Function*
Lactococcus lactis subsp. *lactis* *Lactococcus lactis* subsp. *cremoris*	Acid production
Lactococcus lactis biovar. *diacetylactis*	Acid, gas and flavour production
Leuconostoc mesenteroides subsp. *cremoris* *Leuconostoc mesenteroides* subsp. *dextranicum*	Gas and flavour production
Streptococcus salivarius subsp. *thermophilus* *Lactobacillus delbrueckii* subsp. *bulgaricus* *Lactobacillus helveticus*	Acid production in high-scald cheese
Streptococcus durans *Enterococcus faecalis*	Acid and flavour production in high-scald cheese
Propionibacterium shermanii	Gas and flavour production

Over the years, a number of systems have evolved whereby an active starter culture can be introduced into the cheese vat.

(a) Mother culture and bulk starter inoculation:
  (i) sub-culturing of starters from freeze-dried cultures;
  (ii) sub-culturing of starters from frozen liquid cultures.
(b) Direct bulk starter inoculation:
  (i) using deep frozen concentrated cultures;
  (ii) using freeze-dried concentrated cultures.
(c) Direct vat inoculation (DVI):
  (i) using deep frozen concentrated cultures;
  (ii) using freeze-dried concentrated cultures.

The cultures, as supplied by commercial starter manufacturers, are usually in one of the following forms:

(a) single strains;
(b) defined multiple strains; or
(c) mixed strains (for further details see Tamime (1990)).

Inhibition of starter cultures

The ability of a cheese starter to produce acid at a consistent rate is critical to the cheesemaking process. It may be affected by a number of factors, such as

(a) process conditions, e.g. rate and temperature of scalding, salt addition rate;
(b) milk constituents, e.g. agglutins, enzymes, detergents, antibiotics; and
(c) bacteriophage, i.e. viruses which attack and lyse starter bacteria and are considered the main cause of loss of activity.

Suitable precautions must be taken to reduce the risk of bacteriophage contamination, and these include heat treating the bulk starter medium at a temperature of 85°C for 20 min; using specially formulated phage-inhibitory media containing phosphates which chelate calcium ions—essential for phage adsorption on the walls of the bacteria; and paying particular attention to plant hygiene.

Developments in the cheese starter area include the production of bacteriophage-resistant starter strains using recombinant-DNA technology, while the increasing costs of bulk starter production have led to a greater use of DVI starter systems.

Other cultures

Non-starter organisms often used in the manufacture of soft and semi-soft cheeses are listed in Table III. The most commonly used are the blue and white moulds, *Penicillium roqueforti* and *P. candidum*, respectively. These strains were classically, naturally occurring moulds found in cheese maturation stores, but nowadays they are available as pure strain cultures from commercial suppliers. Both blue and white moulds serve to give the final cheese a pleasing appearance and contribute to the flavour, and various strains are available, with more or less proteolytic and lipolytic activity, resulting in a range of flavour intensities. White moulds are also available which give different surface growth characteristics in relation to time, and blue moulds are available which give a range of colours from a pale green through to dark blue. The other cultures listed are mainly involved in colour and flavour development; *Brevibacterium linens*, for example, is partly responsible for the colour and flavour of such cheeses as Munster, Livarot and Limburger, and the reddish-brown coloration on the surface of some Brie and Camembert.

Generally, it can be said that cheese flavour is a result of the production of a complex mixture of chemical compounds arising from the enzymic interaction of a number of microorganisms present in the curd.

TABLE III
Other organisms used in soft/semi-soft cheese manufacture

Type	*Function*
Penicillium roqueforti	Blue mould veining, flavour production
Penicillium candidum *Penicillium camemberti*	White mould surface growth
Penicillium album	White/blue mould flavour production
Brevibacterium linens *Brevibacterium erythrogenes*	Colour and flavour production
Micrococcus varians	Flavour production
Debaryomyces hansenii *Candida utilis*	Yeasts involved in flavour production
Rhodosporidium infirmominatium (precursor to *B. linens*)	Yeasts involved in surface colour production
Geotrichum candidum	Moulds involved in flavour production
Arthrobacter globiformis	Surface smear formation
Lactobacillus acidophilus *Bifidobacterium* spp.	Claimed health benefits

Thus, it is difficult to credit any individual strain with the development of a particular chemical compound, and the organoleptic qualities of any cheese will result from

(a) the lactic fermentation: acids (lactic, acetic, butyric), ketones (diacetyl), esters;
(b) lipolysis (fat breakdown): acids (butylic, caproic, capric), methylketones; and
(c) proteolysis (protein breakdown): peptides, amino acids, ammonia.

It can be seen, then, that complex flavours can develop in cheeses as a result of the activities of a range of organisms, including bacteria, yeasts and moulds. Blends are available, especially in France, of one or more of these microorganisms for the production of a uniquely flavoured cheese. For example to produce a farmhouse-style Pont l'Evêque with a greyish-white and orange coat, one would use a mixture of *Brevibacterium erythrogenes*, cheese smear micrococci, *Penicillium camemberti* and *Geotrichum candidum*. The cultures are generally added to the milk, and/or sprayed onto the cheese surfaces.

A/B cultures in soft cheeses

A recent cultural development has been in the area of A/B products, that is fermented dairy products containing *Lactobacillus acidophilus* and *Bifidobacterium* spp. These organisms are present in the normal gut microflora of healthy people, and there is much evidence that suggests that ingestion of cultures of these species has beneficial effects on the digestive process.

A number of fresh, soft cheese types, such as Quarg, cottage cheese and Fromage Frais, containing A/B cultures have been marketed in Europe. These cultures have found application also in the control of post-production acidification, so allowing the manufacture of mild-tasting products.

Rennet and Rennet Substitutes

The general conception of cheese curd is that it results from the action of a proteolytic enzyme which cleaves the milk protein, casein, thus rendering it insoluble, so forming a coagulated mass which encloses the other milk components such as the fat.

The most commonly used enzyme is an acid protease (EC 3.4.23.4), designated as rennin or chymosin, which is extracted from the abomasum of the suckling calf. The casein complex in milk has been shown to comprise four moieties: α, β, γ, κ, and it is the κ-casein, which exerts a stabilising influence against coagulation. The rennet enzyme, chymosin, cleaves the phenylalanine–methionine bond (105–106) in the κ-casein molecule which effectively destabilises the casein complex. It is a two-stage process, the first being enzymic:

$$\kappa\text{-casein} \xrightarrow{\text{enzyme}} para\text{-}\kappa\text{-casein} + \text{macropeptide}$$

and the second, non-enzymic stage which occurs concurrently:

$$para\text{-}\kappa\text{-casein} \xrightarrow[\text{pH } 6{\cdot}0\text{–}6{\cdot}4]{Ca^{2+}} \text{dicalcium } para\text{-}\kappa\text{-casein}$$

This stage requires the presence of calcium ions, and is the reason cheesemakers often add calcium chloride to the cheese vat. There is a third stage where proteolysis of the casein continues at a low level during cheese maturation, so making a contribution to flavour development. Some 6% of the chymosin added to the milk will be retained in the curd, while the remainder is found in the whey.

Rennet extracted from a young calf will contain between 88 and 94% chymosin and between 6 and 12% pepsin, while extracts from the older bovine animal will contain 90–94% pepsin and only 6–10% chymosin. Thus, the ratio of chymosin to pepsin depends on the age of the calf at slaughter. With commercial pressures on suppliers to produce adequate quantities of 'Standard Rennet' at a realistic price, there has been a tendency over the years to supply product with a reduced proportion of chymosin. Presently, most standard rennets contain between 75 and 80% chymosin, mixtures of standard and bovine rennet 40–50% chymosin, and bovine rennet 25% chymosin.

It was supply and demand pressure on the suppliers that led to the development of a number of alternative coagulants. These have been screened for cheesemaking in order that they meet the following criteria:

(a) must be produced from raw materials available in sufficient quantities and at an acceptable price;
(b) toxicologically safe;
(c) suited to different cheese types without changes in cheesemaking procedure;
(d) must not adversely affect yield and/or product quality;
(e) must be similar to calf rennet, particularly with regard to chemical composition; and

(f) extraction and manufacture on an industrial scale must be possible to a high microbiological standard.

Some of the sources of alternative coagulants are given in Table IV. It has been found that the bacterial and plant enzymes are too proteolytic for cheesemaking, although they are still used in some parts of the world. Work with pig and bovine pepsins and mixtures of these with standard or high chymosin rennet preparations has produced encouraging results in cheesemaking trials. The pepsins tend to coagulate milk more slowly but exhibit greater proteolytic activity at the lower pH values associated with maturing cheese.

Some of the most successful alternative coagulants are the fungal-derived, microbial rennet enzymes, the majority being extracted from the fungi, *Mucor miehei* and *Mucor pusillus*. The products have been shown to have similar proteolytic specificity to chymosin, and are supplied as standardised activity solutions with recommended usage rates for cheesemaking.

In 1981, more than one-third of all cheese produced worldwide utilised microbial rennets, and they have found exclusive use in the production of vegetarian, Halal and Kosher cheeses.

A problem highlighted some time ago was that of residual coagulant activity in whey, particularly when it was destined for use in formulated dairy products, baby foods and dietary aids. The residual coagulant was shown to cause proteolysis during whey processing, resulting in off-flavour development or even coagulation of milk-containing foods. A

TABLE IV
Sources of alternative coagulants

Group	*Source of enzyme*
Bacteria	*Bacillus polymyxa*
	Bacillus subtilis
	Bacillus mesentericus
Fungi	*Mucor miehei*
	Mucor pusillus
	Endothia parasitica
Plant	Pawpaw (papain)
	Pineapple (bromelain)
Animal	Calf (chymosin)
	Ox (pepsin)
	Pig (pepsin)
	Chicken (pepsin)

technological solution was found with the introduction of thermolabile microbial coagulants whose activity was effectively curtailed following whey pasteurisation.

The most recent milk coagulants are the result of recombinant-DNA technology, where chymosin, the key component in calf rennet, is produced by allowing specific gene expression in carrier organisms, such as *Kluyveromyces marxianus* var. *lactis* or *Aspergillus niger*. The resultant product is 100% chymosin, and various researchers have shown that cheese produced with this coagulant is identical to that manufactured using high-chymosin calf rennet with, perhaps, the added benefits that the curd may be cut earlier for an equal addition rate, and that cheese yields are slightly higher due to the absence of other protease activity.

Various recombinant-DNA products have been approved by statutory bodies and are now available commercially, with significant cost savings over standard rennets.

MANUFACTURING PROCESSES FOR SOFT AND SEMI-SOFT CHEESE

Figures 1 and 2 illustrate a functional classification of soft and semi-soft cheese varieties, and Table V lists some typical chemical compositions of representative cheeses from each category. Certain countries may set compositional standards for specific varieties, and not withstanding these, there will be natural variations in levels of, for example, protein and fat due to seasonal variations in the milk supply. Often the fat in dry matter of a cheese may be adjusted or 'standardised' by varying the fat content of the whole milk to be processed. This may be achieved by

(a) removing milkfat using centrifugal separators;
(b) adding liquid skim-milk;
(c) adding skim-milk powder; or
(d) by using processes such as ultrafiltration.

Variations in moisture, salt content, and pH of the cheese may also be experienced due to differences in processing conditions: for example, the rate of addition of starter culture, scald temperature and profile, and the time/temperature relationship of post-whey-off operations.

Not all cheese is manufactured from cow's milk and, in certain countries, substantial quantities are made from sheep's, goat's and water buffalo's milk; Table VI compares the average chemical composition of

TABLE V
Typical chemical composition of soft and semi-soft cheeses

Category	*Variety*	*Moisture*	*Fat*	*FDM*	*Protein*	*Salt*
Soft, unripened particulated	Cottage, creamed	79·9	4·0	19·0	14·0	1·0
Soft, unripened paste	Quarg	79·0	0·2	1·0	15·0	0·7
Soft, ripened surface moulds	Camembert	50·0	23·0	45·0	20·0	1·5
Soft, ripened surface/internal moulds	Lymeswold	42·0	39·0	68·0	14·0	1·6
Soft, ripened surface bacteria	Limburger	46·0	27·0	50·0	21·0	1·7
Soft, ripened internal bacteria, brine preservation	Feta	58·0	21·0	50·0	20·0	4·0
Semi-soft, surface bacteria	Saint Paulin	48·0	26·0	50·0	20·0	1·7
Semi-soft, internal moulds	Stilton	42·0	31·0	53·0	22·0	1·8
Semi-soft, stretched curd	Mozzarella	58·0	17·0	40·0	21·0	1·5
Whey cheese	Ricotta	72·0	10·0	36·0	12·5	1·5

TABLE VI
Average chemical composition of milk from various mammals

Mammal	*Moisture (%)*	*Fat (%)*	*Protein (%)*	*Lactose (%)*
Sheep	80·6	8·3	5·4	4·8
Goat	87·8	3·8	3·5	4·1
Buffalo	82·4	7·4	4·7	4·6
Cow	87·3	3·7	3·4	4·8

each milk. Both soft and semi-soft varieties may be found, while nowadays it is becoming increasingly common to manufacture product using cow's milk, with the addition of lipase, for example, in an attempt to mimic natural flavours found in cheese made from goat's milk. Varieties made from goat's milk, or mixtures of goat's and sheep's milk, include

Feta, Lightvan, Domiati and a large range of French soft, mould-ripened cheeses, including local delicacies such as Crottins de Chavignol, Sainte Maure, Tome de Romans and Levroux. Sheep's milk is used to manufacture the classic blue cheese, Roquefort, and water buffalo's milk has long been used in southern Italy to produce Mozzarella. Sheep's, goat's and buffalo's milk all lack carotene, and consequently cheese made from them, unless artifically coloured, will be white. The major difference between goat's milk and cow's milk is the higher level of capric, caprylic and caproic fatty acids, which give the cheese a sharp, pungent flavour.

The following sections review the manufacturing processes involved for each category of soft and semi-soft cheese as shown in Figs 1 and 2. Where appropriate, processing variations are included, together with a review of the application of equipment and process developments aimed at mechanising unit operations and for increasing manufacturing efficiency. Figure 4 illustrates the range of steps encountered in soft and semi-soft cheese production, although the manufacture of a particular variety may not involve every operation shown.

SOFT CHEESE MANUFACTURE

Soft Unripened Cheese

Soft unripened fresh cheese may be characterised by the following:

(a) high moisture content (up to 80% but varies with fat content);
(b) mildly acidic to bland flavour (dependent on fat content);
(c) short shelf-life;
(d) little rennet used;
(e) no pressing of curd; and
(f) product ready for consumption immediately (no maturation period).

Cottage cheese contains discrete curd particles as a result of its method of manufacture, while the majority of the unripened cheeses are of the paste-type. Variation is due mainly to the fat content, and a wide range of products, both plain and with sweet or savoury additives, may be encountered. Quarg (or Quark) is of German origin, while in France, the 'Fromage Frais' category includes examples such as Fromage Blanc, Petit Suisse, Gervais and Neufchatel. In the former USSR, Tvorog is found, and in the UK, Bakers, Lactic and cream cheese are all in this category.

Cottage cheese

Manufacturing process

Cottage cheese is manufactured from pasteurised skim-milk by way of an acid coagulation stage, followed by cooking the cut curd particles in whey, washing the curd particles with water, draining and blending the curd with a cream dressing. The unit processes for manufacturing cottage cheese are shown in Fig. 5. The illustrated process is known as a 'short set', in which the coagulum is produced at 31–32°C in about 5 h using a 5% bulk starter culture, and the total process time is between 9 and 10 h. This method is now favoured over the more traditional 'long set', in which the coagulum is produced at 22°C in about 14–16 h using 0·5–1·0% starter addition. Although traditionally an acid set product, it is now usual to add very low rennet levels of 1–3 ml per 1000 litres skim-milk. The rennet addition allows a higher pH at cutting, the curd tends to be more resilient during cooking (scalding), and whey fines are reduced.

The coagulum is cut near to the milk protein isoelectric point of pH 4·6 (usually in the range 4·40–4·85), and the size of the cut is varied depending on whether small, medium or large curd particles are required. After cutting and a quiescent period of some 15–20 min, stirring is commenced, and the whey temperature is progressively raised from 31–32°C to 50–55°C. This cooking process is critical to cheese quality and yield, and in the initial stages, the scald profile must not exceed a 1°C rise every 5 min or curd clumping will be experienced. At a whey temperature of 40°C, the increments may be increased to 2°C every 5 min. At 50–55°C the curd particles have the desired moisture content and firmness. To prevent the curd particles clumping at this temperature, it is necessary to introduce wash waters which also serve to remove lactic acid and lactose from the cheese grains, these losses give rise to the characteristic bland flavour. The chilled water is often introduced into the cheese vat as the whey is drained off and in this way the temperature is reduced to 8–10°C. The procedure is repeated two to three times, and it is essential to have chilled water of very good microbiological quality. To this end, the water is often chlorinated (5–25 ppm) and/or pasteurised, or passed through a UV sterilisation unit. Also the pH of the wash water is often adjusted to the pH value desired in the final product.

After the washing stage is completed, the curd is drained and blended with a formulated cream dressing. The dressing formulation will vary from manufacturer to manufacturer with the fat content adjusted to give 4·0% in the final product. Often, sorbates are added to the dressing at

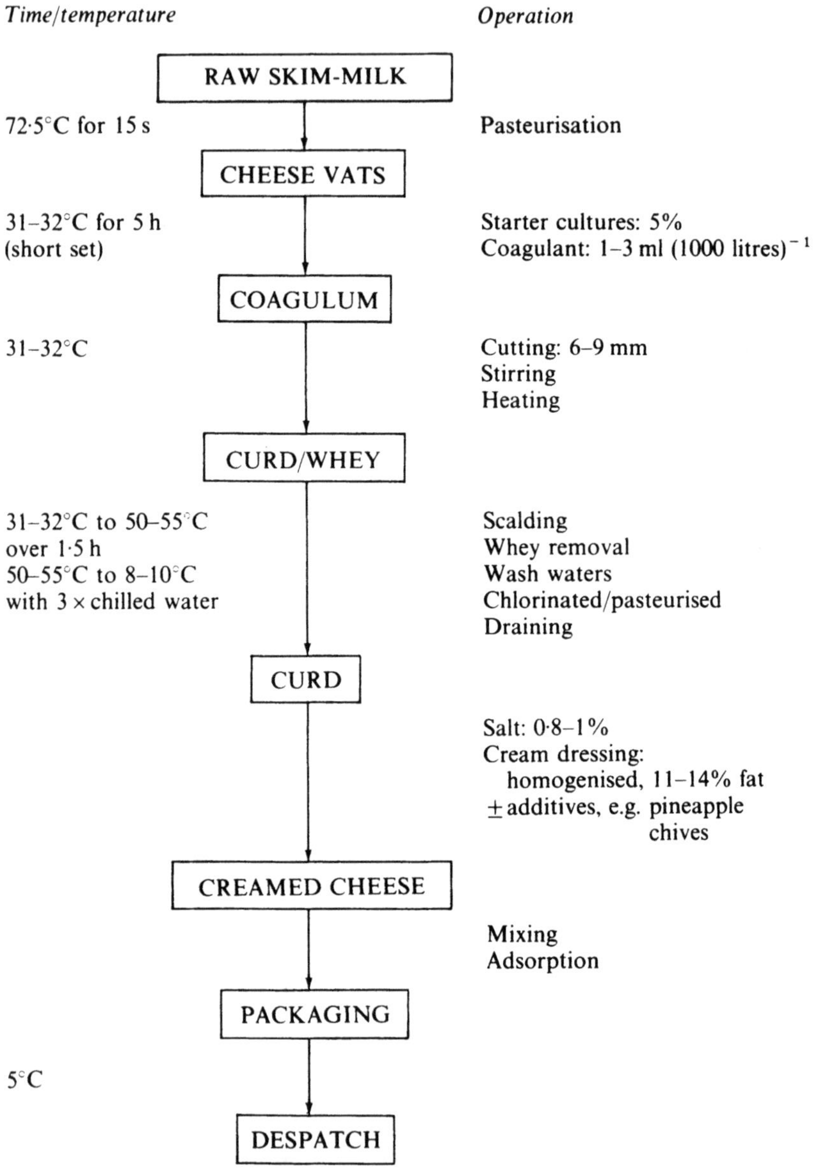

Fig. 5. Steps in cottage cheese manufacture.

0·1% to extend the shelf-life, stabilisers to prevent syneresis occurring, and salt to give 0·8–1·0% in the product. Starters may also be added to the dressing in order to reduce the pH, and in some cases, to produce carbon dioxide, both aimed at extending the shelf-life of the product. Finally, various additives may be used in order to produce a range of products, and these may include pineapple, chives, or pimentoes, all of which must be of good microbiological quality before adding to the cottage cheese.

After the curd and dressing have been blended and some 30–40 min allowed for adsorption, the cottage cheese is ready to be packaged. The packed produce should be immediately transferred to a cold room, and after 6–7 h, the product temperature should be less than 5°C. During this time a further adsorption of free cream will take place. The cheese is now ready for sale with a shelf-life of 2–3 weeks.

Equipment and process developments

Traditionally, cottage cheese has been manufactured in rectangular, open vats, and while some manufacturers still utilise this method, a number of equipment developments have been introduced in order to maximise process efficiency (see Fig. 6).

The majority of equipment developments have evolved in the US where there is a well-developed market for cottage cheese. Various systems are available from manufacturers such as Grace, Stoelting, Rietz and Damrow, who have engineered equipment aimed at mechanising the following process stages.

Curd manufacture. Conventional cottage cheese vats employed steam in the vat walls, but circulated water is now used in order that a gentler, more uniform heating can be achieved. This change was necessary to obviate problems with curd 'burn-on' and differential heat treatment, and this type of vat is often referred to as a 'spray-vat'.

As the smaller, traditional vats of 4000–5000 litres began to be replaced by larger vats upwards of 10 000 litres, the need to achieve a constant heat distribution during the 55–60°C cook became more critical. Thus, the cooking process referred to as 'jet cooking' was introduced. The basis for this method is the introduction of culinary steam into the whey layer formed on the top of the vat after cutting. This development went 'hand-in-hand' with the development of more efficient stirring systems necessary to ensure good heat distribution, due to the fact that the warm whey naturally tended to remain at the vat surface. The Stoelting

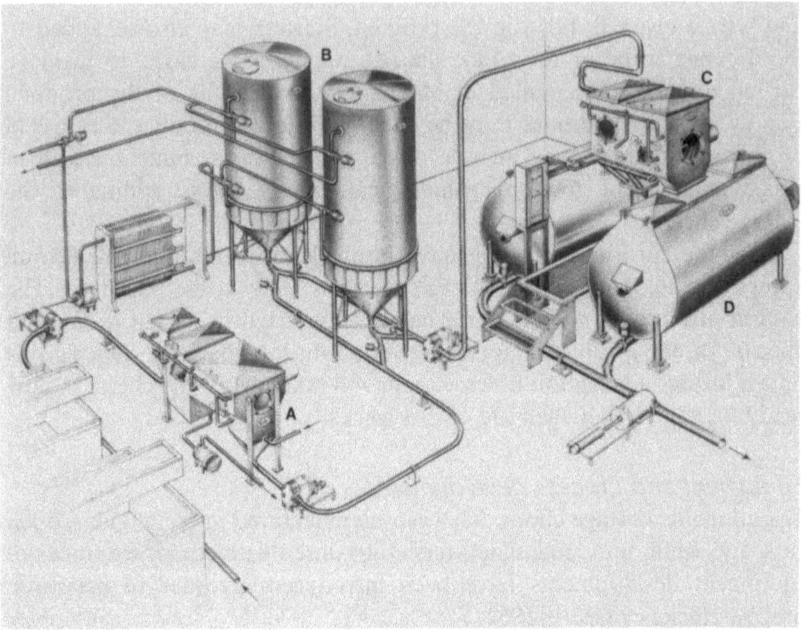

Fig. 6. Equipment for the manufacture of cottage cheese. (A) Whey drainer, (B) curd washers/coolers, (C) curd drainer, (D) creamers. (Reproduced by courtesy of Bepex Corp.)

'verti-stir' system uses agitators with large blades which gently scoop the curd from the bottom of the vat into the warm whey stream, so that all the curd particles receive a uniform heat treatment.

Further development of the 'jet cooking' system has resulted in the use of a concentric tube heat exchanger, which allows the whey to be heated without the dilution effect experienced with steam injection systems.

Washing/draining. Amongst the early cottage cheese equipment developments, the removal of the curd washing, draining and creaming operations from the cheese vat contributed much to overall process efficiency and plant utilisation. A number of manufacturers offer combined drainers/creamers, and the washing process is performed in the cheese vat followed by draining and addition of the cream dressing in a separate, tared tank system. Damrow and Rietz, for example, offer whey drainer systems where undiluted whey can be recovered, followed by separate washer/cooler tanks and curd drainers. Stoelting have developed a high speed, continuous drainer called 'Flexi-press', which separates curd from the final wash water at rates in the order of 100 kg min^{-1}. Damrow also have a belt system which is made up of four sections. In the first, the whey is drained from the curd particles which are spread as a thin layer on the belt. Washing is carried out in the second and third sections, with the chilled water sprayed onto the third section being recovered and sprayed onto the second section—this guaranteeing rapid cooling with an overall low water consumption. Finally, the fourth section permits drainage of the curd and is equipped with pressure rollers which can be adjusted to control the final moisture content of the curd.

Creaming. A number of batch curd creamers are available which are equipped with gentle agitation systems and sprays for the addition of curd dressing. Load cells systems are often used to ensure a correct and accurate ratio of curd to dressing. Other developments which can be applied to the manufacture of cottage cheese include ultrafiltration and direct acidification procedures, (reviewed later), and the use of acidophilus/bifidus cultures.

Quarg, Fromage Frais and other fresh/unripened soft cheeses

Over recent years, demand for this type of cheese has been growing rapidly throughout the EC, with, for example in France, an annual growth rate of 5–7% with a 1987 total production of 381 000 t.

Manufacturing process

The basic manufacturing process involves an acid coagulation assisted by the addition of small amounts of rennet, a coarse cut of the coagulum, separation of the curds and whey by various methods, cooling of the curd, followed by packaging and sale.

Figure 7 illustrates the process steps used in the manufacture of a low-fat Quarg from skim-milk. The illustration is for a 'long set', in which the coagulum is produced at 22–23°C in about 16–18 h using 1–2% of a bulk starter culture. The incubation proceeds until a pH of 4·7 is attained (0·55–0·6% titratable acidity as lactic acid). The incubation time may be reduced to 5–6 h ('short set') using a 5% starter inoculum at 30°C. The incubation time is generally chosen to fit in with production scheduling. The cultures used are generally homofermentative strains of *Lactococcus lactis* subsp. *lactis* and/or *Lac. lactis* subsp. *cremoris*, and the small amount of coagulant added allows the formation of a firm coagulum at pH 4·7–4·8, that is above the protein's isoelectric point of pH 4·6. The

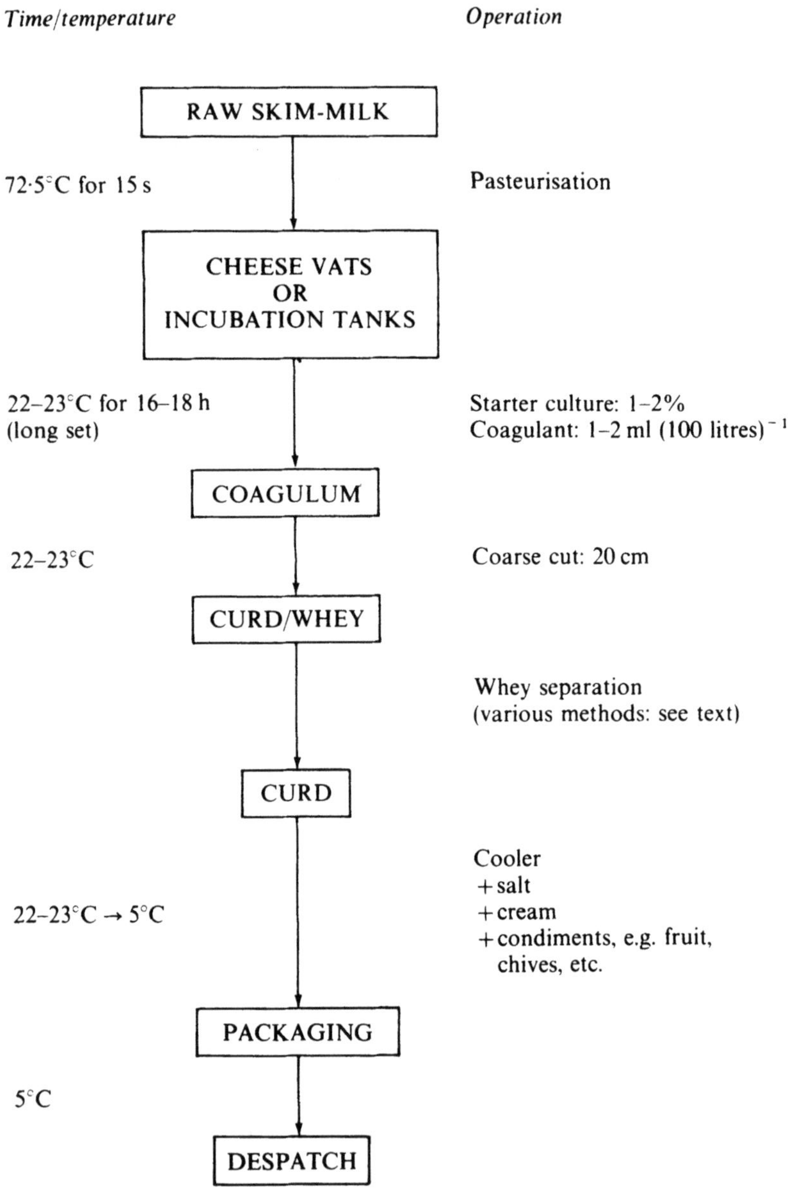

Fig. 7. Steps in Quarg manufacture.

coagulum treatment depends on the methods used for whey separation. Traditionally, curd drainage is carried out in muslin bags, and the coagulum is cut very coarsely into approximately 20 cm cubes. The introduction of centrifugal whey separators led to the coagulum being slurried up using mechanical agitators prior to feeding the separators (see later).

After whey drainage/separation, the curd is cooled either in muslin in cold rooms or, more usually, using a tubular heat exchanger/cooler. The chilled curd is then blended with salt and, in some cases, the addition of cream and/or condiments, such as herbs and fruits, may be carried out. The product is then filled into pots or tubs and distributed immediately. The shelf-life of the product will be in the region of 14 days.

Quarg may be bulk packed and used for the manufacture of other soft cheeses, such as medium-fat and full-fat varieties, where butter or cream is blended in; the blend is heat treated, often up to 80°C, a stabiliser is added to reduce serum separation, and various additives such as herbs may be introduced. The final product has the advantage of an extended shelf-life in the order of 10–12 weeks.

Equipment and process developments

Traditionally, whey separation is achieved by cutting the coagulum and transferring the curd mass to linen or muslin bags, where the whey drainage proceeds until the optimum curd moisture content is reached. This method is cumbersome and labour-intensive, and the product can be subjected to aerial contamination from yeasts and moulds. A further development of this system, which is used extensively in France, is the Berge process, where whey drainage takes place in a series of cloths held in a rack which allows the curd to be slowly compressed over a defined time period. Even with this semi-mechanised process, whey drainage times can still be up to 8 h, and difficulties may be encountered in obtaining consistent cheese quality.

As the demand for Quarg and Quarg-based products increased, it became obvious that the cloth drainage systems were not suitable for large-scale, sanitary operations. This commercial pressure led to the development of centrifugal separators and ancillary equipment in the late 1950s and early 1960s. The two major manufacturers of these separator systems are Westfalia and Alfa-Laval.

In a typical centrifugal Quarg separator, such as the Westfalia model (KDA), Lehmann *et al.* (1991) coagulated skim-milk flows through a central feed tube into a set of discs revolving at about 5500 rpm. Whey flows to the top of the separator bowl and is discharged foam-free under pressure. The cheese curd in ejected from the separator bowl by nozzles, and the hood, cooled by chilled water circulation, guides the Quarg vertically into a curd collecting trough. Four, slowly rotating, scraper blades then move the Quarg to the discharge outlet from which it slides into a Quarg catcher. Separator capacities are typically up to 2500 kg curd h^{-1}. The cheese curd is transferred from the Quarg catcher by a variable speed, positive displacement pump through a tubular cooler and a Quarg mixer into a holding tank. The Quarg mixer is used for the manufacture of creamed Quarg (medium-fat and full-fat products). Mixing takes place in a cylinder fitted with three pairs of vanes, and a piston proportioning pump controls the amount of cream added to the Quarg. Blended Quarg is passed to a silo tank where a positive displacement pump delivers the chilled Quarg to the packing machine, Fig. 8 illustrates a typical Quarg production line. In order to meet the increasing demands for hygienic manufacture, Westfalia have developed the KDC model, the first, steam sterilisable soft cheese separator.

Separators can also be used for the manufacture of full-fat cheese, although the bowl design is significantly different from that in the Quarg separator. Typically, full-fat cheese is made from pasteurised, standardised milk with a butterfat content of 11%. This milk is homogenised at 45–80°C, and the resultant fat–protein complex will be less dense than whey (0·93 compared to whey density of 1·025), and it is this fact that dictates the design of the bowl (as in the Westfalia KSA separator, for example), which continuously discharges the lighter, full-fat cheese solids concentrate.

A range of products of various fat contents may be obtained by blending different proportions of Quarg (0·2% fat) and full-fat cheese (33% fat) from the two designs of separators.

Following on from the introduction of complete separator production lines for Quarg, there have been a number of developments aimed at increasing the yield of Quarg from a unit volume of skim-milk. Currently, the two most successful processes are as follows.

The Thermo-Quarg process. Using a standard, separator-based Quarg production line, the whey has a protein content in the order of 0·8%; these valuable whey proteins are lost from the Quarg and reduce yield. Westfalia have developed a Thermo-process which allows the recovery of approximately 50% of the whey proteins, resulting in an increased Quarg yield of some 10%. The process is as follows: skim-milk is heated to between 82 and 90°C for 5–6 min, followed by cooling to normal ripening temperature when the additions of starter and rennet are

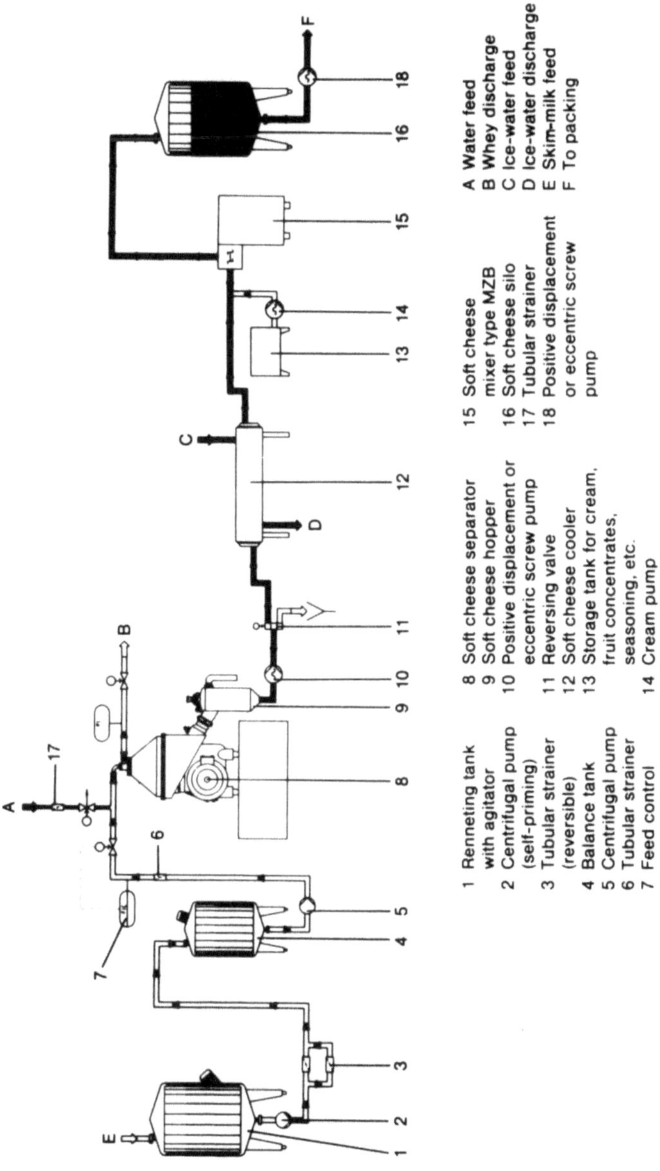

Fig. 8. Production of soft cheese based on skim-milk by the standard process. (Reproduced by courtesy of Westfalia Separator, UK.)

made. Following formation of the coagulum, the bulk is further heat treated at, typically, 60°C for 3 min to allow further whey protein denaturation and complexing with the casein. The coagulum is then cooled to separation temperature, and the curd discharging from the separator will contain the whey proteins, thereby increasing the yield. Figure 9 illustrates a typical thermo-Quarg production line and Fig. 10 shows a typical installation based on a KDA 30 separator.

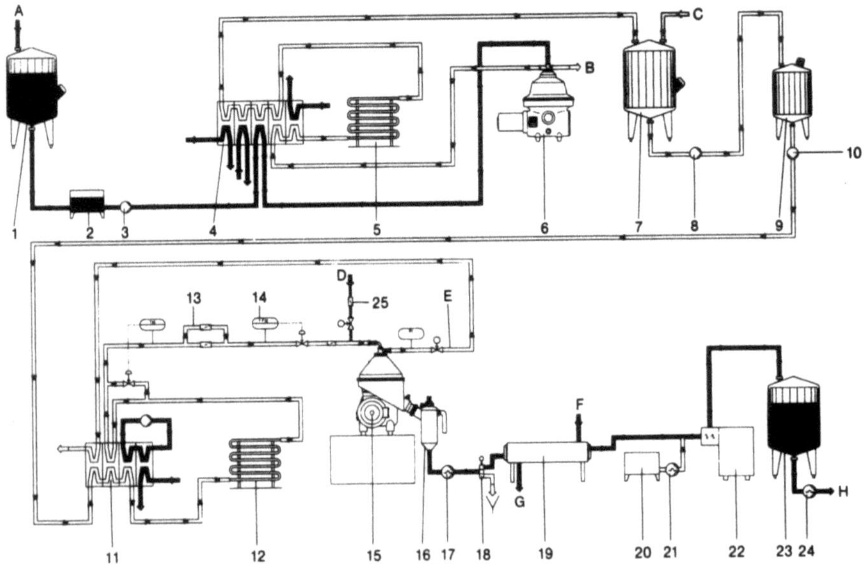

1 Raw milk tank with agitator
2 Balance tank
3 Centrifugal pump
4 Plate heat exchanger
5 Holding tube
6 Skimming separator
7 Renneting tank for cheese milk with agitator
8 Centrifugal pump
9 Balance tank with agitator
10 Centrifugal pump
11 Plate heat exchanger
12 Holding tube
13 Double strainer (reversible)
14 Feed control
15 Soft cheese separator
16 Soft cheese hopper
17 Positive displacement pump
18 Reversing valve
19 Soft cheese cooler
20 Cream tank
21 Mono pump
22 Soft cheese mixer
23 Soft cheese silo
24 Positive displacement pump
25 Tubular strainer
A Raw milk feed
B Cream discharge
C Culture and rennet inoculation
D Water inlet
E Whey discharge
F Ice water inlet
G Ice water outlet
H to packing

Fig. 9. Production of soft cheese by the Thermo-soft cheese process. (Reproduced by courtesy of Westfalia Separator, UK.)

Ultrafiltration. The use of membrane processing, and in particular ultrafiltration, for the manufacture of soft cheeses has received much attention, and ultrafiltration may be used for Quarg manufacture. Skim-milk is heated to 95°C from 5 min, and then cooled to the ripening

Fig. 10. Thermo-soft cheese line with KDA 30 separator and two tubular coolers. (Reproduced by courtesy of Westfalia Separator, UK.)

temperature of typically 28°C. Acidification proceeds until a pH of 4·4 is reached, when the cheese milk is heated to 60–63°C for 3 min. The fermented skim-milk is then cooled to 40–42°C before being passed through a continuous, multi-stage ultrafiltration plant, (see later for further details concerning ultrafiltration for soft cheese manufacture). The soft cheese leaving the ultrafiltration plant can be further processed as for other production processes. Very little protein is lost during ultrafiltration and the need for a separator is eliminated, since the de-watering is achieved in the ultrafiltration plant where it is lost as permeate. Increases in cheese yield of up to 40% have been reported, although, at high levels of whey protein incorporation, the product may have different organoleptic qualities from conventional Quarg.

Quarg and Fromage Frais are used as a base to produce a range of fresh and extended shelf-life value-added products. Typically, the base

Fig. 11. A Stephan Universal UMM1sk machine. (Reproduced by courtesy of Stephan Machinery (UK) Ltd.)

cheese is blended in a batch cooker, (Fig. 11 illustrates a typical cooker/blender as manufactured by Stephan GmbH), with a suitable stabiliser (usually proprietary brands based on guar gum, locust bean gum or carboxymethyl cellulose) and heated by indirect or direct steam injection to between 60 and 80°C for 2–5 min with continuous stirring. Additives, such as fruits and herbs, may also be mixed in at this stage prior to product packing. Hot filling and/or the addition of preservatives, such as potassium sorbate, also assist in extending shelf-life up to 12 weeks. Products are often aerated or whipped by incorporating nitrogen gas under pressure to produce stable foams with up to 200% overrun, and these may be used in layered products with fruit jellies and/or cream resulting in a range of attractive, value added, retail products.

Soft cheeses manufactured using specific starter cultures have been introduced offering the benefits of a milder taste, perceived health benefits and extended product shelf-life.

A number of cultures are now available which have a limited acidification rate below pH 4·8–5·0. For example, Biogarde® (Germany), produce a culture containing *Stretococcus salivarius* subsp. *thermophilus*, *Lactococcus lactis* subsp. *lactis*, *Lac. lactis* biovar. *diacetylactis*, *Leuconostoc mesenteroides* subsp. *cremoris*, *Lactobacillus acidophilus*, *Bifidobacterium bifidum* for the manufacture of fresh, unripened cheeses. It is the presence of the bifidobacteria which has the major effect on product stabilisation against over-acidification, and shelf-life in a cold chain can be usefully extended from two to three weeks at least.

Soft Ripened Cheeses

Figure 1 shows the range of ripened soft cheese classified according to the major ripening agent, that is surface or surface/internal moulds and surface or internal bacteria.

Surface mould-ripened cheese

Manufacturing process

Surface mould-ripened soft cheeses, or pâtes molles as they are known in France, are characterised as cheeses that have undergone other fermentations as well as a lactic fermentation. The curd is not scalded and the cheeses are not pressed, but they are matured. The French recognise two sub-categories of the mould-ripened cheese; those with a 'rind in bloom' (croute fleurie) such as Camembert, Brie and Melbury, and those with 'washed rinds' (croute lavée), such as Pont l-Evêque and Livarot.

Of the 'croute fleurie' or surface mould types, Camembert and Brie are undoubtedly the best known, and the 'rind in bloom' characteristic is due to the growth of a white mould species, *Penicillium candidum*. During the production of the 'croute lavée' varieties, such as Pont l'Evêque and Livarot, the white mould growth is removed during cheese maturation by 'smearing' or washing, and the surface growth of a red-pigmented bacterium, *Brevibacterium linens*, is encouraged.

Over the years, a number of methods have evolved for the manufacture of soft mould-ripened cheese, and the technological development of various mechanised systems will be discussed later. As the systems have evolved, slight changes have been made to the manufacturing regime, and various 'recipes' may be found showing variations in time/temperature relationships at different stages, amount and type of starter culture, and/or coagulant addition.

Fundamentally, two types of manufacture exist: the traditional and industrial, although in practice a spectrum of systems can be found with varying degrees of technological innovation offering good examples of how a traditional method has evolved into highly sophisticated, mechanised plants having a large milk throughput. Figure 12 shows the typical steps for manufacturing Camembert by an industrial process.

Traditional manufacture

The essential points of the traditional manufacturing process can be summarised as follows.

(a) Raw milk is used for manufacture, and acidification results from the natural bacterial flora. In cases where this limits the rate of acidity development, a mesophilic, mixed strain starter culture may be added at reduced levels (0·05%).
(b) The raw milk is filled into 100 litre tanks or 'bassines' after temperature adjustment to 30–32°C. Calcium chloride may be added at the rate of 6–10 g (100 litres)$^{-1}$ milk to aid coagulum formation.
(c) Acidity development proceeds until a level of 0·20–0·25% lactic acid is achieved.
(d) Rennet, or other suitable coagulant, is added at the rate of 15–20 ml (100 litres)$^{-1}$ milk.
(e) The coagulum is ready to be moulded approximately 70–90 min after renneting.
(f) No cutting of the curd occurs, and it is ladled by hand from the bassine to a series of moulds, which are open-ended cylinders with whey drainage holes, placed on drainage mats in trays which allow turning of the filled moulds.
(g) Whey drainage is allowed to proceed in the moulds for some 24–28 h in a room having a temperature of 28°C and relative humidity (RH) of 95–100%. To assist whey drainage, the moulds are turned approximately every 5 h.
(h) The cheeses, after removal from the moulds, are salted by sprinkling the surfaces with dry salt.
(i) The cheese is ripened or matured at a temperature of between 11 and 13°C and relative humidity of 90–95% for from 3 weeks to 1 month.
(j) The cheese is then packaged and stored at between 4 and 8°C prior to distribution.

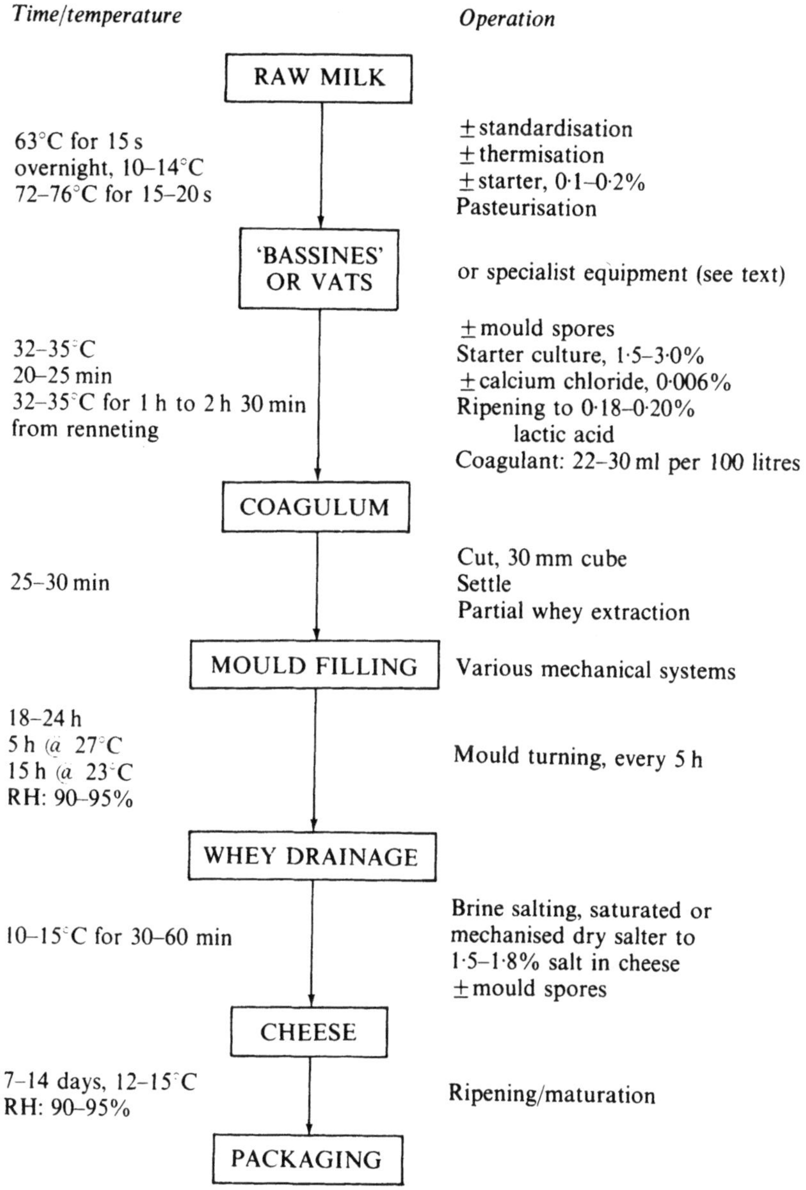

Fig. 12. Steps in Camembert manufacture (industrial).

Industrial manufacture

The essential points of the industrial manufacturing process can be summarised as follows.

(a) Milk is often standardised for fat content to give a cheese of known 'fat in dry matter'. Milk is pasteurised at 72–76°C for 15–20 s, and between 1·5 and 3% of a mesophilic mixed strain starter culture is added. Heterofermentative cultures are generally used comprising strains of *Lactococcus lactis* subsp. *lactic Lac. lactis* subsp. *cremoris*, *Lac. lactis* biovar. *diacetylactis* and *Leuconostoc mesenteroides* subsp. *cremoris*. In certain situations, thermisation of the milk may be carried out on reception (63°C for 15 s) followed by an overnight prematuration of the milk by inoculating the milk with 0·1–0·2% starter culture at between 10 and 14°C. Addition of calcium chloride is generally carried out at between 5 and 20 g (100 litres)$^{-1}$ milk.

(b) The pasteurised, standardised (and perhaps thermised and pre-ripened) milk, after dosing with starter, is filled into 100 litre bassines (or specially constructed vats or equipment) with the temperature being adjusted to 32–35°C.

(c) Acidity development proceeds until a level of some 0·18–0·20% lactic acid is achieved (20–25 min).

(d) Rennet addition at the rate of 22–30 ml (100 litres)$^{-1}$ milk is carried out.

(e) Coagulum forms and is ready for moulding at approximately 60–150 min after renneting.

(f) The curd is cut into 30 mm cubes and may be allowed to settle in the vat for some 10–15 min followed by partial extraction of whey using a pump. The cut curd is they transferred manually or mechanically to various types of mould systems where it is distributed as evenly as possible amongst the moulds.

(g) Whey drainage is allowed to proceed in the moulds for approximately 18–24 h: 5 h at 27°C followed by 15 h at 23°C with relative humidity between 90 and 95%. To assist whey drainage, the moulds are turned approximately every 5 h.

(h) The cheeses, after removal from the moulds, are salted either in a saturated brine solution at 10–15°C for 30–60 min, or by means of a mechanical dry salter. Salt content of between 1·5 and 1·8% is considered acceptable in the finished cheese.

(i) The cheese is ripened or matured at a temperature of between 12 and 15°C and relative humidity of 90–95% for from 7 to 14 days.

(j) The cheese is then packaged and stored at between 4 and 8°C prior to distribution.

The characteristic white mould coat found on Camembert and other soft ripened cheeses was traditionally formed by a naturally occurring mould strain found in cheese ripening rooms. Nowadays this surface mould growth is controlled by the use of pure cultures of *Penicillium candidum*, a strain having pure white coat forming ability and characterised by its high salt tolerance and highly aerobic nature.

The cultures of *P. candidum* may be inoculated in three ways:

(a) by addition of the mould culture to the milk;
(b) by spraying a solution of the mould culture onto the cheese surfaces; and
(c) by applying dry mould spores along with the salt to the cheese surfaces.

Ripening or maturation of the cheese is brought about by the action of both the lactic bacteria and the *P. candidum*. Camembert characteristically ripens from the outer surface as the *P. candidum* releases proteolytic enzymes. As the hydrolysis continues, casein is progressively broken down to ammonia, the body becomes smooth, and the breakdown of fat gives characteristic flavours. The central white, pasty layer gradually diminishes as ripening continues, and eventually the cheese will become overripe and liquefy with a characteristic aroma of ammonia. Traditional Camembert ripens differently from the industrial varieties, and this is attributed mainly to differences in the bacterial flora in the cheese.

Other varieties of soft ripened cheese are manufactured using similar methodology, buy by varying the speed of whey drainage, ripening conditions and mould size, a different finished product results. For example, Brie manufacture is very similar to that for Camembert. However, slight differences in production and the dimensions of the cheese mean that the internal ripening, and hence characteristic flavour and aroma, is different.

Equipment and process development

In view of the high proportion of the production cost attributable to raw materials (that is, the milk), and the fact that the cheeses are sold as small units (usually 250 g pieces), it is essential to minimise the variation observed in unit weight, and attempt to reduce the mean milk volume used per cheese. In traditional systems, 90–100 litre bassines or vats are used for producing the curd which is subsequently filled into a constant number of moulds. Consequently, irregularities of filling the bassines will result in variations in cheese weights, and certain lines of equipment development have been aimed at reducing this variability. Distribution of curd into the moulds has also been a problem with a variation of weights being observed.

The net effect of the above two points is shown in Fig. 13, which illustrates the influence of mechanisation on cheese weight dispersion and volume of milk per cheese. Essentially, during traditional manufacture, the problem of cheese weight scatter results in the manufacture of cheeses of average weight 270 g from an average volume of 2·0 litres of milk per cheese, in order that the minimum number of cheeses is downgraded because of short weight. By introducing mechanical processes, the weight scatter observed has been substantially reduced, and now cheeses of average weight 250 g can be manufactured from an average volume of 1·85 litres of milk without fear of producing an excessive number of underweight cheeses.

Thus, the primary objective of mechanisation of soft cheese manufacturing processes has been the reduction of weight dispersion due to irregularities in vat filling and distribution the curds among the moulds,

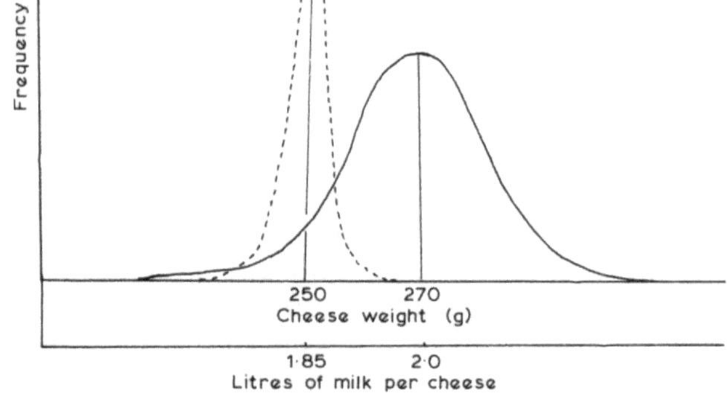

Fig. 13. Influence of mechanisation on cheese weight dispersion and volume of milk per cheese.

together with the improvement in productivity. Other economies have been made in the handling and packaging stages, as these two items, traditionally of a very labour-intensive nature, were also prime candidates for mechanisation.

In some traditional factories still using small volume bassine manufacturing methods and hand moulding, a certain amount of mechanisation has been introduced in the form of bassine emptying, mould stacking, turning and conveying. Various manufacturers, such as Pierre Guerin, Alpma, Waldner, Burton Corblin and Cartier, produce such equipment which can considerably reduce the amount of labour required.

Several mechanised soft cheese manufacturing systems have evolved over the years and the various lines of development will now be considered.

Several systems, aimed at eliminating cheese weight dispersion due to irregular filling of small vats of bassines, utilise large capacity vats from 1000 to 5000 litres. The vats, usually of semi-circular section with a curd discharge opening at one end, are equipped with stainless-steel partitions which allow the vat contents to be divided into sections for subsequent transfer to multi-mould systems. Pierre Guerin, Waldner, Burton Corblin, Alpma and Steinkecker manufacture soft cheese vats of this design.

Cheeses manufactured using this method are still subject to some weight dispersion due to variations in distributing the curd into the multi-mould systems. This problem has been partially resolved by the use of mechanical curd draining equipment prior to mould filling. Two such systems are manufactured by Alpma and Waldner where curd, after production in vats, is fed onto a mechanical whey drainage belt and finally into vertical moulding tubes, where portions of cheese are cut-off at the bottom by pneumatically operated knives. Each portion is automatically transferred to a mould for further drainage and handling.

The second line of development, also aimed at eliminating cheese weight dispersion, involves coagulating milk in small capacity 'vats'. A quantity of milk, corresponding to the volume required for a single cheese, is processed in a 'micro-vat' or 'micro-bassine' which is part of a mechanised handling system. Coagulation and whey drainage operations are carried out in the micro-bassines, and very good cheese weight dispersion can be achieved.

Two systems have evolved in France—the Rematom and the Hugonnet processes, but they have not gained much popularity due to processing constraints. Problems are experienced in ensuring good distribution of manufacturing ingredients such as starter, mould culture and rennet, in controlling uniformly the temperature, in cleaning the micro-bassine system, and in general mechanical control.

A third line of development has proceeded with milk coagulation being carried out in medium capacity units with uniform distribution of the curd at the moulding stage. Pierre Guerin have a system based on shallow rectangular bassines of between 30 and 80 litre milk capacity (see Fig. 14). Milk, to which rennet, starter culture and mould culture are metered, is filled into the small rectangular bassine and coagulation occurs while moving on a conveyor. After curd cutting, a divider is placed into the bassine which serves to segregate a quantity of curd sufficient to produce one unit of cheese. The curd is effectively transferred to a block mould system by placing a block mould with corresponding mat and whey drainage tray on top of the bassine and divider. Bassine and moulding system are turned mechanically and the curd transferred to moulds. The bassine and divider are then cleaned and re-used, while the block moulds and drainage trays are stacked, regularly turned to allow good whey drainage, and conveyed mechanically to the store (Fig. 15).

A fourth line of development is the Alpma coagulator which consists of a sectioned, slow moving belt of semi-circular cross-section on which all the different sequential manufacturing steps involved in traditional manufacture are carried out. Milk, to which rennet, starter and mould culture

Fig. 14. Cartier system for the manufacture of soft, mould-ripened cheese using small bassines or vats (30 litre). (Reproduced by courtesy of Pierre Guerin.)

Fig. 15. Mechanised handling of the block moulds used for the production of soft, mould-ripened cheese—the turning facilitates drainage of the whey. (Reproduced by courtesy of Alpma, UK.)

have been metered is filled into a section of rubberised flexible belt which is formed into a semi-circular cross-section by the stainless-steel walls. Milk is compartmentalised by the lowering of a stainless-steel spacing plate, and coagulation proceeds while the belt slowly moves forward. The spacing plates are removed prior to cutting of the coagulum and return, via an automatic cleaning system, to the beginning of the belt for further use. The belt advances the curd through automatic cutting devices which give uniform curd cubes, which are heated while being stirred (see Fig. 16) to allow whey drainage to proceed, before being filled into block moulds on a tubular filler, (see Fig. 17). The block moulds are then conveyed, automatically stacked, turned and passed through acclimatisation chambers, (see Fig. 18), where temperature and humidity are controlled to allow consistent curd drainage. The block moulds are then forwarded for demoulding, salting and maturation.

In 1992 there were 39 Alpma coagulators worldwide, up to 71 m in length and with a maximum throughput of 35 000 litres h^{-1}. Products manufactured include Camembert, Brie, Coulommier, Feta, Mozzarella, Lymeswold and Gorgonzola from standardised milk and from recombined and ultrafiltered milks (see later for further details).

A fifth line of development involved a method of continuous milk coagulation and curd draining. The Stenne–Hutin process, for example,

Fig. 16. The Alpma Continuous Coagulator showing curd stirring and whey drainage stages. (Reproduced by courtesy of Alpma GB.)

Fig. 17. The Alpma Continuous Coagulator showing end of curd drainage and block mould-filling stages. (Reproduced by courtesy of Alpma GB.)

Fig. 18. Block moulds passing through an acclimatisation chamber for controlled whey drainage. (Reproduced by courtesy of Alpma GB.)

consists of concentrating milk to a total solids content of about 36% by evaporation, and then cooling, ripening with bacterial starters, and renneting. The cold, renneted, concentrated milk is then dosed with hot water which has the dual purpose of reconstituting the milk to the original total solids level and heating the milk to the coagulation temperature of 32°C. Coagulation follows in approximately 45–60 s, when the curd is cut and tipped conventionally into block moulds. The APV Co. Ltd, in conjunction with Stenne–Hutin, developed a piece of equipment to allow continuous coagulation of the cold renneted milk (the Paracurd machine). Other processes, for example, Berridge, Nicoma and Multitube Schulz, all make use of this cold hydrolysis of casein, but none have passed the stage of experimental use as they rely greatly on homogeneity of raw materials and very precise control of the manufacturing parameters.

Finally, a sixth line of development is the application of ultrafiltration to milk for cheesemaking. Using the process of membrane filtration, it is possible to remove the necessary amount of water, lactose and minerals before coagulation and acidification of the cheese milk. A retentate or 'pre-cheese' can be produced having a similar chemical composition to a drained cheese. Good control of the cheesemaking process can be achieved, and the procedure has the additional advantage of increasing cheese yield as the soluble whey proteins, normally lost in the drained whey, are retained in the concentrated 'pre-cheese'.

A further process development has been that of 'curd stabilisation', which utilises a particular mix of starter cultures. The development has arisen due to a consumer preference for more mild soft cheeses with a longer shelf-life. The products generally have the same characteristics in terms of dry matter, size and surface appearance, but their pH after salting is higher, the fat content may be higher (e.g. 60–70% FDM), and the shelf-life is longer.

The technique involves using partial or total replacement of the mesophilic starter strains with a thermophilic species, such as *Streptococcus salivarius* subsp. *thermophilus*. Modifications to the cheesemaking procedure are necessary, in that the set temperature is increased from 30–32°C to 37–40°C. Thus, acidification proceeds at an elevated temperature, but after moulding and cooling, the acidification rate is decreased and the curd pH stabilises in the region of 4·9–5·1 after salting. This results in a relatively stable product whose textural characteristics remain fairly constant throughout a shelf-life of 12–14 weeks.

Surface/internal mould-ripened soft cheese

A number of cheeses in this category have been developed over the last 10 years in Europe, and these include Lymeswold, Blue Brie, Cambazola, Saga, Bavarian Blue and Opus 84. Production methods are generally as outlined in the previous section for soft mould-ripened cheese, and the products are further characterised by the presence of blue veining or pockets throughout the body of the cheese, as well as a white mould surface coat. The products are generally of comparatively high fat content, (60–70% FDM), and possess a mild flavour due to the use of *Penicillium roqueforti* (blue mould) strains with low proteolytic and lipolytic activity. An extra processing stage is necessary after cheese brining, where piercing with needles is carried out to allow growth of the aerobic blue mould spores added to the vat milk. Some manufacturers use curd stabilisation techniques to prolong product shelf-life, particularly for export.

Surface bacterial-ripened soft cheese

Some surface mould-ripened cheeses, such as Pont l'Evêque and Livarot, are known as 'washed rind' cheeses, and these are produced by removing any white mould growth from the cheese surface and encouraging the growth of the red-pigmented bacterium, *Brevibacterium linens*. Other bacterial-ripened cheeses include Limburger, Liederkranz, and the semi-soft varieties Saint Paulin, Brick, Munster and Port Salut.

Generally, the manufacturing steps for the different soft varieties are similar to those used for mould-ripened varieties, but the flavour intensities can vary from weak to strong. Strong flavours result from increasing moisture levels, increasing the period and temperature of maturation, not removing the surface growth, and/or increasing the surface area of the cheese in relation to its volume.

The flavour-producing bacterium *Brevibacterium linens* produces a characteristic reddish-brown surface to the cheese, and the essential process step is known as 'smearing'. This involves rubbing the surface with warm salt water either by hand or using a mechanical brushing device. The 'smearing' does not introduce the *B. linens* onto the cheese surface, as the bacterium is generally engrained into the wooden shelves in the ripening room. At the commencement of manufacture of these varieties, it is customary to apply a culture of *B. linens* to the shelves. It is essential to maintain a high relative humidity environment for the growth of *B. linens*, and growth is limited to the cheese surfaces due to the highly aerobic nature of the organism.

In the initial stages of ripening, lactose fermenting yeasts establish themselves, growing well in the high salt, low pH (*c.* 5·2) environment. As the yeasts increase the pH of the cheese surface to the order of 5·9, the *B. linens* begins to grow, and further control of the growth is achieved by washing and brushing techniques.

Internal bacterial-ripened soft cheese

The cheeses in this category may be considered as soft mould-ripened cheeses without the white mould coat. Some Coulommier is manufactured by this route, as are the traditional English varieties—York and Cambridge. Because these varieties lack the stabilising influence of the mould coat and the moisture content is high, the shelf-life is limited to 2 weeks at refrigeration temperatures. Its manufacture is thus restricted to the small producer or farmhouse situation, where cheese can be marketed locally.

Internal bacterial-ripened and brine preserved cheeses

Manufacturing process

Cheeses included in this category include Feta, Domiati and Halloumi. This group is often referred to as pickled cheese, as they are matured in brine. Feta, traditionally made from sheep's milk or a mixture of sheep and goat's milk, is now more commonly made from bovine milk with the addition of a bleaching agent, and lipase enzyme to enhance flavour production. A wide range of Feta types are on the market today with differences primarily in texture and flavour. It is typically of soft crumbly texture, with a subtle aromatic flavour, and a very white colour. Figure 19 shows a typical manufacturing procedure for a traditional Feta.

Milk is pasteurised, standardised and, if using cow's milk, bleached by addition of a chlorophyll decoloriser; a starter culture (2%) is added and milk is ripened. Rennet, and optionally, calcium chloride and/or kid or lamb lipase are added, and once set, the curd is cut. Curd is ladled into moulds where whey drainage proceeds. These are turned and sometimes subjected to slight pressure for an overnight period. The following day, the cheese is removed from the moulds and cut into blocks, followed by dry salting and/or immersion in brine solution. The cheese is then usually transferred to sealed tins containing brine, for storage. Cheese may be consumed fresh, or held for up to 6 months.

Equipment and process developments

Although Feta cheese may be produced on a wide range of traditional and modern equipment, the majority of cheese is now made from ultrafiltered milk, and two types known as 'unstructured' and 'structured' are produced. The difference between these is that the unstructured type has a dense smooth texture, while the structured tends to resemble the traditional, crumbly product. Ultrafiltration-based systems are available from a number of equipment suppliers, and these include the Pasilac 5600 coagulator, and the Alfa-Laval Alcurd continuous coagulator.

The ultrafiltration-based process involves the following steps:

(a) milk standardisation, pasteurisation, homogenisation;
(b) milk concentration (usually 5:1) by ultrafiltration at 50°C;
(c) pasteurisation and homogenisation of the ultrafiltration retentate.
(d) addition of starter and lipase;
(e) addition of rennet to the ripened retentate and transfer to the coagulator system—tubular in Alcurd and small batch tanks in the Pasilac 5600;
(f) the coagulum may be cut and subjected to a holding/heating stage usually in the permeate stream from the ultrafiltration plant, followed by dosing into moulds. This effectively texturises the final product which is then known as 'structured' Feta.

Time/temperature		*Operation*
	RAW WHOLE MILK	
72°C for 15 s	↓	± standardisation ± pasteurisation ± bleaching agent (decoloriser)
	CHEESE VATS	
30–32°C (ripen) 30–32°C for 45–120 min (set)	↓	Starter cultures: 1·5–2·0% ± lipase enzyme Coagulant: 20–30 ml per 100 litres ± calcium chloride: 0·02%
	COAGULUM	
30–32°C 30 min–1 h	↓	Cut: 2–3 cm cubes Stand/occasional stir
	CURD/WHEY	
16–20°C	↓	(1) Ladle into moulds, or (2) ladle into cloths ± weak pressure Overnight drainage
	CHEESE	
pH: 4·3–4·7	↓	Cut: 1 kg pieces (1) Dry salt, or (2) brine salt: 16–23% NaCl solution
	PACKAGING	
3–4 weeks at 8–10°C then 5°C until despatch	↓	(1) Cans with brine, or (2) vacuum pack
	DESPATCH	

Fig. 19. Steps in Feta manufacture (traditional).

(g) Alternatively, the retentate may be dosed into moulds or directly into storage tins, to which brine is added. This is 'unstructured' Feta.

Cheese yield, as a result of UF manufacture, is in the order of 5500 litres milk per tonne of product, compared to 7300 litres milk per tonne product of traditional manufacture. This increase in yield is due to incorporation of whey proteins into the cheese, and also a more efficient recovery of milk fat.

The technique of direct acidification may be used for the manufacture of Feta instead of starter cultures, and this has led to a recent development whereby cooled, ultrafiltered milk is directly acidified and then dosed as a liquid pre-cheese into cartons on a vertical form–fill–seal packaging machine. The pre-cheese then coagulates in the carton to produce an unstructured type Feta.

SEMI-SOFT CHEESE MANUFACTURE

Figure 2 illustrates a functional classification of semi-soft cheeses into 3 categories, namely, surface/internal bacterial-ripened, internal mould-ripened and 'stretched curd' cheeses.

Surface/internal Bacterial-ripened Semi-soft Cheese

Cheeses in this category include Saint Paulin, Munster, Brick and Port Salut, and the surface flora contribute to the flavour development, although recently varieties, such as Saint Paulin, have been manufactured without the surface organisms, which are replaced by a dye, to satisfy consumer demand for a smooth, clean-flavoured product. The establishment of the reddish-brown coloured, surface bacterium, *Brevibacterium linens*, has been outlined above.

Saint Paulin is a French cheese derived from Port Salut made in a monastery near Laval. It has a creamy white, soft but sliceable texture, with a mild, slightly acid flavour and delicate, aromatic aroma. The rind is yellow to orange in colour, and this is usually achieved nowadays with the use of coloured plastic coatings, although some product may be found with the natural rind dyed with annatto, or even with a surface growth of the reddish-brown bacterium *B. linens*. Figure 20 illustrates the process steps involved in Saint Paulin manufacture.

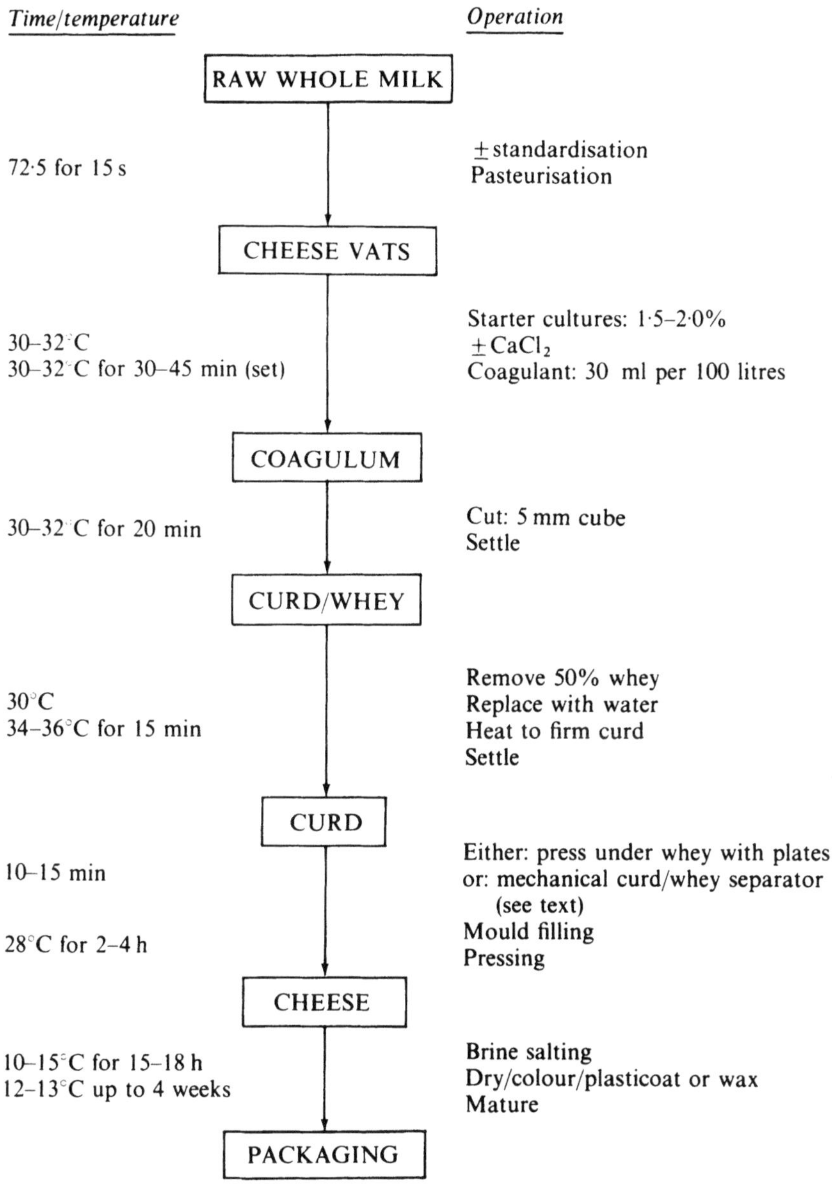

Fig. 20. Steps in Saint Paulin manufacture.

The manufacturing process is characterised by the removal of 50% of the whey and replacement with water (or brine solution in some cases), this effectively reduces the acidification rate and results in mild flavour development. The curd is subjected to a cooking temperature of between 34 and 36°C, followed by a slight pressing, and the cheeses are salted by immersion in a saturated brine solution.

Equipment and process developments

Curd manufacture is carried out in a range of traditional and modern vats with the appropriate development of systems for carrying out the curd washing stage, i.e. removal of a volume of whey and replacement with water. This is generally achieved with the use of a perforated boom, which can be lowered into the whey during the settling stage. Whey is then removed by vacuum and water introduced, either through the boom arrangement or from jets installed in a pipe running along the internal surfaces of the vat. The most significant development has been the introduction of curd drainage and moulding machines, available from Stork-Friesland, Koopmans, Alfa-Laval and Pasilac, for example. These consist basically of tubular drainage cylinders where curd is separated from the whey, and the curd is subsequently dosed into cheese moulds (see Chapter 2 for further details).

Other developments include single cheese, multi-row, automated tunnel press systems, automatic deep brining systems, mechanised equipment for plasticoating cheese, and automated cheese maturation stores. More recently, work has been carried out on the application of ultrafiltration to Saint Paulin manufacture. This approach has resulted from the development of mineral membranes, which have allowed milk to be concentrated to the higher solids levels necessary for semi-soft cheese manufacture.

Internal Mould-ripened Semi-soft Cheese

Cheeses in this category include Stilton, Roquefort, Gorgonzola and Danish Blue. Stilton manufacture is officially limited to the three English counties: Derbyshire, Leicestershire and Nottinghamshire, while Roquefort must be made from sheep's milk in the Roqueforte region of southern France.

The blue cheeses are characterised by the internal veining resulting from the growth of the aerobic mould species, *Penicillium roqueforti*, which is available exhibiting a range of colours from green to deep blue, and varying proteolytic and lipolytic activities. The blue cheeses are generally produced in round-wheel forms, containing 3–5% salt and possessing a spicy, piquant flavour associated with certain free fatty acids and ketones resulting from the mould's lipolytic activity. The textures are generally moist, with a slight stickiness and crumbliness.

Manufacturing process

The steps involved in manufacturing a semi-soft blue cheese are outlined in Fig. 21. The process is characterised by the homogenisation step which increases the susceptibility of the fat to lipolysis, and consequent flavour development, while also assisting in the production of a porous curd and open-textured cheese required for good blue mould growth. The *P. roqueforti* is either added to the milk, or at a later stage, such as onto the curd at salting. The curd is not pressed, but knits together under its own weight in open-ended moulds which are turned frequently.

During ripening, the cheese must be pierced to allow the escape of carbon dioxide and the entry of oxygen which promotes mould growth.

Stilton manufacture differs from the typical blue cheese process, as illustrated, in that

(a) there is no homogenisation;
(b) starter inoculation is very low (0·01%);
(c) there is no cooking stage;
(d) the curd is left overnight in coolers/drainers;
(e) the curd is cut into chunks, salted and milled twice before filling into moulds;
(f) cheese drainage occurs in moulds for 4–5 days; and
(g) the cheese is pierced fairly late in ripening, at 6 and 8 weeks.

Equipment and process developments

The manufacture of Stilton and other blue cheeses is considered very much a traditional process, and consequently much production is still carried out as a labour-intensive operation using basic cheese vats and curd drainers. However, some cheese is now manufactured using modern, mechanised vats, such as the Damrow 'Double O', curd drainage belts, and mechanised mould-filling equipment. Some curd has been manufactured using the Alpma continuous coagulator and, as with other categories of cheese, the use of ultrafiltration has been investigated. Significant advances have been made in the storage of blue cheeses with environmentally controlled, fully automated stores being available from manufacturers, such as Elten and Koopmans, and automated piercing, cheese

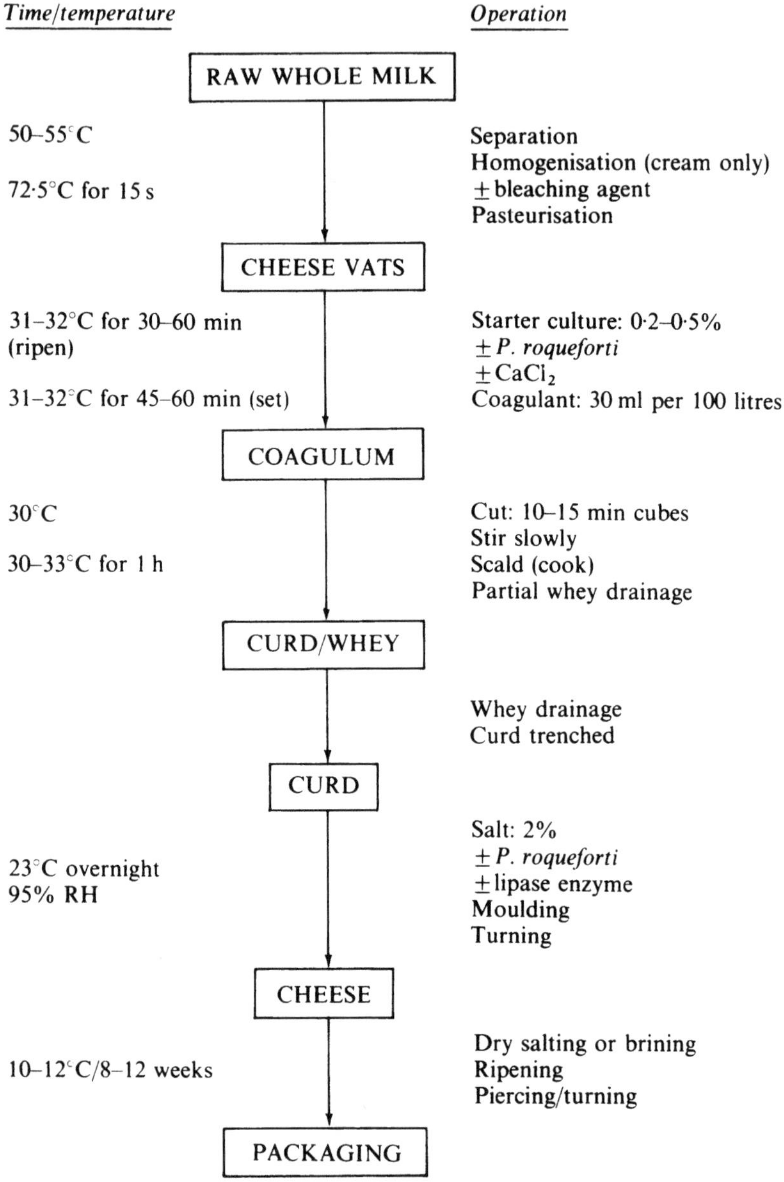

Fig. 21. Steps in semi-soft blue cheese manufacture.

washing and packaging operations are being introduced. Many manufacturers have gone away from the long maturing, less salty types of blues, in favour of high yielding, quick maturing types capable of being manufactured on modern, mechanised equipment with reduced labour involvement.

Stretched Curd Cheese

The major cheese types included in this category are Mozzarella and Provolone, which are also referred to as 'pasta filata' or 'plastic curd' types. Mozzarella was originally manufactured from high-fat buffalo milk, but it is now made all over Italy, in other European countries and the US from cow's milk. Some cheese is made from partially skimmed milk, especially for use in catering as a pizza cheese. It is an unripened semi-soft cheese which may be consumed shortly after manufacture. It has a soft, waxy body with a bland, but mildly acidic, flavour.

Manufacturing process

The steps involved in the manufacture of Mozzarella are illustrated in Fig. 22. Although traditionally manufactured from raw milk, it is more common to pasteurise nowadays. Often a bleaching agent, such as titanium dioxide, is added to mask the natural β-carotene yellow colour, and high temperature, or thermophilic starter cultures, such as *Lactobacillus delbrueckii* subsp. *bulgaricus* are used together with *Streptococcus durans* for flavour production. The use of high-temperature starters is necessary because of the high cook temperature used during processing, and the need to produce curd of optimal pH for the stretching operation.

Lipases from kid, lamb, or calf are often added in order to stimulate the production of piquant flavours resulting from free fatty acid production. At a critical pH of 5·1, the curd is heated under hot water, stretched and moulded into shape, and finally filled into moulds.

The chemistry of the stretching operation involves conversion, by lactic acid, of the dicalcium *para*caseinate, produced as a result of rennet action, to monocalcium *para*caseinate, which, when heated to 54°C or higher, becomes smooth, pliable, stringy and retains fat:

$$\text{calcium caseinate} + \text{rennin} \xrightarrow{\text{pH } 6{\cdot}2} \text{dicalcium } para\text{caseinate}$$

$$\text{dicalcium } para\text{caseinate} + \text{lactic acid} \longrightarrow \text{monocalcium } para\text{caseinate}$$

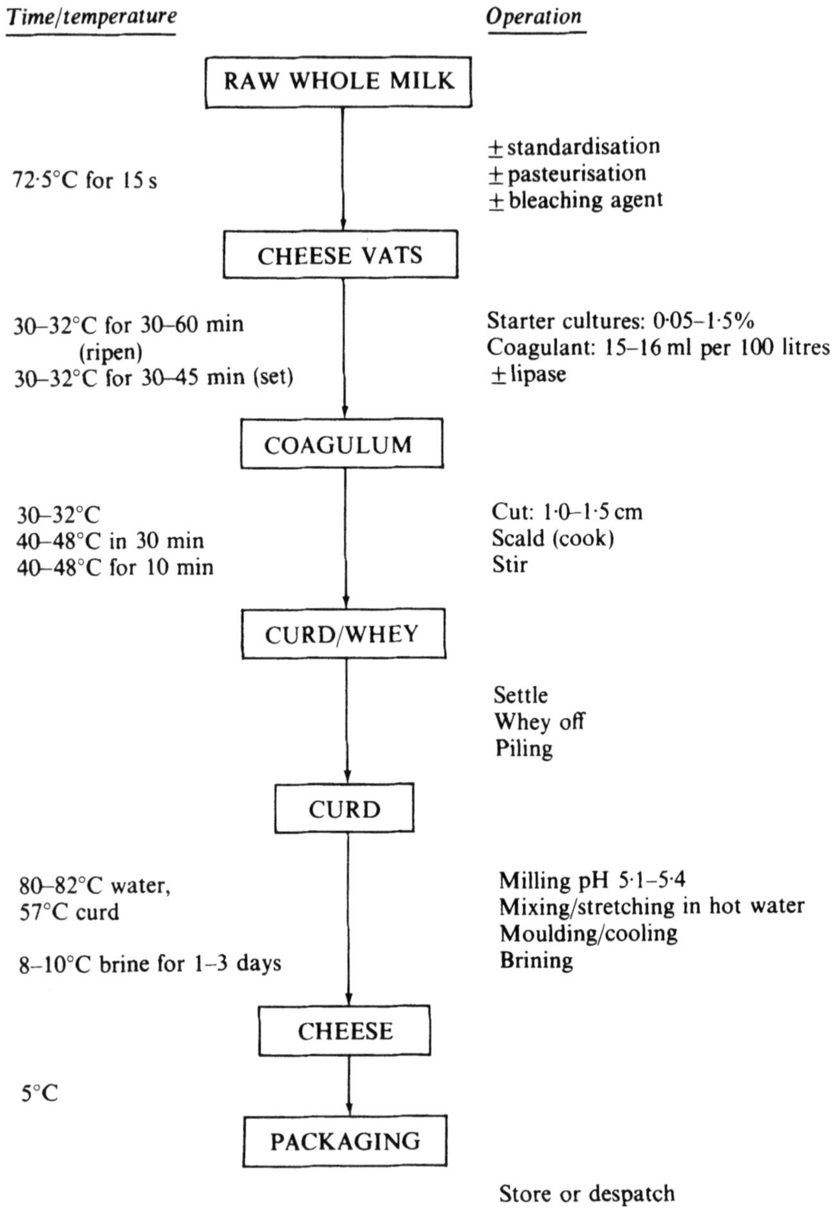

Fig. 22. Steps in Mozzarella manufacture.

Equipment and process development

Traditionally, Mozzarella manufacture was carried out in wooden vats with the curd mass being manually manipulated under hot water. Now the curd is produced in modern automated vats and curd drainage equipment (see Chapter 2), and the downstream operations of milling, heating, stretching, moulding, cooling and brining have all been mechanised and automated by various equipment manufacturers, such as Modena and Alfa-Laval. The ripened curd blocks are reduced to smaller size particles using a mill which may be separate or part of a complete line. The milled curd is then heated with hot water or steam, and subjected to a gentle pulling or stretching into a smooth plastic mass, which is then transferred to a moulding module which forms specific weights into desired shapes and units by weight. Finally, the moulded cheese is passed through a cooler into an automatic brining section.

Other developments include the use of direct acidification techniques instead of, or as well as, the use of starter cultures, and the application of ultrafiltration technology.

Product developments include the introduction of shredded admixtures of Mozzarella with Cheddar and Samsoe for pizza manufacture, and 'string' cheese: a 'finger' of Mozzarella targetted at children as a 'snacking' product.

Other Soft/Semi-soft Cheeses

Figure 2 refers to three other categories of cheese which may contain soft and semi-soft types: whey cheese, processed cheese and cheese substitutes.

Whey cheese

Soft cheeses made from whey include Ricotta, originating from Italy, and Mysost and Gjetost made from cow's milk and goat's milk whey, respectively, and originating from Norway. The Ricotta, although originally made from the whey from Provolone and other cheese manufacture, is now made from mixtures of whey and milk, or from whole milk alone. The cheese resembles cottage cheese curd, is soft and creamy, with a light delicate texture and slight caramel flavour. The Mysost and Gjetost are light tan in colour, with a smooth creamy body and a caramelised flavour.

Manufacturing process

Ricotta is made by heating whey to 85°C by direct steam injection followed by the addition of citric or acetic acid. The whey proteins will precipitate out, and the curd can be transferred into drainage moulds for 4–6 h. Following drainage, the cheese is ready for packaging and consumption.

Mysost and Gjetost are manufactured by evaporating whey to 60% total solids. The cheese is a condensed mass of caramelised lactose, protein, fat and minerals. The mass (60% total solids) is transferred to a vacuum cooker and further concentrated to 80–85% total solids. The brown concentrate is then further heated until the required flavour and brown colour is reached. The product is kneaded to produce a butter-like texture, and is then moulded into blocks and packaged. It has a good shelf-life due to the low moisture content (15–20%), and it can be stored for up to 6 months at 5°C. The product does not undergo ripening.

Processed cheese

A number of soft/semi-soft processed cheeses may be found in the market place in the form of spreads, dips and dressings.

Processed cheeses are manufactured by blending and heating one or more base cheeses with a suitable emulsifying salt until a homogeneous mass results. Processed cheese spreads may also contain other dairy ingredients, such as skim-milk, cream, butter and whey powder. Many 'value-added' spreads are available containing such additives as shrimp, crab, pepper, horseradish, nuts, mushrooms, garlic, herbs, port or kirsch.

Manufacturing process

Processed cheese spreads are made by selecting suitable cheeses according to age, flavour, body and texture. The cheese is then ground and mixed with emulsifying salts and water, followed by heat treatment in a kettle at 70–85°C for 5–15 min, dependent on the product being made. Heating can be indirect, or direct by steam injection. The hot, plastic cheese mass, to which additives such as herbs, nuts or fruit may be added, is transferred to filling machines, where it is dosed into the final package. The emulsifying salts used are citrates, monophosphates and polyphosphates, and these cause an increase in pH which, in turn, solubilises the protein and results in a smooth, homogeneous mass.

Equipment and process developments

Many large production facilities now exist in the US and Europe for processed cheese manufacture. There have been many types of batch and continuous cookers developed. In Europe, round-bottomed, stainless-steel vessels equipped with direct or indirect heating and vacuum operation are widespread. In the US, stainless-steel horizontal cookers with direct steam injection are popular, and more recently, continuous cookers, such as the swept-surface, heat exchanger (e.g. the Votator or Kombinator), have been introduced.

The use of enzyme modified cheeses (EMC) for processed cheese manufacture has recently become widespread. EMCs are made by treating cheese with a mixture of proteolytic and lipolytic enzymes, which results in the production of an intensely flavoured material which can substitute in a processed cheese formulation for a proportion of the mature cheese; it thus introduces economies to the processed cheese operation. Many EMCs are available commercially from flavour houses in the form of pastes or powders.

Cheese substitutes

The cost of producing dairy products has risen considerably over the years, and this has given an impetus to the development of a range of dairy product substitutes, including cheese in the form of blocks, slices, spreads and dips. The major application for cheese substitutes (or imitation, analogues, artificial) is in formulated foods manufactured by catering or industrial establishments, e.g. a Mozzarella substitute for use in pizza manufacture.

Manufacturing process

Substitutes may be manufactured by two routes, and these involve the use of fat and/or protein sources other than those native to milk, together with a suitable flavour system. The one route uses a liquid 'milk', whether skim-milk plus vegetable oil, or totally synthetic, and involves conventional in-vat cheesemaking methods, the products often being referred to as 'filled' cheeses. The other route involves blending various raw materials together using techniques similar to those for processed cheese manufacture. The raw materials used can be various vegetable proteins and oils, or alternatively partial-dairy ingredients, such as casein and caseinates, together with a hydrogenated vegetable oil, such as soya bean, palm, cotton seed, coconut or corn.

The flavour system might be artificial, or of natural origin using EMC for example. As well as savings in the manufacturing process, raw materials are considerably cheaper per tonne of product than milk, with vegetable oil some 75% cheaper than butterfat, and casein up to 50% cheaper than skim-milk powder, if calculated on an equivalent protein content basis.

Methods of manufacture vary but, generally, processed cheese equipment can be utilised, for example batch cooker/mixers or continuous scraped-surface cookers.

The future of cheese substitutes depends on the relative prices of vegetable fats and proteins compared with milk ingredients, and perhaps more importantly on a greater consumer acceptance of the products.

GENERAL DEVELOPMENTS

Over the last 10 years a wide range of cheese products have been finding their way onto the supermarket shelves. In the UK for instance, while there has been a static Cheddar and territorial market, there has been considerable growth in the soft and semi-soft cheese sector. Imported products, such as Brie, Camembert and Fromage Frais have proved popular and this has led to the development of home produced cheeses. Changes in dietary fashion have also resulted in diversity of product development, with a number of low-fat or reduced-fat cheeses now being produced, as well as higher fat 'luxury' products for special occasions. Export opportunities for products, such as Feta and mould-ripened cheeses, have also stimulated production demand. Consideration of the high labour and raw material cost have led to a need to increase process efficiency, and this in turn has stimulated equipment manufacturers to develop mechanised, automated lines capable of producing good-quality high-yielding cheeses at minimal unit cost. Many of the equipment and process developments have been reviewed in the sections relating to each category of cheese, but the following are worthy of further consideration.

Ultrafiltration

The technique of ultrafiltration has been applied to the manufacture of many soft cheeses, including Camembert, Quarg, cottage cheese, Feta, Mozzarella and Danish Blue, Renner & Abd El-Salam (1991).

Ultrafiltration is a membrane separation process operating at a molecular level. The membrane has extremely fine pores (about 1–20 nm in diameter), and small molecules such as water, lactose and dissolved salts can be forced through the pores under pressure, while larger molecules, such as proteins and fats cannot go through and can, therefore, be progressively concentrated. Ultrafiltration membranes are based on cellulose acetate (first generation), polycarbonates or polysulphonates (second generation), and zirconium oxide (third generation), and ultrafiltration systems are available in various forms: hollow fibre, tubular or plate. The advantages of using ultrafiltration for cheesemaking are as follows:

(a) Higher protein recovery due to incorporation of whey protein, (over 20% of total protein in milk);
(b) higher fat recovery due to reduced loss into whey;
(c) reduced volume of fluid to be processed;
(d) less rennet and starter required;
(e) smaller plants required; and
(f) continuous processing is made easier.

Ultrafiltration can be utilised in four ways.

(a) Pre-concentration of milk up to two times (i.e. half volume of original milk). Concentrate is processed using conventional cheesemaking equipment, and the major advantage is increased plant utilisation. There may be a slight increase in yield due to increased recovery efficiency of fat and protein.
(b) Production of a 'pre-cheese'; here, the milk is effectively concentrated to the final solids content of the cheese type with the concentrate or retentate being dosed, after addition of starter and rennet, into moulds. As there is only limited whey drainage, increase in cheese yield can be significant. Some Feta (unstructured), soft unripened cheese, and Camembert are produced by this method. The ultrafiltration concentrate cannot be processed in conventional vats and, therefore, specialist equipment has been developed for ultrafiltration cheesemaking.
(c) Production of a concentrate, followed by cutting of the coagulum produced after addition of rennet and starter. This method allows some texturisation of the final product as in the case of structured Feta, Mozzarella and Danish Blue manufacture. Again yield increase can be significant, but specialist equipment is required for processing the concentrate into cheese.
(d) Ultrafiltration of whey from cheesemaking to concentrate whey proteins which are subsequently returned to the cheese vat as a slurry. Only relatively small increases in yield can be achieved by this method without affecting product quality.

A number of continuous systems based on ultrafiltration have been developed recently, and these include the Camatic system from Alfa-Laval for the manufacture of soft ripened cheeses, such as Camembert, the Alcurd system from Alfa-Laval for production of structured Feta and Mozzarella, and the CC2500 coagulator from Pasilac, again for the production of structured Feta and Mozzarella.

Direct Acidification

Acidification of milk during cheesemaking is generally as a result of lactic acid production by lactic bacteria (starter cultures) fermenting milk lactose.

However, direct acidification has been traditionally practised in the manufacture of a number of unripened cheeses, such as Ricotta and Queso Blanco, where lactic, acetic or citric acid is used to adjust the pH of the milk to 5·0.

More recently, work has been carried out on the application of direct acidification techniques to the manufacture of cottage, Mozzarella and Feta cheese.

The advantages of direct acidification or the direct set method are

(a) elimination of problems associated with starter cultures;
(b) decreased production times;
(c) improved product consistency; and
(d) greater control of process.

The process involves acidification of cold (3–5°C), milk with a mineral acid such as phosphoric, and this is usually achieved in-line using a helical mixer coupled to an acid dosing pump. The system is linked to a pH controller, which monitors and holds the pH in the range 4·90–4·95 (for cottage cheese manufacture). Following this operation, the acidified milk is heated in the vat to 30–32°C and an acidogen, D-glucono-δ-lactone is added. This hydrolyses producing gluconic acid at a controlled rate (*cf.* the production of lactic acid by starter cultures), until pH 4·70–4·75 is attained. A small amount of coagulant may be added to assist coagulum formation, and cheesemaking proceeds as for production using starter cultures.

As no enzymes are formed with this method, compared to the use of starter bacteria, cheese flavour is bland, and often a starter distillate is added to the cottage cream dressing in order to give a 'creamy' note to the flavours. Presently, direct acidification has only found applications in the manufacture of soft unripened cheeses, where flavour development is somewhat irrelevant, but no doubt developments in the field of enzyme technology will result in a greater use of the technique in the future.

Automation, Mechanisation and Continuous Cheesemaking

Reference to the specific sections reviewing equipment developments illustrates how many technical solutions have been found for automating and mechanising the various stages of cheesemaking. Cost considerations have been the major impetus for these developments, and nowadays the labour requirement has been minimised with large throughput, process-efficient plants controlled by centralised computer systems being available for the manufacture of many cheese varieties. Modern cheese vat design achieving good milk solids conversion coupled with efficient curd/whey separation, and cheese handling equipment allows traditional cheesemaking procedures to be mechanised and automated, while the development of new equipment for handling ultrafiltration concentrates and the introduction of continuous curd production on equipment such as the Alpma and Alfa-Laval Alcurd coagulators has opened the door to a new era of cheesemaking technology.

FURTHER READING

Davis, J. G. (1966). *Cheese, Vol. III: Manufacturing Methods*, Churchill Livingstone, Edinburgh, UK.

Eck, A. (1987). *Cheese making Science & Technology*. (Lavoisier, Paris, France).

Fox, P. F. (1987). *Cheese: Chemistry, Physics and Microbiology (Vol 2) Major Cheese Groups*. Elsevier Applied Science Publishers, London, UK.

Guerault, A. M. (1966). *La Fromagerie devant les Techniques Nouvelles*, Editions Sep., Paris, France.

Kosikowski, F. (1977). *Cheese and Fermented Milk Foods* (2nd edn). Edwards Brothers Inc., Ann Arbor, MI, USA.

Scott, R. (1981). *Cheesemaking Practice*, Elsevier Applied Science Publishers, London, UK.

UK Cheese Regulations, S.I. 1970 No. 94 as amended by S.I. 1974 No. 1122, S.I. 1975 No. 1486, S.I. 1976 No. 2086 and S.I. 1984 No. 649, HMSO, London, UK.

Van Slyke, L. L. and Price, W. V. (1979). *Cheese*. Ridgeview Publishing Co., Reseda, CA, USA.

REFERENCES

Bundesgesegzblatt, (1986), German Federal Cheese Ordenance, No. 15, 23 April 1986, pp 412. (Amended 3/12/87, 16/12/88, 23/6/89, 12/11/90 & 29/10/91.)

Campden, R. A. (1987). Technical Manual No. 19. Guidelines to the establishment of Hazard Analysis Critical Control Point (HACCP), Campden Food Research Association, Chipping Campden, Gloucestershire.

FAO/WHO Code of Principles concerning milk and milk products, International Standards for milk products and International Individual Standards for cheeses, Codex Alimentarius Commission, Vol XVI, 1st edn, Rome.

ICMSF (1988). Microorganisms in Food 4. Application of the Hazard Analysis Critical Control Point (HACCP) system to ensure microbiological safety and quality. International Commission on Microbiological Specifications for Foods. Blackwell Scientific, Oxford.

Lehmann, H. R., Dolle, E. & Bücker, H. (1991). *Processing Lines for the Production of Soft Cheese.* Westfalia Separator AG, Germany.

Renner, E. and Abd El-Salam, M. H. (1991). *Application of Ultrafiltration in the Dairy Industry.* Elsevier Applied Science Publishers, London, UK.

Tamime, A. Y. (1990). In: *Dairy Microbiology* (2nd edn), ed. R. K. Robinson. Elsevier Science Publishers, London, UK.

Chapter 4

Developments in Frozen-Products Manufacture

H. L. Mitten*

Formerly APV Crepaco, Inc., Chicago, USA

and

J. M. Neirinckx

APV Pasilac AS, Pasteursvej, DK-8600 Silkeborg, Denmark

Frozen dairy products are well known in most of the world, but there is a wide range of composition and physical characteristics from area to area. In those areas where frozen dairy products are made commercially, several types are defined either legally or by the industry.

CLASSIFICATIONS

There are two broad categories: hardened products and soft-serve products. Hardened products are packaged in the semi-frozen state as extruded from an ice cream freezer, then frozen in a hardening room, tunnel or other device, and stored prior to distribution. Soft-serve products are those which are served directly to the consumer in the semi-frozen state as extruded from the ice cream freezer.

Frozen dairy products may also be classified as ice cream, dairy ice cream, ice milk, frozen yoghurt, sherbet, sorbet and water ice. All of these except sorbet and water ice contain some milk solids, and all of them may be aerated to some extent.

Frozen novelties are variations in the form in which the frozen products are marketed. They are usually stick and stickless lollies or bars, extruded bars, small cups and ice cream cakes or pies. Many

*Present address: 400 W. Butterfield Rd., Suite 726, Elmhurst, Illinois 60126, USA.

of these have chocolate or other coatings, and some are elaborately decorated.

Ice cream usually contains 8–18% fat, 9–12% milk solids-not-fat (MSNF), 13–16% sugar other than the lactose of the MSNF, and small amounts of stabiliser and emulsifier. In the UK, ice cream may contain milk fat, vegetable oils or animal fats other than milk fat or combinations of these fats. If the fat is all milk fat, the product is known as dairy ice cream. This is the practice in much of the world, and because of relative costs of the fats or oils, only about 20–25% of the ice cream is dairy ice cream. This is not the case in the US where the legal standard of identity requires that the product contain at least 10% milk fat if it is to be called ice cream. When vegetable oils or other animal fats are used, the product is called mellorine or artificial ice cream.

Ice milk is similar to ice cream, but has a fat content of about half that of ice cream, usually 4–6%. It also has more MSNF and sugar. Ice milk generally has a lower calorie content than ice cream, and this has made it attractive to diet-conscious consumers in recent years.

Ice cream and ice milk are aerated during freezing so that the overrun (percentage volume over volume of non-aerated mix) is from 25% for the new super-premium products to 120% or more for standard products.

Frozen yoghurt is generally similar to ice milk, and is sold both as a soft-serve product and as a packaged, hardened product. Since about 1970, it is usually made in two parts which are combined for the final mix. The first part, 80% of the total, contains some of the milk fat and MSNF and all of the added sugars and stabiliser-emulsifier. The second part is a skim-milk or low-fat yoghurt. These parts are processed separately, the first according to the procedure for ice cream mix, and the second as yoghurt. When the acidity of the yoghurt is 1·2–1·4%, it is cooled to 4°C and combined with the first part for freezing. In the proportion of 80 to 20, the acidity of the yoghurt dilutes to 0·35–0·4% in the combined mix. Frozen yoghurt mix is formulated so that the composition of the combined product is 0–4% fat, 12–12·5% MSNF, 13–14% sugar and 0·4% stabiliser-emulsifier. Although there is no official standard in the US for frozen yoghurt, it is generally agreed that the final product should contain viable yoghurt bacteria and have a titraitable acidity as lactic acid of not less than 0·35%.

Sherbets vary greatly in composition, taste, colour, and eating characteristics. Generally, they contain fruit or fruit juice, some added acid for greater tartness, about twice as much sugar as ice cream, 1–2% milk fat, from 2–5% by weight of total milk solids and from 0·2 to 0·5% stabiliser, depending upon the type and strength. Sherbets are aerated to give overruns ranging from 20 to 35%.

Water ices are similar to sherbets except they do not contain any milk solids.

Sorbets are often considered to be water ices, but in some areas small amounts of milk solids may be added and whipping agents may be used to permit somewhat higher overrun (30–35%) and to control brittleness and smoothness of body.

These definitions or descriptions are very general. The manufacturer of frozen dairy products should familiarise himself with the legal regulations in his marketing areas.

HISTORY OF FROZEN DESSERTS

The history of frozen dairy products is given briefly by many writers, but a publication of the International Association of Ice Cream Manufacturers (1966) is interesting and comprehensive. It traces the beginnings of ice cream and frozen desserts to iced drinks enjoyed by Alexander the Great in the 4th century before Christ down through the years of Nero in the first century AD, Marco Polo in the 13th century, through Italy and France to England in 1643–1649, and to America around 1700.

In 1846, an American woman, Nancy Johnson, invented the hand-turned ice cream freezer which was widely copied, leading the way to a new industry.

The first wholesale ice cream business in the US was started in Baltimore in 1851 by Jacob Fussell. After the introduction of mechanical refrigeration around 1890–1900, the frozen dairy products industry grew rapidly throughout America and Europe.

The first continuous ice cream freezer was developed by The Creamery Package Mfg Company of Chicago about 1908. It was a crude machine with refrigerated discs rotating in an ice cream mix flowing through a vat-like trough so that the discs were partially submerged in the mix. As the mix froze to the discs, it was scraped off and blended into the liquid. Very softly frozen product flowed out one end and into containers.

This was followed shortly by a batch-type freezer invented by Miller. It had an improved, brine refrigerated, freezing cylinder and a rotating dasher with a beater and scraper blades. Batch freezers were improved by direct expansion and full-flooded refrigerating jackets, and were soon manufactured by several equipment makers in America and Europe. This type of freezer really started the era of industrially made ice cream, and was the predominate type of freezer until around 1938.

In 1926, Clarence Vogt invented the first truly successful continuous ice cream freezer. It aerated and froze the product in a closed cylinder under pressure with the product in continuous flow. Within the next decade, others developed similar machines and by 1940, about half the ice cream was made with continuous freezers.

High-speed packaging equipment and rapid hardening equipment developed along with ice cream freezers led to the mass production of frozen dairy foods, taking them from a high-cost luxury dessert to a readily available, nutritious, and nearly universally desired food.

The manufacture of frozen dairy products is an economically important segment of the dairy foods industry. Information published by the International Ice Cream Association (1991) indicates that the total production of ice cream and related frozen products in 46 selected countries was approximately 12 287 862 700 litres in 1990. Of this total the US made 5 463 270 700 litres or 44·5%. Japan ranked second with 910 091 700 litres, about 7·4% of the total, and the UK ranked seventh with 301 175 280 litres, approximately 2·5%. The trend is worldwide growth, but the increase in the US and the UK has slowed to around 1% annually.

MANUFACTURING METHODS

Ice cream and several of the other frozen dairy products are very complex in their chemical and physical properties. Readers are referred to books or papers on the more scientific aspects of the products, such as those of Arbuckle (1986), Berger (1976) and Keeney (1972).

Raw Ingredients

The raw ingredients necessary to provide the components of ice cream or ice milk must contain sufficient fat and milk solids in proportions that can be combined to make a mix of the desired composition. The usual sources of milk fat are milk, cream of 30–35% fat, butter and sometimes concentrated whole milk. The MSNF of the milk and cream alone are seldom sufficient for the mix. Therefore, concentrated whole or skim-milk or dried skim-milk solids are included in the necessary raw ingredients. Sugar is required for sweetness, and for additional solids desired for the characteristic body and texture of the finished ice cream. The amount of sugar depends upon market preferences to a large extent. Corn syrup of moderate dextrose equivalent (DE) can povide solids to build body and give chewiness to the ice cream without making it overly sweet. Both sucrose and corn syrup solids are available as syrups or in granulated or crystalline form. Other ingredients, such as the stabilisers and emulsifiers, are usually in dry form and are used in small amounts compared to the others. Water may also be considered as one of the ingredients, especially if dried milk and granulated sugar are used.

The choice of raw ingredients depends upon such influencing factors as quality of ice cream to be made, availability of the ingredient, relative costs, storage equipment and space available, the blending or batching methods to be used and the size of the operation. Nearly all modern ice cream makers having a yearly production of 2 000 000 litres mix or more, use liquid raw ingredients including sugar for most of the year. They may also use dry ingredients some of the time. Small ice cream makers may find that ease of storage and lower costs make dry ingredients more attractive, and will choose processing methods and equipment designed primarily for convenience in accommodating these ingredients. Occasionally, very large ice cream makers design their operations around dry ingredients, because of the location of the plant and availability of the ingredients. When there is no great economic advantage of one type of ingredient over another, most ice cream makers prefer liquid sources which can be pumped through closed systems with relatively lower labour costs for handling and for equipment cleaning.

Receiving

Receiving operations are similar to those of any other dairy processing plant. Milk, cream and concentrated milk or concentrated skim-milk are delivered by transport tanker, weighed, and pumped into storage tanks. The storage tanks may be of the horizontal type or the vertical or silo type. They should be insulated sufficiently to allow holding the cold ingredient for 2 days or more without a temperature rise of more than 4°C, when the temperature difference from ambient to product is 27°C or less. Air agitation may be used in milk storage tanks, but it may cause the incorporation of air and foaming when cream and concentrated milk are involved. Mechanical agitators of special design are required for these products.

Liquid sugar and syrups are also received from tankers and pumped to storage tanks designed especially for syrups. Because liquid sugars and syrups are very viscous at lower temperatures, they are usually transported and stored at temperatures of 35–45°C, where viscosity is low enough for ease in pumping and for good drainage characteristics. Care must be exercised to prevent the temperature of these syrups from exceeding 45°C where changes in flavour and colour may occur, and result in shortening the storage time and quality of the final product.

Storage tanks for liquid sugar and syrups should have stainless-steel product contact surfaces or the equivalent in cleanability and corrosion resistance, should be insulated and should have means for warming the syrup or keeping it warm. As an alternative, single shell, uninsulated tanks located in an insulated, heated room are often used. The temperature of the room is easily maintained by ordinary steam-heated blower units with thermostatic controllers.

The concentration of liquid sugar and syrups is sufficient to prevent or minimise growth of microorganisms within the liquid; however, there may be some growth of yeasts and moulds on the surface due to contamination of air in the headspace of the tank. Such contamination can be prevented by a UV lamp in the air vent line combined with an absolute filter on the air entry to this line. Sometimes a UV lamp is installed within the headspace of the tank. While this is effective in destroying yeasts and moulds, it must be removed for cleaning the tank and is a hazard in case it breaks.

Granulated sugar when used in large quantities may be received in bulk and conveyed into bins or silos. More often, granulated sugar and corn sugar solids are received in bags and stored in a dry area near processing equipment.

The emulsifiers, stabilisers, cocoa and other ingredients such as fruits and nuts are usually received in drums and stored in appropriate areas near their point of use.

MAKING THE MIX

Putting the ingredients together in the correct proportions to make a mix of the desired composition, and preparing it for further processing, is known as blending or batching the mix. The procedures for making the mix are influenced by the size of operation, and whether pasteurisation is to be by batch (vat or long-hold) method or by continuous high-temperature, short-time (HTST) pasteurisation.

In small volume operations, where less than 2000 litres of mix per hour for about 5 h per day is made, pasteurisation is often by the batch method. The mix ingredients are usually blended in the pasteurising vat, with the liquid ingredients added by pumping and dry ingredients dumped by hand directly into the liquid portion in the pasteuriser tanks. The amount of each ingredient required for the recipe is determined prior to the beginning of the blending operation. The amounts of liquid ingredients may be added by measuring the volume by calibrated dipstick, pumping through a flow-meter or by counting cans of a given volume and dumping the contents into the blending vessel. Dry ingredients are usually measured by the bag with part amounts weighed in a separate container. The ingredients which are used in small amounts, such as emulsifiers and stabilisers, are measured or weighed separately, and usually added after the other ingredients are in the blending tank.

Sugars, dried milk solids and stabilisers are much easier to blend when the liquid portion of the mix is warm. Some of the stabilisers disperse and hydrate best with the mix blend at 60°C or higher.

With batch pasteurisation, mixes are usually blended warm. Heating can start with the addition of the first liquid ingredient, and by the time the batch is completed, the temperature is at or near the pasteurising temperature. Little or no time is lost.

As homogenising and cooling following batch pasteurisation are continuous processes, two or three blending/pasteurisation tanks are often employed to provide an uninterrupted flow of mix.

MIX BLENDING SYSTEMS

Those ice cream makers who produce more than 25 000 litres of mix per day usually employ mechanical or automated mix blending systems and HTST continuous pasteurisers. The larger the operation, the more sophisticated the mix making system, and when more than 8000 litres per hour of mix is made, multiple lines are often employed. For example, in a factory which makes 20 000 000 litres or more of mix in a year, 120 000 litres per day might be made during the four months of peak production and 60 000 litres per day during the other eight months. With a 12-hour mix making day during the peak season, the average rate is 10 000 litres per hour. Since two 5000 litre batches can be made each hour, a line of a 5000 litre batching system and a 10 000 litres per hour pasteuriser can handle the production requirements. Most ice cream makers find it necessary to make some of the mixes in small volumes, and therefore, in order to maintain a high average production rate, prefer to have two mix making lines. In the example here, two 4000 litre batches per hour feeding an 8000 litres per hour pasteuriser, plus another line capable of about half the capacity of the main system, would provide greater flexibility in production and keep the production day within the desired time restraints.

Blending of ingredients to a recipe for the proper mix composition can be done in many ways. If all the ingredients are liquids, volumetric or mass flow-meters can be used. Metering systems with a meter on each ingredient line are the most rapid of batching systems, for all of the ingredients can be metered into the blending tank simultaneously. Weighing systems require adding ingredients individually, one after the other.

Meters can be set manually or electronically to activate valves or pumps or both for predetermined amounts, so that flow is stopped when the desired quantity of ingredient has passed through the meter. Metering systems can be fully automated using computers or microprocessors to compute the recipe, to set the quantity of each ingredient, to start and stop flow when the quantity has been satisfied, and to total and record data desired for inventory and production control.

Mass flow-meters are probably more accurate than volumetric meters and can be used to measure ingredients of different densities, but they are generally more expensive than volumetric meters. Properly designed volumetric metering systems can be very accurate, provided that the meters are maintained in excellent mechanical condition and are calibrated after a change in piping or a change in the ingredient to be metered. In smaller systems, one volumetric meter may be used for two or more ingredients if corrections are made for the different densities.

When it is necessary to add dry ingredients or if dry ingredients are used regularly, they can be weighed into hoppers above the blending tank and dropped into the liquid ingredients, or they may be fed by air or screw conveyors to a continuous weighing device and dropped into the blending tank. The blending tank should have a turbine agitator which will produce a deep vortex into which the dry ingredients can fall. This will assure rapid wetting and dispersion of the powders. The computerised control system can be designed to integrate the data from the meters with data from the dried solids weighing equipment.

Some mix makers prefer to make up a concentrate or slurry of the dry ingredients which can be stored and then metered to the blending system. When this is done, care must be exercised to assure that the composition of the slurry or concentrate is the same from batch to batch, and that the slurry or concentrate is cooled to 7°C or lower if it is to be held longer than 4 h before being used.

There are some worthwhile advantages to pre-liquefying dry ingredients. Air or foam tends to come out of the mixture during holding, and holding allows time for the dried skim-milk solids, whey solids, etc., to hydrate for a better mix later on.

Dry ingredient incorporators, high shear mixers, funnels and similar equipment may be used in mixing the dry ingredients with water.

Weighing systems are generally preferred over metering systems since weighing is a direct or basic process without the need for converting volume to weight. Weights are not influenced by differences in density due to incorporated air or composition of the ingredient, and properly designed weighing systems are very accurate. The batching tank may be suspended from beam scales or from a strain gauge-type load cell, or it may be mounted on multiple load cells. The suspended types are often located above blending tanks so that after the first liquid is in that tank, subsequent additions can be subjected to mixing for a considerable part of the batching time. Figure 1 shows a load cell batching system.

Load cell weighing systems are well suited to computer operation and control, and fully automated systems are possible. The fully automated systems may be equipped with a very small separate unit for the micro-ingredients, such as the stabilisers and emulsifiers. These in dry or liquid form are weighed by the micro-system for greater accuracy with very small increments, and dropped into the main weigh tank. The stabilisers and emulsifiers are more often weighed manually and added by hand to the main load cell tank where they are included in the total weight of the batch. This small manual operation requires little time and is nearly always preferred over the expensive and delicate automated micro-load systems. After batching the mix with either the metering or weighing system, it may be pumped directly to the surge or balance tank of the HTST pasteuriser or to a larger surge tank depending upon the overall design of the plant.

Mix batching systems and their controls can be simple or very complex and expensive. With computer or microprocessor control, almost every operating desire can be engineered into the system if the budget permits. Much of the time, simplicity in both the ingredient handling and in the batching system is prudent, less costly and results in greater flexibility for future operations.

In recent years, microprocessors have replaced mini-computers as the choice for automated mix-making systems. Microprocessors are small, can be expanded when operations are changed, and can be located in the

Fig. 1. Automated mix batching system. The weigh tank with its load cell is suspended above two blending tanks. After each ingredient is weighed, the control system indicates or records the weight and opens a valve to allow the ingredient to flow into one or the other blending tank. After a batch is completed, the next is weighed and dropped into the other blending tank. (Reproduced by courtesy of APV International, Ltd.)

processing room near or on the equipment to be controlled. Individual microprocessors located adjacent to specific operations may be linked to a central control room.

A recent innovation is the use of the personal computer (PC) along with programmable logic controllers (PLC) to operate and control mix assembly, processing operations and cleaning of equipment and pipe lines. The PCs with programs on hard disks or floppy diskettes do all the computing, determine the routing or ingredient flow, etc. Figure 2 shows such a system. D. A. Sleiberling (1991, pers. comm.) has written the following:

> Control of processing and cleaning operations in both new dairy facilities and for major renovation and expansion projects is now accom-

Fig. 2. Computer control room showing a desk-top personal computer used with a programmable logic controller for formulating mixes, assembling mixes, operating all processing equipment in the plant and controlling the cleaning and sanitising of equipment and pipe lines. The sub-panels on the wall at the left are those instruments required (in the US) for legal control. (Reproduced by courtesy of Seiberling Associates, Inc.)

plished through replacement of the hard wired system previously used with a combination of a PC and PLC. Only the required instruments for legal controls (dairy products pasteurisation and CIP operations) is retained. All other sensing and control of level flow, temperature and pressure is by means of PLC hardware and software, and the operator interface is through the PC monitor for observation and the keyboard for set point adjustment. This combination system combines the ruggedness, simplicity, and easy maintenance of the industrial PLC with the number crunching, reporting, calculation and data storage capability of the modern PC. The PC may be a desk-top system [as shown in Fig. 2] or may be installed as an industrially hardened workstation in the process area, or in combinations of these two components. In addition to providing the operator interface to the PLC, and thence to the process, the PC is also capable of communicating with mainframe computers for purposes of receiving or up-loading inventory data, production requirements, and other information as required. Most of these systems are installed [as shown in the photograph, Fig. 2], in a comfortable air-conditioned control room in the heart of the processing area, permitting a single operator to handle complex and extensive processes including receiving and raw product storage operations, mix formulation, automatically controlled batching and pasteurising, mix flavouring en route to the production centre, and automatic cleaning of the complete process. The system may provide substantial information such as processing variables, inventory reports, production reports and CIP reports.

Aids for incorporating dry ingredients

When dry ingredients such as powdered milk, whey solids, and granulated and crystalline sugars are used, mix blending may be slowed, and incorporation and dispersion incomplete in the time available. There are items of equipment which can assist the mix maker in improving his operations.

The simplest mechanical aid to speed up the addition and dispersion of dry ingredients into liquid is a funnel and centrifugal pump combination. A simple funnel or hopper is attached to a tee in a pipeline just upstream from a centrifugal pump. Liquid from the pasteurising tank or blending vessel flows by the tee to the pump and is returned to the tank. Dry ingredients are drawn from the funnel through the tee into the liquid by the partial vacuum developed by the centrifugal pump, and are wetted and dispersed very quickly by the turbulent flow in the line and the pump. A valve may be installed between the funnel and the tee so that after the dry ingredients have been added, recirculation can be continued for additional mixing without drawing air into the product. A more sophisticated funnel and pump incorporator or mixer is available as a unit, complete with the funnel, valve, and special pump which has a mixing chamber. These aids are especially useful for small volume, manual operations. For greater volumes and more rapid mixing, two other aids are available, and are especially helpful with high-volume mix blending systems when dry ingredients are used. One is a high shear mixer, and the other is a dry ingredient incorporator. In use, a portion of the liquid ingredients are put into one or the other of these mixers, the agitator turned on, and the pre-weighed dry ingredient added. Mixing is very rapid, sugars are dissolved quickly and dried milk solids dispersed almost instantaneously. The dry ingredient incorporator has a high-speed turbine type agitator which forms a deep vortex in the liquid. Dried ingredients flowing into the vortex are wetted and dispersed quickly. Air is not intimately mixed, but forms large bubbles that break out at the top surface as the mix circulates outward along the bottom, away from the turbine and up the side walls of the tank. A turbine agitator with a closed bottom is especially effective in dispersing powders of lighter density than the liquid portion of the mix. A turbine with a partly open bottom lifts heavy ingredients off the bottom of the tank and disperses them effectively, and easily handles the low-density ingredients as well. The open bottom turbine agitator is especially useful in rapidly liquefying granulated sugar and heavy syrups which tend to accumulate on the bottom of the tank.

Other considerations in mix blending

Some stabilisers and emulsifiers are not readily incorporated or dispersed with cold blending. The choice of stabiliser should be based upon the desired effect on the product, and then upon how it can be blended. As a compromise which may satisfy both requirements, ice cream makers often experiment from time-to-time to optimise their recipes and to evaluate new commercial stabilisers. Generally, sodium alginate must be added when mix temperature is 70°C or higher. Gelatine, propylene glycol alginate, carboxymethylcellulose and mixtures of carrageenan with locust bean gum or guar gum can be dispersed into a cold mix, and are suitable for pasteurising through an HTST system.

Emulsifiers are usually not a problem in blending either hot or cold. Sometimes emulsifiers are combined with stabilisers, but many ice cream makers prefer separate emulsifiers because they permit variations in the amount used for different products.

The choice between blending hot or cold is an important consideration when continuous pasteurisation is to follow blending. As seen earlier, when vat or batch pasteurisation is employed, heating during blending is an advantage and no choice is necessary.

When continuous HTST or ultra-high-temperature (UHT) pasteurisation is used, much greater economies in energy utilisation are realised when cold rather than warm mix enters the pasteuriser, for it is possible for regenerative heating and cooling to act over a greater product temperature range. For example, in an HTST pasteuriser designed for 80% regeneration, pasteurising at 80°C, and final cooling to 4°C, mix entering at 16°C after blending is heated to 67·2°C by regeneration, to 80°C by steam, cooled to 28·8°C by regeneration and to 4°C by refrigeration. If the mix is blended warm and enters the HTST at 45°C, it would be heated to 73°C by regeneration, to 80°C by steam, cooled to 52°C by regeneration and to 4°C by refrigeration. Summarising the net energy requirements, it is seen that with cold blending, the steam heating range is 12·8° compared to 7° with warm mix, and the refrigeration range is 24·8° with cold mix compared to 48° in the case of warm mix. At first it appears that the warm mix offers savings in the steam requirements, but heating of the raw ingredients for warm blending must be added to that of the HTST unit. This example assumes that raw dairy ingredients at 6°C and liquid sugar at 43°C are used in about a 74% to 26% proportion to blend at approximately 16°C. For warm blending at 45°C, the 29° difference requires additional steam and this added to the 7° in the HTST total 36° net steam heating range for the warm blended mix. As steam and refrigeration requirements are directly proportional to the temperature ranges involved, the energy savings for cold blending compared to warm blending is 35·6% in steam and 51·7% in refrigeration.

There are other factors to consider in deciding the temperature for blending. Sugars, especially heavy syrups, dissolve more slowly in cool liquids, and dried milk solids are difficult to wet, disperse and hydrate at temperatures below 35°C. Air entering the mix with dried milk solids and with the cream and concentrated milk, if used, tends to remain in the mix until heated to 35–40°C. When this occurs in a closed, continuous pasteuriser, the resulting mix and air may cause fouling of the heat exchange plates or surfaces, reducing the heat exchange rates and shortening the pasteurisation run. Hydration or holding tanks used between blending and the balance tank of the pasteuriser may permit enough of the incorporated air to escape to prevent problems downstream, but blending equipment which minimises the incorporation of air should be employed.

PASTEURISING, HOMOGENISING AND AGING

Pasteurisation

Both batch or long-hold pasteurisers and continuous HTST pasteurisers were discussed briefly, and are generally understood by processors of dairy foods, but some additional comments on continuous pasteurising equipment are in order.

While the great preponderance of continuous pasteurisers are of the plate type, other types are sometimes used. These are of the tube-within-a-tube, triple tube, and small diameter, high-velocity tubular types. The reasons for selecting these in preference to plate-type units are various and usually not related to operating costs, for plate pasteurisers with a high degree of regenerative heating and cooling offer low operating costs seldom equalled by other types. Tubular units are sometimes chosen because of limits on floor space, for they can be mounted along a wall or suspended from the ceiling.

Small diameter, high-velocity tubular types are usually limited to 6000 litres h^{-1} or less capacity with most makes having a maximum capacity of half that. These are generally available as a complete processing unit with controls, heater, pumps, surge tanks, holder, and homogeniser all mounted on a base. This type of pasteurising–homogenising system is especially attractive for ice cream makers in remote areas where installation skills are lacking, or when a smaller, complete line is required which can be installed with minimum disturbance to existing operations. These systems are also used for UHT pasteurisation or sterilisation of mixes. The latter is occasionally used to give long shelf-life to mixes for soft-serve products, or for small-volume freezing plants not making mix on site.

UHT pasteurisation or sterilisation of mixes is also done with direct contact steam heaters combined with a flash chamber to remove the added steam, and with plate heat exchangers for pre-heating, cooling and regenerative heat exchange. The direct steam heaters are of two types—injector and infusor. With injector heaters, steam is injected through a venturi into the flowing liquid mix; with infusion, the liquid mix flows into an atmosphere of steam. Both heat the product to almost the steam temperature nearly instantaneously, and by controlling pressure, very high temperatures—150°C or higher—can be obtained. A short holding tube providing 0·2–6 s or more residence time, as desired, assures complete heating of all the mix.

The flash removal of the vapour added is usually done in an evaporation/vapour separation chamber with controls to assure that vapour removed is equal to that added, so that no change in the composition of the mix occurs. Of course, the controls can be adjusted to concentrate or dilute the product, if that is desired. Vapour removal also removes some volatile odours associated with weeds and feeds ingested by the cattle, and much of the air and oxygen is also removed.

Cooked or heated flavour associated with UHT treatment is minimised by the direct heating pasteurising and sterilising systems; Fig. 3 shows a typical system.

Homogenisation

Homogenisation of ice cream mix reduces fat globule size to prevent churning in the ice cream freezer, improve the smoothness of the ice cream, and permit more of the milk proteins to adsorb onto the fat globule, which increases the viscosity of the mix and produces a smoother body and texture in the frozen ice cream.

Homogenisers are specialised reciprocating pumps having from three to seven pistons and an homogenising valve or valves. Mix is pumped through the homogenising valve, usually with quite a high pressure which is then released suddenly, dissipating the pumping energy into the product as heat.

How the fat globule reduction is accomplished has been studied since homogenisation was introduced, and even today there is no general agreement. Probably hydraulic shear forces, developed as the mix flows through the homogenising valve, distorts the liquid fat globule beyond the limits where the surface forces can hold the fat in globule form. When these forces are exceeded, the fat is attenuated, and during turbulent flow, separated into several new, smaller globules. When the average diameter of the fat globules is halved, the number of globules is increased by eight times and the total surface is doubled, allowing more proteins to adhere to the surface.

Homogenisation occurs only when the fat is liquid. This means that the mix for frozen desserts must be at about 50°C or higher for homogenisation, and the higher the temperature, the easier it is to produce good homogenisation with smaller and more uniformly sized globules.

Fat globules in ice cream mix can be made so small that the amount of natural phospholipids available to form the globule membrane is insufficient to cover the increased surface. In this case, the globules may coalesce to form larger globules or clumps which have the effect of larger globules.

Fig. 3. Ice cream mix UHT pasteurising/sterilising system of the direct steam infusion heating type. The upright vessel in the left foreground is a jacketed steam infusion heater. The vessel in the centre is the steam flash (removal) chamber. The plate regenerator and homogeniser are beyond these two vessels. (Reproduced by courtesy of APV Crepaco, Inc.)

Over-homogenisation may also affect the fat globule surface phenomenon to allow clustering or the incomplete dispersion of individual globules; clusters act physically much the same as larger globules.

In practical terms, large fat globules, clumps and clusters indicate instability in the fat-in-water emulsion of the mix, which may lead to partial reversal of phase during freezing due to the severe agitation of the dasher or mutator. The partial reversal of phase shows up as churning or smearing of the fat with buttery deposits on the scraper blades, the inside surfaces of bends, the edges of extrusion nozzles, and as visible particles in the product. When this occurs, the body and texture of the finished product is inferior.

The purpose of homogenising mixes containing fat is to produce a good frozen product without evidence of churning or smearing. This does not necessarily mean that the smallest fat globule size is necessary. A practical and economical approach to accomplishing satisfactory homogenisation is to place the homogeniser in the HTST pasteurising system at a point where the temperature is highest. This allows excellent homogenisation at lowest homogenising pressures. Further, the pressure should be adjusted to the lowest value which prevents churning in the ice cream freezer. It is important to keep homogenising pressures low to conserve energy and reduce production costs.

Another factor affecting homogenisation is the fat content of the mix. In general, the greater the fat content, the lower the homogenising pressure required. It is easy to over-homogenise or use too high pressure for high fat mixes.

Ageing of Mixes

Following pasteurisation and homogenisation, mixes are cooled to 4°C or lower and held in a holding or storage tank until they are required at the freezer. The tanks may be insulated and, in addition, may be refrigerated to maintain the low mix temperature. This prevents the growth of microorganisms, and promotes crystallisation of fat and other changes which improve freezing, air incorporation, smoothness in body and texture, and resistance to melting. Holding time combined with a low mix temperature is called ageing, and may be deliberate or incidental to the production procedures.

The changes which occur during ageing are more complete hydration of stabilisers, increase in adsorption of protein to fat globules and continuation of fat crystallisation. The effects of ageing are most pronounced with gelatine stabilisers, and improvements in ice cream and freezing performance are more pronounced as the ageing time increases from about 4 to 12 or more hours. Because gelatine is expensive and must be used in greater amounts than other stabilisers, it is no longer used extensively.

Mixes containing most of the other types of stabilisers require much less ageing time, and improvements in the mix may be more the result of increased protein adsorption to fat globules and fat crystallisation than a change in the stabiliser. Those stabilisers which hydrate completely during cold blending, or as the mix is heated during HTST pasteurisation, are affected very little by ageing. CMC (sodium carboxymethylcellulose) and guar gum are in this category, but a small amount of carrageenan (Irish Moss) must be blended with them to prevent serum separation (wheying-off) in the mix during holding, especially if the cold mix is held longer than 4 h.

Many progressive ice cream makers have observed that ageing can be shortened by cooling the mix to temperatures of 0–2°C. This probably increases the crystallisation of the fat and improves adsorption of milk proteins to the fat surface. A more certain result is an increase in the ice cream freezer capacity, for the freezing point in the freezer is reached much sooner when the additional cooling load is done outside the freezer. To cool the mix to such low temperatures efficiently and conveniently, a direct expansion chilled scraped-surface heat exchanger is most often used.

FREEZING ICE CREAM

Ice cream is a very complex food. The mix is nearly always more than 60% water. The water dissolves the sugars, both natural lactose and the added sugars, and also dissolves a portion of the salts from the milk solids. Then there is a colloidal system suspended in the water. The milk proteins and proteins from stabilisers and the insoluble salts make up the colloids here. Another system which depends upon the water is the fat-in-water emulsion from milk fat sources or from vegetable oils. These systems coexist. When the mix is passed through an ice cream freezer, it is chilled to the congealing point where air is incorporated, and then chilled to freeze out some of the water. All of this is accomplished under the rather severe agitation of the scraper blades, dashers and beaters, if any.

As ice is removed from the mix system, the sugar in water solution is concentrated, as are all the other systems, and the freezing point is depressed or lowered further by the increased sugar concentration in the remaining water. As ice is frozen, the ice crystals are suspended in the water, and very small air cells are incorporated into the mixture. When the drawing temperature is −5.6°C about 50% of the water, in most mixes, is frozen. This means that in a mix having 38% total solids, the semi-frozen ice cream extruded will have 69% of its contents as solids suspended or dissolved in the remaining water, which amounts to only 31% of the whole. If that mix started with 12% fat or 19·4% fat in water, it would be extruded with 38·7% fat in water. If freezing continued to a drawing temperature of −9·4°C where 67% of the water is frozen out, the fat in water content would be 58·7%. It is not unreasonable to expect such a concentrated emulsion to break or reverse phase to result in partial churning or a greasy texture. This is why some mixes, regardless of how well they are homogenised, cannot be frozen into the 'low temperature' range in an ice cream freezer with its severe agitation.

In addition, as ice crystals are formed, they add to the solids already in the mix, and this increases viscosity and motor load requirements.

In continuous-flow, industrial ice cream freezers, the air for overrun has very little effect in the freezer cylinder because it is compressed. In a freeze operating with 5 atm (~505 kPa) absolute cylinder pressure, the air required to give 100% overrun occupies only one-sixth the volume of the total mix and air. Thus, the density of the mixture in the freezer is not affected enough by the air to interfere with rapid internal heat flow to the cylinder walls. When the semi-frozen ice cream is extruded, it expands as the pressure is lowered to atmospheric and only when this expansion is complete is overrun fully realised.

Most textbooks deal with air incorporation as it occurs in batch-type ice cream freezers at atmospheric pressure. In this case, the mix takes on air when chilled to its 'congealing' point where it changes from a liquid to a more plastic mass. This, of course, starts at the initial freezing point. At atmospheric pressure, air is incorporated readily between the initial freezing point and −5°C to −5.3°C.

In continuous freezers, the cylinders are under pressure and very high overruns at very low drawing temperatures can be obtained. Generally, air for overruns up to 130% at drawing temperatures to −7°C is easily incorporated with cylinder pressures of 3·5–5·5 atmospheres, depending upon the dasher and blade design and the condition of the blades. For overruns in excess of 130% additional cylinder pressures may be required. When the drawing temperature is lower than −7°C, cylinder pressures may have to be increased by 2–3 atm (~202–303 kPa).

The temperature of the mix as it enters the continuous ice cream freezer is very important in ice cream freezer performance. If the temperature of the mix is uniform throughout the run, the overrun control and freezing rate are predictable, provided that the refrigerant supply and suction conditions are also uniform. If such a mix is also supplied to the freezer's mix pump at a constant pressure, operating controls can be adjusted at the beginning of a run to give the overrun and temperature desired with no significant changes required throughout the day.

A mix temperature of 0° to −1°C will optimise freezer performance. It will also assure that the fat or vegetable oils are crystalline, and will practically eliminate the necessity of ageing where formulation has not already done so.

The consistency of ice cream as it is drawn from the freezer is often referred to as 'wet' or 'dry' or 'stiff'. This consistency is influenced more by formulation than by any other factor, and while related to the drawing temperature, it is not directly so related. Some quite cold ice creams can still have some wetness and be flowable, and some stiff, dry ice creams can be quite warm when drawn. The design of the freezer dasher and the volume of product in relation to throughput does have some effect on this quality of product, but mix that produces a characteristic wet ice cream can be reformulated to give a drier product at the same drawing temperature. On the other hand, if the freezer itself causes the dryness and stiffness, the drawing temperature must be raised or an extensive reformulation made in order to obtain a wetter, more flowable product.

Freezers with displacement dashers combined with relatively high rotative speeds tend to cause a shearing action that may produce a physical change in the structure of the protein–fat–sugar complex. This is more pronounced as the drawing temperature is lowered, but it is definite at −5°C. The ice cream produced this way is stiffer and drier at a given drawing temperature than that made from the same mix but with a more open, slower dasher. This ice cream is slower melting after hardening, and is characterised as a 'warmer eating' product. This eating quality is a result of slower melting in the mouth, and some would say that this product is less refreshing than the wetter-drawn, faster-melting ice cream. Flavour is not released so readily with the warm eating product, and melt down may be flaky in appearance. The 'warmer eating' ice cream, when its characteristics are not carried to extremes, is preferred by about 25% of the consumers and producers in some markets.

From an engineering viewpoint, 'stiff, dry' ice cream is less than ideal, for it has poorer flowability, causes more pressure drop in pipelines and has poorer container filling characteristics than a 'wetter appearing' ice cream at the same temperature; stiff, dry ice cream also requires more power input from the dasher motor and the pump drives. The term 'wetter appearing' ice cream as used here means one which has a sheen or gloss indicating some unbound moisture, but a product that is stiff enough to stay in the container during packaging operations and which, if drawn colder, is stiff enough for extrusions for some of the stickless products.

Freezing point curves for ice cream and ice milk of various compositions are very similar (Fig. 4). They differ according to the depressing effects of the dissolved sugars and salts, but a curve for one mix composition falls slightly above or slightly below that of another. A freezing point curve is a practical reference for any ice cream maker, because it shows important general facts. For example, from earlier discussions, it was seen that because of the nature of ice cream, the most practical extrusion temperature is between −5.5 and −6°C. The amount of ice formed is from 52 to 55% of the water in the mix. If ice cream

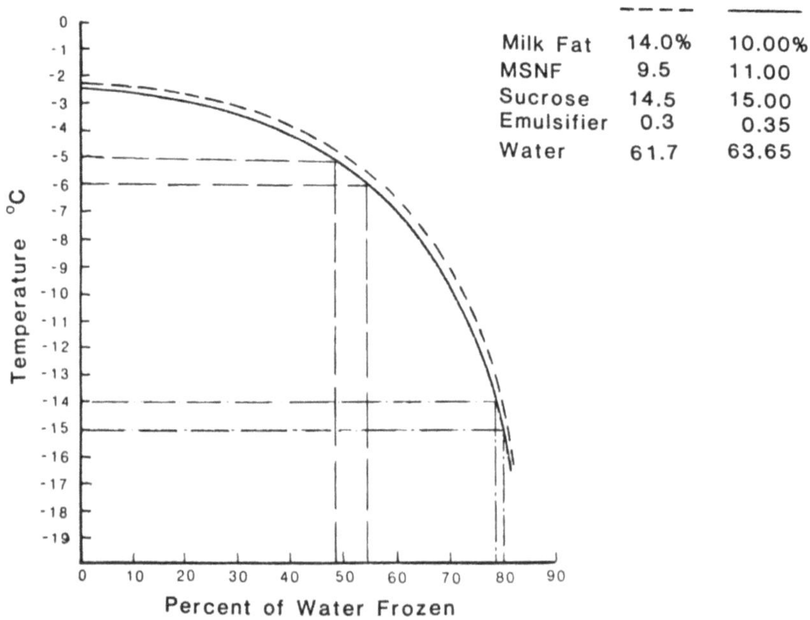

Fig. 4. Freezing point curves for two ice cream mixes. The horizontal and vertical lines indicate percentage of water frozen from the mix at a given temperature for the mix having 10% milk fat. It is apparent that a temperature change of 1°C at higher ranges affects more of the ice crystals than a change of 1°C at lower ranges.

extruded at −6°C were held long enough between extrusion and hardening to permit its temperature to rise to −5°C throughout the mass, the amount of water frozen is reduced to 48·5 from 54·5%. This means that 6% of the water has been affected or 11% of the ice formed has melted; when refrozen, larger ice crystals result. To prevent coarse, icy ice cream, the product temperature should not be permitted to rise, but should be lowered continuously until well hardened. When hardened to −15°C, about 80% of the water is frozen. If the temperature goes up to −14°C, about 78·5% of the water is still frozen, but has only about 1·9% of the ice has melted, refreezing will not produce as much coarseness as in the case of the same 1°C change from extrusion temperature. The freezing point curve shows that temperatures should be lowered continuously, that fluctuations cause iciness, and that at lower temperatures, there is a diminished effect. It can be deduced that repeated cycles of temperature fluctuations will accelerate this problem. This information is important in the proper management of packaging, hardening and cold storage of the product.

ICE CREAM FREEZERS

There are two types of ice cream freezers: the batch type and the continuous flow type. Batch freezers are used in very small ice cream factories, or for small volume speciality lines and for pilot plant research in larger factories. They are also used rather extensively in ice cream stores which freeze and harden ice cream in-house.

Batch Freezers

Batch freezers made today are generally of the horizontal cylinder type with the freezing cylinder refrigerated with one of the halocarbon refrigerants, most often R22 or R502. There is a mix supply tank located above the freezing cylinder, so that mix will flow by gravity into the cylinder when a valve is opened. The usual procedure is to charge the freezer, turn on the dasher and start the refrigeration. When freezing has increased the power required to turn the dasher to a value which indicates sufficient freezing, the refrigeration is turned-off, and air is whipped in until the desired overrun is obtained; this is determined by drawing samples for overrun determination. At this point, any fruits, purées or flavours may be added directly into the open port of the front door and mixed into the ice cream. A slide or pivot valve on the bottom portion of the front door allows the ice cream to be drawn into containers or bulk cans. The dasher is designed to propel the product toward the discharge port. Subsequent batches are made in the same manner. A new charge is put into the mix supply tank as soon as the previous charge is dropped into the freezing cylinder. A batch of ice cream can be made every 6–7 min by experienced operators.

The sizes of batch freezers vary according to the manufacturers' designs, but 18–20 litres and 36–40 litres of 100% overrun ice cream are the two sizes found most often. If greater than 100% is desired, the mix charge must be reduced enough to prevent overflow of ice cream during whipping.

As freezing in a batch freezer is at atmospheric pressure, the whipping ability of the mix is important and must be considered in formulation. The dashers of the freezers are designed with heaters to promote whipping, but when freezing to temperatures below $-5°C$, air is less readily incorporated, and some may even be expelled.

Generally, ice cream made with batch freezers has both larger ice crystals and bigger air cells than ice cream made with the same mix on continuous freezers. Overrun control to close tolerances is difficult with batch freezers, and it may vary by 10–15% from the beginning of drawing the batch to completely emptying the barrel. This occurs because whipping continues all during the drawing time. Where most of the ice cream is hand packed and where it is sold by weight, close overrun control is regarded as unnecessary.

Modern batch freezers are not much different from the earlier models except in better hygienic design, more durable materials, and better refrigeration systems.

Continuous Ice Cream Freezers

At present, there are approximately 10 manufacturers of continuous ice cream freezers in the world. These are in Europe, Japan and the US. Three of these supply between 60 and 70% of world requirements in numbers of units, and about 80% in terms of ice cream capacity.

All, but a few, very small, continuous freezers, have horizontal freezing cylinders, and all freeze with the cylinder pressurised. What happens during freezing was discussed earlier, and applies regardless of make of freezer.

Some makes have one cylinder or freezer to a unit, others have one, two or three freezers in a single housing, and one make uses pairs of cylinders in series to provide sufficient heat exchange surface for its larger models.

The capacity of continuous ice cream freezers is difficult to rate, since products around the world have greatly different characteristics which affect refrigerating requirements. Then, the different refrigerants are not thermodynamically equivalent. While there is no generally adopted standard for capacity rating, most manufacturers list very similar conditions for their nominal maximum rated capacity. Some, but not all, have a margin of safety included in these ratings. The conditions for rating include the following:

(a) machine is new or in excellent operating condition;
(b) refrigerant is clean, free of oil (if R717) and free of non-condensible gases; ammonia (R717) is the refrigerant used for rating:
(c) mix is an ice cream mix of approximately 10% fat, 15–16% sugar and 37–38% total solids;
(d) mix enters freezing cylinder at 4·4–4·5°C and ice cream is drawn at −5·6C; one major manufacturer rates at a drawing temperature of −5°C;
(e) refrigerant evaporating temperature is at −30·6°C (saturated conditions); and
(f) rated capacity is either in terms of litres of 100% overrun ice cream, or in litres of mix input.

The manufacturers' rating is a nominal rating which can reasonably be expected under the conditions stated for rating, but it does not assure that the ice cream maker will be able to operate at that rating. The types of sugars and some ingredients of specific formulations influence the viscosity of the freezing ice cream and its internal heat flow characteristics to such an extent that the maximum nominal capacity may not be attainable.

In 1991, the sizes of industrial continuous ice cream freezers available ranged from 300 to 3000 litres per hour per cylinder in terms of 100% overrun ice cream. Small pilot-plant continuous freezers down to 100 litres per hour were also offered.

Pumps and Overrun Systems

Pumps on ice cream freezers are usually of the rotary type with the capability to pump against pressures of 7–14 $kg\,cm^{-2}$ (~690–1380 kPa) with reasonable volumetric efficiency.

There are two general pumping arrangements, both designed as a part of the overrun system. The first employs a pump (or a pair of pumps or a compound pump), to pump or meter the mix into the freezing cylinder, plus a hold-back valve at the ice cream discharge port. The hold-back valve may be spring loaded with manual adjustment, or it may have an air operator with adustable air pressure supplying the operating power. The hold-back valve permits imposing a pressure on the cylinder during freezing which compresses the air admitted with the mix for overrun. Cylinder pressure of 3·5–4 atm (~354–404 kPa) keeps the volume of air in the freezing cylinder sufficiently small so that it does not significantly slow the internal heat transfer out from and through the mix, and that pressure is sufficient for good air dispersion and small air cell size. Higher pressures may be imposed on the cylinder, but in most cases, the improvement of heat transfer and air cell size is not great enough to offset the disadvantages of increased pumping costs.

The earlier continuous freezers using the pump and hold-back valve arrangement had two pumps in close proximity, both powered from a common drive, but with the second operating at about three times the volume of the first. As mix was pumped, a partial vacuum was produced between the pumps. Air for overrun was allowed to flow into the partial vacuum so that the difference in pumping volume between the pumps was made up with air. An adjustable snifter valve on the air intake allowed control of the amount of air to give the overrun desired.

Current freezers using this system make use of a compressed air source with an air regulator, sometimes combined with an air flow-meter, to control air from a slight vacuum to moderate pressure at entrance to the mix-air pump. This allows better control as the pumps wear and permits greater versatility in overrun control.

The other general pumping arrangement has a mix pump or mix-air pumps metering the mix into the freezer cylinder, and another rotary pump at the discharge of the cylinder. The mix-air pumps operate in the same way as in the pump and hold-back valve arrangement, but when a single mix pump is employed, compressed air is admitted directly into the freezer cylinder or to the mix line between the mix pump and cylinder. The ice cream discharge pump operates in a ratio to the volume of the mix pump that will develop cylinder pressure. For example, if the ice cream pump is operated at 1·23 times the volume of the mix pump, it will cause a cylinder pressure of 4·4 atm (~444 kPa) with 100% overrun ice cream, and 5·2 atm (~525 kPa) with 120% overrun. Low cylinder pressures result with low overrun products, and if the pumping ratio is not changed or a restriction is not placed at the inlet to the discharge pump, the cylinder may be emptied when overruns are less than 25%. One current model using this two-pump system has a hydraulic pump drive which, along with a cylinder pressure sensor and speed controller, permits a continuously variable ratio of pumping volumes between the mix and ice cream pump to maintain any preset cylinder pressure from 1 atm (~101 kPa) for products without overrun, to in excess of 12 atm (~1212 kPa) for very high overrun products drawn at cold temperatures.

With the two-pump arrangements, the mix or mix-air pump works against only the cylinder pressure, while the discharge or ice cream pump works against the difference between the downstream line pressures and the cylinder pressure. With the mix or mix-air pump and hold-back valve arrangement, the mix-air pump has to work against the imposed cylinder pressure plus the downstream line pressures. This may total 10–13 atm (~1010–1313 kPa), near or exceeding the design limits of some of the rotary pumps found on freezers. In the case of such high pressures, pump and pump rotor life is short and pump slip is relatively great.

Both systems are capable of the uniformity of pumping necessary for excellent overrun control, when the pressure on the pump inlet and downstream line pressure is uniform throughout the run. This is an ideal which seldom prevails. Levels in the mix supply tanks affect the pressure at the inlet even when the mix is pumped to the freezer, and there are variations in pressure drop through the lines downstream of the freezer. Such changes may be caused by slight fluctuations in product viscosity resulting from variations in extrusion temperature and changes in heat flow into the lines as frost thickness changes. Any changes in downstream pressures are transmitted hydraulically to the cylinder through a hold-back an old-back valve. On the other hand, the cylinder is effectively isolated from downstream pressure changes with a product discharge pump. Thus, when pressures outside the freezer are unsteady, the two-pump system yields more uniform overrun.

Ice cream freezer pumps are driven by various means, but all of these provide for varying the pump speed. Usually the set of pumps for each cylinder is powered by one drive. Drives are of three types:

(a) electric motor powering a mechanical variable speed drive;
(b) frequency inverters with electronic speed control for standard electric motors; a gear reducer is nearly always used between motor and pump; and

(c) hydraulic pumping systems connected to hydraulic motors on the pumps; the hydraulic pumping units may be located within the freezer housing or remotely outside the production room.

Automated Overrun Control

Automated overrun control systems which measure the density of the extruded ice cream and, by feedback, adjust the air supply to attain and maintain the desired overrun are not yet available. This type of system has been sought since about 1948, but the very nature of ice cream and the way it is produced introduces so many variables, that such feedback systems are not successful.

One of the problems is choosing the point at which to measure the density. Ice cream and related products containing air for overrun are compressible and full overrun is not attained until the product has expanded to atmospheric pressure. This requires some time and, in a continuous flow, is not realised until the product is in its package. Measuring weight in the package rather than density is probably an easier and more accurate control element, but this is subject to the variations in filler performance and container size, as well as differences in the cut-off of product from the extrusion nozzle.

The second major problem is the time lag between density or weight measurement and the change in air input. In a typical continuous freezer with a dasher displacement of 35–50%, the flow through the cylinder at 80% of rated maximum capacity varies from 0·9 to 1·2 min (as air is compressed within the cylinder, mix flow rates are the governing values). This is a considerable lag, and the tendency is for the instrumentation and controls to overcorrect resulting in a hunt and seek cycling of overrun which may be greater than with manual control.

The major manufacturers of ice cream freezers offer automated overrun systems which use microprocessors to regulate air input in relation to mix input. These provide for pre-setting the desired overrun. After operation is underway, the overrun is carefully adjusted manually, and the set points noted. These can be used for subsequent runs of the same products with excellent reproduction. Once overrun has been adjusted, the microprocessor will maintain the flow rates, pressures and other conditions to maintain accurate overrun control.

How accurately do these systems control overrun? With 100% overrun ice cream, and with a properly designed installation of equipment, standard deviations in the order of 1·5–3% in terms of overrun, or

1–1·5% in terms of weight can be expected. These standard deviations are about half those with standard freezers.

A word of caution is in order here. If the mix has an excess of air incorporated in it from the blending operations, from a leaky seal on the suction side of a pump, or from unmelted rerun (refreeze) in the mix, no amount of automation will control the overrun until these undesired air sources are eliminated. Automation is not a replacement for good management practices.

With good management practices and proper operator skills, manual overrun control can be within the standard deviations expected for automated systems. Good practices include proper maintenance of all equipment and pipelines in the system, proper blending of mixes with sufficient hydration and ageing time, minimising air incorporation, air removal, complete reprocessing of refreeze, keeping mix temperature low and constant throughout the day, supply of mix under uniform pressure to the freezer mix pump, assuring that there is nearly constant pressure in the suction mains of the refrigeration system, and keeping frozen product lines between freezer and packaging point as short as practical. Good production management will also provide for long operating runs of one product any one day to avoid unnecessary changes in products where freezing must be interrupted and restarted.

Dasher Design

Dasher types and their influence on ice cream was discussed earlier in the section on freezing ice cream. Most manufacturers of industrial ice cream freezers offer at least two different types of dashers or mutators for their machines, in order to produce the wide variety of frozen products being made around the world. One set of specifications for a freezer cannot be expected to produce optimum results for all of the products.

Dashers were originally designed for batch freezers as rotating carriers for the scraper blades, and for beating or whipping air into the congealing, partially frozen product. The first continuous freezers had a small diameter cylinder and a solid mutator to which scraper blades were attached. This solid dasher displaced almost 80% of the volume of the cylinder, and having a small annular space for the product and a high rotative speed, the mix was subjected to rather severe shear. These conditions, combined with rapid freezing, produced a very smooth textured ice cream which, with many mixes, tended to be greasy, warm heating and slow melting. In operation, this type of freezer was sensitive to changes in refrigerant temperature and froze-up quickly.

The next continuous freezer to be introduced had a much larger diameter cylinder and an open-type dasher with a beater similar in design to that of batch freezers. When ice creams from this freezer were compared to those from the other with the solid dasher, when drawn at the same temperature, the ice cream appeared wetter, melted faster and seemed to be more refreshing to the taste. It was also somewhat coarser and more nearly like that made with batch freezers.

During the years following the introduction of these first continuous freezers, designs of the cylinders and dashers of both makes changed, becoming similar in several ways. The first make increased its cylinder diameter and reduced the displacement of its dasher, making the freezer less sensitive to refrigerant changes and producing a very smooth ice cream with less tendency to be greasy. Dasher speeds were also reduced which produced better melt-down in the ice cream.

Later on, other dashers were designed for different products. Open dashers with beaters operating with greater cylinder volume are the least sensitive to operator inattention, as they respond slowly to both intentional and unintentional changes in air, refrigeration or flow rates. Dashers with 30–40% displacement, with or without beaters, are nearly universal in respect to the types of products handled, and are excellent for the currently popular low overrun, premium ice creams, and for mixes having a relatively high content of whey solids. Solid dashers with 65–80% displacement are preferred for cold-drawn extrusions. They require less power than the more open dashers and enhance quick freezing. They are very sensitive to changes in refrigeration, freezing-up rather easily unless monitored closely by the operator or by automatic controls.

Dasher types and rotative speeds may be matched to characteristics desired for special products. For example, ice crystal size is especially important in certain types of delicate water ices, and the proper choice of dasher and its speed can produce just the crystal structure desired.

Operating Controls and Automation

All continuous ice cream freezers have controls for operation which include on–off switches for power to pump and dasher motors, and for air compressor motors (when these are part of the freezer), for solenoid valves on hot gas defrost lines, air lines, and refrigerant supply lines, speed regulation of pumps, refrigeration supply and back pressure, pressure gauges for the refrigeration system and cylinder or air pressure, and dasher motor ammeter, wattmeter or motor load indicator.

In addition, the newer, more sophisticated machines may have a viscosity meter and controller, and a programmable controller or microprocessor to operate and control most functions of the ice cream freezer.

The viscosity controller takes a power signal from the dasher motor power line and, through a transducer and controller, adjusts the refrigerant back pressure to raise or lower the evaporating temperature which lessens or increases the degree of freezing. As evaporator temperature is lowered, the extrusion temperature of the product is also lowered, and more ice is frozen from it. This increases the viscosity or stiffness of the product within the freezing cylinder, thus requiring more power to turn the dasher. The dasher power for the desired stiffness or viscosity of product can be preset and changed any time during operation.

Typical of the state of the art is an ice cream freezer with a microprocessor programmed to control all the functions of operation including overrun, viscosity of product, cylinder pressure, all operating steps such as start up, routine or emergency shutdown, resumption of operation after an automatic shutdown when the cause for shutdown has been corrected, preparation for cleaning, and the valve and pump bypass required for automated cleaning (see Fig. 5).

Microprocessor-programmed operation assures that all functions are performed in the proper sequences, and under the conditions envisioned by the designer of the freezer. This is especially beneficial to the ice cream maker in preventing damage to the freezer in emergency situations, thus avoiding the incidental, unplanned down-time in production.

NEW ICE CREAM FREEZERS

New or up-dated ice cream freezer models for 1992 were introduced in the US in October, 1991 by the world's four major manufacturers of ice cream freezers, two from Europe and two from the US. All of these freezers were equipped with microprocessors for operating the freezer and for control of the functions as described above. Three of the makes had the two-pump system, and three had automatic cylinder pressure control through variation of the ratio of ice cream pump to mix pump speed. For overrun control, three makes used air mass flow-meters and mix magnetic flow-meters with computer control of air input in proportion to the mix

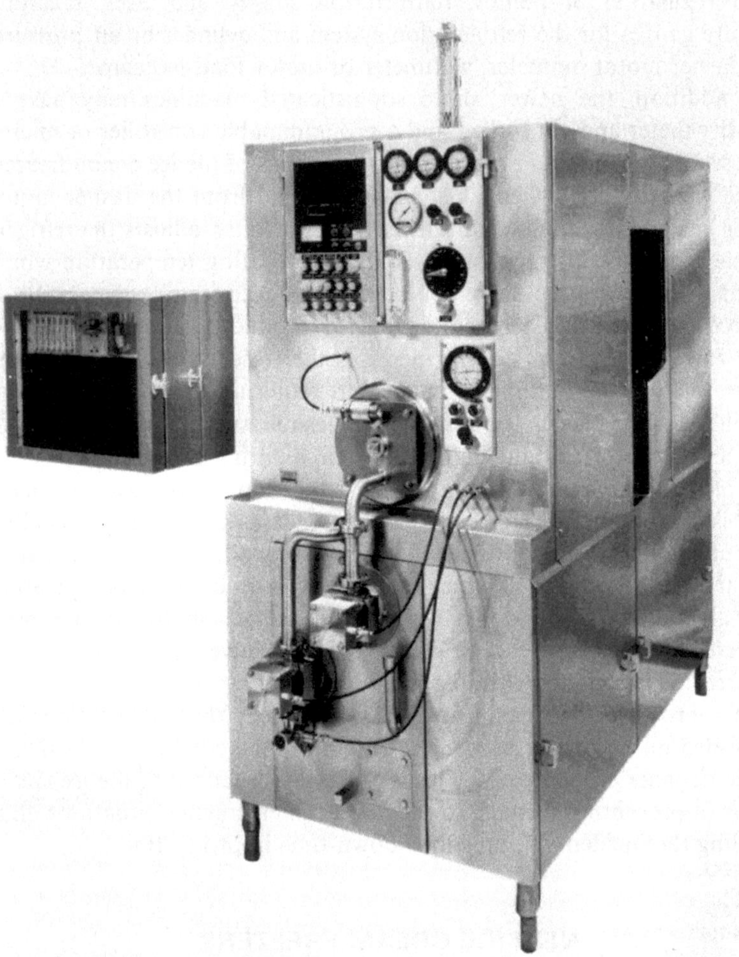

Fig. 5. A modern continuous ice cream freezer with microprocessor control. The specific machine shown here is a single-cylinder model. The miroprocessor, shown at the left of the freezer, can operate up to three cylinders and can be mounted on the side or rear of the freezer, or it may be located in a control cabinet at a distance from the freezer. (Reproduced by courtesy of APV Crepaco, Inc.)

flow as measured by the mix flow-meter. One used the principle of critical flow to control the air for overrun. Critical flow occurs when the velocity of flow in the throat of an orifice is at or greater than the speed of sound.

This occurs when the ratio of absolute pressure immediately downstream the orifice to the absolute pressure immediately upstream the orifice is at or less than approximately 0·5 for air. Any further decrease in pressure downstream the orifice has no influence on the flow rate. The quantity of air passing through the orifice, then, is dependent only on the upstream air pressure which is regulated through the microprocessor and a transducer to proportion the air to the mix input.

Pre-freezer Aeration of Mix

The new freezers also offer an optional pre-freezer aerator, either as a separate unit or built into the housing of the freezer. Pre-freezer aeration is a new technology which was described by Andreasen (1987) and introduced late in 1987 on prototype equipment. Pre-freezer aeration appears to offer benefits in air incorporation, slower melting, increased stability in storage, and increased creaminess of product.

The pre-freezer aerator is a high-speed, high-shear machine similar to that used in the baking industry to plasticise and reduce the density of shortenings. It is placed in the mix feed line after the mix pump and flow-meter. Air, instead of being added at the cylinder, is directed into the mix in the aerator. The aerated mix then flows into the freezing cylinder and is frozen in the normal manner. Such pre-freezer aeration apparently results in the release of greater amounts of free fat which is believed to aid in producing smaller air cells and to result in the benefits listed above. In one US installation, variations in overrun when using the pre-freezer aerator are routinely one-third to one-half the magnitude of the overrun variations without the aerator when using the same mix and make of freezer.

The pre-freezer aerators are costly, and whether or not application of this new technology is worth the added cost is yet to be proven, for at the beginning of 1992, there were very few commercial installations.

Special Ice Cream Freezers

Two special types of ice cream freezers which were popular in the past are receiving renewed attention. These are the recirculation freezer and the low-temperature freezer.

The recirculation freezer, which was invented in 1961, consisted of a standard freezer with a third pump which recirculated a portion of the partially frozen ice cream from a point between the discharge port of the freezer and the product discharge pump and discharged it into the mix line between the mix pump and the inlet to the freezing cylinder. This recirculation pump was driven by the common pump drive, but it was sized to pump approximately 30–35% of the freezer throughput.This produced a product that was smoother textured and slower to melt than that not recirculated when using the same mix and drawing temperature.

The recirculation freezer was first used for making a warm-drawn, very fluid ice cream or ice milk for stick bar novelties (lollies), for the product had superior retention of its overrun through the hopper and into the moulds of the stick bar machine, yet it flowed readily into the moulds. Because there was less loss of overrun prior to mould filling, the saving in mix was substantial, often enough to pay for a new freezer in a few weeks of operation. Sales for this freezer were very good for about 10 years, but dwindled as the market reached near saturation. At present three manufacturers offer this type of freezer.

Some ice cream makers used the recirculation freezer for ice cream extruded at normal temperatures (−5 to −6·6°C), and found that after hardening, the product was smoother textured and slower melting than that made without recirculation. The additional mechanical working of the product may produce some of the benefits claimed for the pre-freezer aeration treatment.

Low-temperature freezers are those capable of extruding ice cream products at temperatures as low as −9 to −10°C where almost 70% of the water of the mix is frozen.

The original low-temperature freezers had two freezing cylinders in one housing with series flow of product through them. The first cylinder to receive the mix froze the product just as if it were a normal freezer. The second cylinder, usually smaller in diameter and equipped with a dasher that turned at about one-third the speed of that in the first cylinder, did the additional freezing.

The product extruded at such low temperature was very stiff, and therefore difficult to package. It was excellent for extrusion of ribbons of product which could be hardened, cut to desired length and then enrobed with chocolate.

With the recent growth in popularity of high-quality, stickless novelties, such as ice cream candy bars, logs, etc., the interest in low-temperature freezers has increased even though standard freezers produce acceptable extrusions at −7 to −8°C.

The four major manufacturers of ice cream freezers now offer auxiliary freezers for this application. The auxiliary freezer is much like a refrigerated scraped-surface heat exchanger with a mutator (dasher) carrying scraper blades and turning at 80–100 rpm. Ice cream is pumped from the standard freezer through the auxiliary freezer where a further freezing occurs. The auxiliary freezer has no pumps or provision for adding air, and the only controls required are those for refrigeration.

Adding Ingredients and Flavours

Flavouring materials, other than chocolate, are nearly always added after the mix has been made. These may be added at the ageing or holding tanks, or in flavour tanks located just upstream of the ice cream freezer. Fruit juices, flavour extracts, colours and similar materials are added at these points. Pieces of fruit and purées should not be added to the mix prior to freezing in continuous ice cream freezers, for they tend to settle out in the tank with subsequent poor distribution in the frozen ice cream. Further, seeds in fruit or other gritty content harms the close-fitting pumps, the dasher bearings and seals and dulls the scraper blades.

Ingredients such as pieces of fruit, purées, nuts and candies, are inserted into the ice cream after extrusion from the freezer. When the ingredients are purées, syrups or conserves, they may be pumped into the product carrying line by a small pump often called a ripple pump. A device within the line at the entry from the ripple pump may be used to produce a variegated effect, or to distribute the ingredients more uniformly throughout the ice cream.

Ingredient feeders, often referred to as fruit feeders, have a hopper for the ingredient, an auger or other means for metering or proportioning the fruit, a rotor or plunger for inserting the ingredient, a mixing chamber or blender, and a variable speed drive. A typical ingredient feeder is shown in Fig. 6. An auxiliary unit is a vibratory dry ingredient feeder used to feed nuts, marshmallows, or other dry ingredients into the enrobing throat of the ingredient feeder. Dry ingredients may be fed simultaeously with the fruits for a greater variety of flavoured ice creams. Ingredient feeders are capable of adding ingredients continuously at combined ice cream and ingredient rates up to 11 000 litres h^{-1}, and with ingredients constituting 10–12% of the product.

Fruits and other ingredients of similar composition, fresh, frozen or canned should be sugared to prevent iciness in the final product. The amount of sugar required varies with the fruit from none for banana and grape to as much as 50% for raspberries. There is an excellent discussion

Fig. 6. Ice cream freezer installation. From the left is a three-cylinder ice cream freezer, an ingredient feeder, and a vibratory dry ingredient feeder, (not in its operating location). The long radius, sweep bends in the lines which carry ice cream reduce pressure drops and minimise damage to the produce. (Reproduced by courtesy of APV Crepaco, Inc.)

of fruit preparation and recommended fruit to sugar ratios in Arbuckle (1987).

PACKAGING

Packaging should be done as close to the ice cream freezers as is practical, for the longer the pipelines, the greater the possible damage to the ice cream. In the design of the ice cream pipelines, every effort should be made to not only keep the lines short, but to eliminate as many bends and other fittings as possible. Long radius, sweep ends are preferred over standard elbows for all lines carrying semi-frozen products.

Packaging Equipment

Packaging equipment is available for nearly all types of containers. These machines have only a few features in common: an extrusion head, some with a spreader plate, and a means for holding the container, moving it to the point of filling, and moving it away from the filling point.

Packaging–filling machines designed to handle paperboard cartons may store the flat folded carton in a feed cartridge, set it up for filling, fill it and close it. Filling may be on a time-cycle basis or on a volume/weight basis. Closure depends upon the carton design which may be of the interlocking-flap type, or a type which is held closed by an adhesive applied upon closure.

Carton fillers are available for various sizes from 250 ml to 4 litres. In the US, the pint and half-gallon sizes are most popular; the quart-size, rectangular carton is seldom used. In Canada, the litre and 2-litre sized rectagular cartons are popular, and in the UK and some areas of Europe, the 1-litre size is fairly common. A typical 1-litre, home-pack carton and its set-up and filling machine are shown in Fig. 7.

Other containers are the cylindrical paperboard containers used extensively in the US in the pint, quart and half-gallon size for the expensive premium ice creams; plastic cups and tubs in various sizes; and form-seal plastic containers formed in a web, filled, sealed and cut into separate packs on the same machine.

There are also bulk containers of 2, 4 and 6 litre size in Europe, and 2, $2\frac{1}{2}$, 3 and 5 gallon size in the US. These are often hand-filled, but filling and lidding machines are available. In addition, there are speciality machines which use extrusions of ice cream in varous forms, such as slices and sandwiches, and packaged in cartons containing six, eight and 12 pieces.

In the filling of cartons or other containers, those machines which use weight of contents to trigger the cut-off of filling generally give greater accuracy of fill than the time-fill type.

After the containers are filled, they may be bundled into large units by means of a shrink-wrap machine before being hardened, or they may go directly to the hardener.

HARDENING ICE CREAM

Ice cream as drawn from the freezer and packaged is only partially frozen. It was indicated earlier that the most practical temperature of ice cream as it is extruded from the ice cream freezer is between $-5{\cdot}5$ and -6°C,

Fig. 7. Carton set-up and filling machine. The machine shown here is for 1-litre home-pack cartons used primarily in the UK. (Reproduced by courtesy of APV Anderson, Inc.)

where the amount of ice formed is 45–55% of the water in the mix. The product is relatively soft as it is extruded, and can be pumped easily with moderate pressure drops. In filling, it flows into the corners of the containers and does not leave air voids as happens when extruded at lower tempeatures. If the product is to be rigid enough to store properly, to transport and to maintain its overrun, body and texture, it must have additional freezing. This freezing is done quiescently after the product is in its final container or package. This further freezing is known as hardening.

Hardening should be done as quickly as possible from the drawing temperature to about −18°C core temperature. Generally, the faster this can be done the smaller the ice crystals and the better the quality of the ice cream. However, if the temperature progression is continuously downward, hardening over a period of 10–12 h will cause no significant increase in ice crystal size. Faster rates of hardening are desirable for better space utilisation in cold stores, and for faster turnover of inventory.

Hardening Rooms

Cold cells or hardening rooms within, or separate from, the cold stores are still common, especially for smaller volume operations. If the hardening cell is located within the cold stores it should be partitioned-off from the storage area, so that air can be circulated at greater velocity with better control of flow. The hardening room may be equipped with shelves or racks on which the product is placed in bundles of open-type baskets. For rapid hardening, the product must be arranged to allow air to pass on all sides. The air temperature should be −25°C or lower, and air should be circulated with enough velocity to give good turbulence. Generally, there should be enough space in the hardening area of 24 h of peak production.

Hardening in a room of the type described will usually produce a core temperaure of −18°C in 10–12 h in bulk or bundled packages, and less time with 1 litre or smaller packages.

Hardening Tunnels

Hardening tunnels are used by most of the larger ice cream makers. Some are custom made, but most are standard models available from a number of manufacturers, and nearly all employ cold air at −30°C to −35°C circulated at a velocity in the order of 180 m min^{-1}. Hardening is accomplished in about 2 h for package sizes up to 2 litres, and around 5 h for 4 and 6 litre containers. Air cooler and blower units are arranged to circulate the air across the packages as they are carried by conveyors or trays through the length of the tunnel.

Spiral Tunnels

Spiral tunnels are one of the types of hardening tunnels used for ice cream. This type consists of a spiral conveyor, usually motivated by a centre drive, which carries the product up and around the tunnel for discharge to an exit conveyor. Large capacity units may have two spirals with one carring the product to the top of the tower and passing it to the second spiral, which conveys it down for discharge to the palletising operations or to the storage area.

Spiral tunnels may have their own enclosures, or they may be located in a cold room with walls to separate the conveyors from the storage area, and to control the flow of air.

Recently, a new oval path, spiral tunnel has been introduced in the UK. It has a unique edge-driven belt, and its mechanical drive is located outside the tunnel. It offers advantages over other spiral tunnels in space savings, access to product and ease of maintenance.

Straight-Through Tunnels

Straight-through tunnels are available with a variety of conveyor and moving tray or belt arrangements. While some of these are made up of standard modules, others are custom designed using some standard components. The three most efficient types available are of the hanging-tray type, with a number of shelves in each tray. The conveyor carries the trays vertically at the ends, and makes a double pass through the air stream. The trays move in short strokes, being indexed a shelf at a time to the loading-unloading position, where an automatic pusher moves fresh packages onto an empty shelf and simultaneously pushes hardened packages off the adjacent shelf onto an exit conveyor.

Contact Plate Freezers

Contact plate freezers (Fig. 8) with automatic loading and unloading for continuous operation are excellent hardening devices, freezing very rapidly, requiring a minimum of floor space, and operating at low maintenance costs. They are more efficient in the use of refrigeration than any other hardeners, for there are no fans with their mechanical heat added to the refrigeration load as in the case of hardening cells and tunnels. With plate hardeners, packages of product are moved into the space between two plates; the plates carry refrigerant through internal passages. When the entire space is filled with packages to be hardened, the plates close until they contact the packages on both top and bottom. Heat flows by conduction from the product through the container walls directly to the cold plates. Hardening is completed in 2 h or less at ratings up to 6800 litres of ice cream per hour.

Fig. 8. Contact plate hardener. This type of hardener is the most efficient of hardening systems for ice cream in rectangular cartons. It will accommodate round containers which have large area flat tops and bottoms, but efficiency is reduced because package configuration on the plate prevents contact with as much container surface as in the case of rectangular cartons which fully fill the plate area. (Reproduced by courtesy of APV Crepaco, Inc.).

Contact plate hardeners can be used for products in rectagular containers of one size, or of two sizes which have a common dimension. Where an ice cream maker produces a variety of packaged products but has a large volume in rectangular containers, the plate hardener is especially attractive because of savings in space, refrigeration, power, and ease of maintenance. Further, it is a small enough unit to allow easy relocation should that be necessary.

Contact plate hardeners are becoming quite popular in the US, where a considerable portion of the ice cream is packed in half-gallon rectangular cartons.

Special Tunnels

The special tunnels used for Eskimo Pie, Glacier and Polarmatic systems are familiar to most makers of stickless novelties. Product is extruded onto plates which are conveyed through the air blast tunnel or cabinet for hardening. After hardening, the product is carried to a discharge point where a mechanical hammer blow to the plate loosens the hard extrusion. It is then swept to another carrier for coating, drying and packaging.

Another special tunnel freezing system is that which has a continuous, stainless-steel belt conveyor travelling through a controlled air blast. Product from the ice cream freezer is forced through extursion nozzles onto the moving conveyor belt outside the tunnel to form ribbons of ice cream. The design of the extrusion nozzles gives the cross-sectional shape to the ribbons, including grooves into which nuts, broken biscuit pieces, caramel and other materials may be added. These ribbons are hardened as they are carried through the freezing tunnel, then as they emerge, they are cut to desired lengths, transferred to a wire-type conveyor running at a longitudinal speed greater than that of the belt. This separates the pieces which are then carried to the enrober where they are coated with chocolate (Fig. 9). The coated pieces go on to another chilling tunnel to set and harden the coating before they are packaged.

While this type of freezing system has been used for many years, it has only recently been available in the large sizes (up to 24 ribbons wide) which make it more economical and suitable for large-scale production of this type of novelty. A typical 24-wide system handles 4000 pieces of 250–300 ml size per hour.

COLD STORES AND DISTRIBUTION

Cold stores design, management practice, and distribution methods are beyond the scope of this chapter; however several considerations are important here.

Fig. 9. Enrobing ice cream bars made by extruding ribbons of product onto a stainless-steel belt conveyor and freezing them in a blast tunnel. Bars emerge from the tunnel and cutting station from the right and are conveyed to the enrober at the left. The unit shown is a Promco system. (Reproduced by courtesy of APV Crepaco Ltd, Paigton, UK.)

Temperature changes in ice cream and similar products causes ice crystals to melt as the temperature rises, and to refreeze when the temperature is lowered again. The stabilisers in the mix act to reduce migration of the free water, but there is a tendency for a portion of this water to collect and refreeze into larger crystals, or to refreeze onto existing crystals when the temperature is lowered. When this melting and refreezing of some of the ice crystals is repeated a number of times, the product gets coarse and icy. Fortunately, frozen dairy products, especially those with overrun are poor conductors of heat, so there can be considerable change in the ambient temperature cycle before much of the product is affected. However, the product in contact with the container walls is much more subject to the adverse effects of temperature cycling, and such temperature fluctuations should, therefore, be avoided or kept as small as possible. These effects cause more damage when they occur at higher temperaures, but affect very little of the ice at low temperature.

A general rule is to keep cold store temperatures as low as practical and consistent with the expected storage time (−18 to −25°C is desirable), and keep temperature fluctuations at a minimum. This is a good rule to apply to conditions of distribution, as well, but since frozen product is held in retail cabinets for much shorter periods, one or two weeks instead of up to about four months, higher temperatures in the range of −13 to −18°C can be tolerated.

NOVELTY PRODUCTS

Novelty products include ice lollies, ice cream bars, stickless bars, sandwiches, small single servings in slices or small cups, sundaes, fancy moulded novelties, pudding and gelatine bars, fruit and juice bars, ice cream cakes and pies and others.

Novelties without Overrun

A major portion of the novelties are ice lollies and similar products, which are made from an appropriate mixture that is measured into moulds without incorporating air for overrun and frozen quiescently. Sticks are inserted when the product is frozen enough to hold the sticks in position. The frozen pieces are then removed from the moulds, packaged and transferred to cold stores.

Ice lolly mix is made up of water, sugar, acid (usually citric), flavourings, colours and stabilisers. Sugars may total 17–20% and stabilisers, 0·3–0·5%. The amount and type of sugar is chosen to produce the degree of sweetness desired, and to yield a relatively high freezing point. A high freezing point reduces the refrigeration required in manufacture, and makes the product less subject to melt and refreeze cycles in distribution.

Stabilisers must be compatible with the acids of the mix and be of the type which produces the body structure and flavour release desired in the frozen product. Gelatine, pectin and xanthan gum are often used. Pectin produces a short structure and a very quick flavour release, while the other two produce similar results but with slightly diminished flavour release. CMC and guar gum are also used, but flavour release is still somewhat slower.

Although citric acid is the most commonly used acid in water ices, tartaric, lactic, malic, ascorbic and phosphoric acids are also satisfactory. The amount of acid to be used in water ices depends on the sugar content, amount and type of fruit juice, and the degree of tartness desired. Generally, acid content increases as sugar content increases; the range is usually from 0·3 to 0·4% by weight.

Other stick bar products with little or no overrun are pudding bars which may contain some starch and milk solids, fudge bars and fruit and fruit juice bars.

Novelties with Overrun

Novelties with overrun are numerous, but most of them have ice cream or ice milk as the base. Those made on stick bar machines are handled in the same manner as the ice lollies, except that the ice cream or ice milk mix is partially frozen and overrun produced on an ice cream freezer before being transferred to the moulds for quiescent freezing. After extraction from the moulds, the bars are usually coated with chocolate or other candy coating. Sometimes ground nuts, crumbs or small bits of special candies are added after the main coating.

Other novelties with overrun are the stickless bars, sandwiches, cones and cups enumerated earlier. The stickless bars, such as made with the Eskimo Pie, Polarmatic and Glacier systems, are most popular and have many shapes. Some are highly decorated and sell at quite high prices.

A comprehensive discussion of novelties, their manufacture and the special equipment available is not possible in this chapter, but a few general remarks about trends and developments follow.

Trends in Novelty Manufacture

Because of the great variety of novelties produced and included in the sales lists, an ice cream manufacturer can make only a few different items in sufficient volume to be profitable. The trend, especially in the US, is towards factories which specialise in the production of novelties and package them in private label for other ice cream makers and marketing firms. These specialists may also produce a brand which is advertised and distributed on a national basis. Such novelty specialists may make only one type of novelty or they may make several types. Small cups, decorated ice cream cakes or pies and similar products made of ice cream are usually made by the regular ice cream makers.

This trend is likely to continue, for such specialisation makes better use of high-cost capital equipment.

New Novelties

During recent years, there has been a proliferation of new novelties probably reflecting the higher consumer expenditure of two-income families. In Europe the fancy decorated pieces made on the Glacier and similar machines have increased rapidly, and this trend is now seen in the US.

In the US, there have been a number of larger ice cream bars and sandwiches come on the market. These are in 90–120 ml sizes, coated with chocolate, or sandwiched between wafers or various types of cookies. As the intended consumers are young business men and women who make these a major part of their luncheon, these products are often referred to as 'Adult Novelties'.

For a number of years, orange juice, with or without added sugar, has been frozen on ice lolly machines and used in school lunch programmes in southern California and a few other areas of the US. These orange lollies provide a vitamin-C-rich product in a form that is especially attractive to children.

Very recently, juice bars and fruit bars have been introduced. These, too, are stick novelties, quiescently frozen. Various juices and mixtures of juice and pieces of fruit are the major ingredients. Strawberries and grapes are especially popular. These are nutritious alternatives to ice lollies for children, and are appealing to adults as snack foods.

Pudding bars and, more recently, gelatine bars have entered the novelty market. Both are stick bars, but gelatine bars are made by partially freezing and adding air for overrun with an ice cream freezer before freezing in the mould machine. Pudding bars and gelatine bars are not as cold to the mouth as ice lollies and have a smaller ice crystal structure which is slower to melt. These are characteristics favourable to young consumers.

The introduction of new types of flavours and combinations of products maintains consumer appeal and seems to increase sales without reducing the sales of ice cream.

Equipment for Novelties

There have been no outstanding innovations in equipment for making novelties for some time. The straight-line and rotary types of stick bar machines are improved from time to time making them more efficient, adding accessories for special products, and increasing their trouble-free operating life.

The extruded product and freezing tunnel type machines have been improved, and new extrusion nozzels have been introduced (see Fig. 10), but many consider them to be very costly for their production rates.

Other developments are in the form of improvements rather than innovations, and these trends are likely to predominate over the next decade.

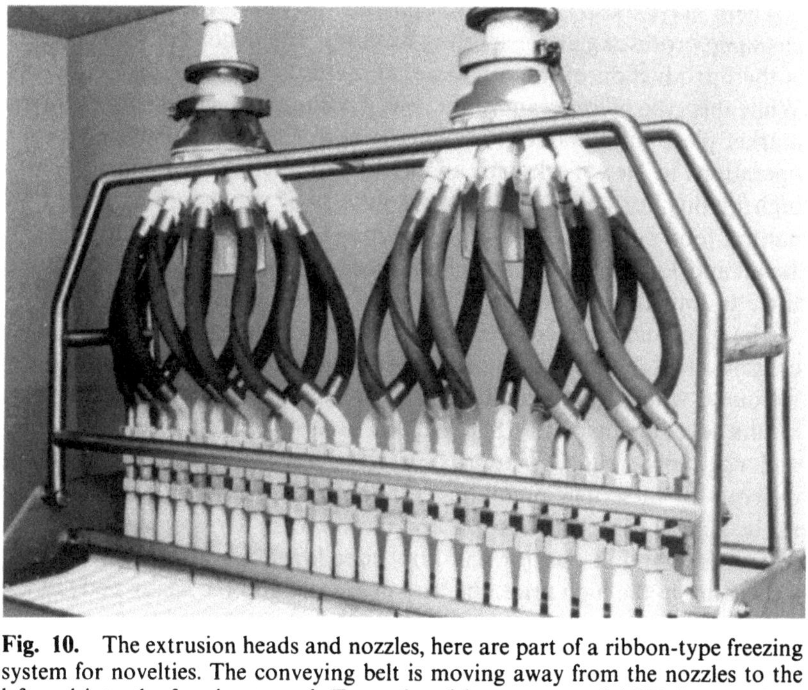

Fig. 10. The extrusion heads and nozzles, here are part of a ribbon-type freezing system for novelties. The conveying belt is moving away from the nozzles to the left and into the freezing tunnel. (Reproduced by courtesy of APV Crepaco Ltd, Paigton, UK.)

RECENT TRENDS AND POSSIBLE FUTURE DEVELOPMENTS

Ice cream technology as it applies to processing methods and equipment for manufacturing frozen dairy products has been rather static for some time, and it appears that developments in the near future will involve improvements rather than innovative technological changes.

Equipment developments during the past 10 years have increased capacity, improved efficiency, reduced maintenance, assured better and safer operation, and reduced labour input, but with the exception of microprocessor control, there have been few really new developments. Possible future developments in processing and processing equipment will likely deal more with energy conservation and recovery systems, improved packaging systems, and more attention to refreeze (rerun) recovery systems.

Automation and microprocessor control of processing and individual machines has been accepted. Developments in this area will continue and bring about significant changes in processing.

There have been some developments in new products which are changing processing and marketing methods. The major development here is the introduction of the premium, all natural, low overrun ice creams. While this type of ice cream is not new, it did not have much success in the market place until about 1978, when a newcomer to the industry began operations in New York with an effective marketing programme, a very high quality ice cream, and the advantage of the attention gained by natural food activists. Now there are many ice cream makers in the US, Japan and Europe producing similar products, and selling them at about three to four times the price of ordinary, good-quality ice cream.

The premium ice creams are made from natural dairy products, contain no stabilisers or emulsifiers other than egg yolk solids, and are flavoured with natural vanilla or juices and fruits. These ice creams have a milk fat content of 13–18%, MSNF of 9·5–10%, sucrose of 15–16%, and egg yolk solids of approximately 0·8–1·4%. The most common overrun for these new premium ice creams is 25%, although a few have up to 50 or 60%. They are packed in bulk cans for dipping stores and restaurants, and in pint (or half-litre) sizes for retail.

Because of the high price, the very cold taste and the heavy body of the product, conventional ice cream makers at first considered these premium products to be a short-term fad, but now after more than six years of continuous and rapid market gains, it appears that the appeal to affluent consumers is real and enduring.

CONCLUSION

The frozen dairy products industry is a dynamic segment of the broader dairy products industry with its own special technology and marketing methods. The per capita consumption statistics for much of the world indicates opportunity for considerable growth in the use of the more nutritious of the frozen dairy products.

REFERENCES

Andreasen, T. (1987). *Grinsted System for Stick Novelties*, TP214-1e, Grindsted Products A/S, Denmark.

Arbuckle, W. S. (1986). *Ice Cream* (4th edn). AVI Publishing Co., Westport, CT, USA.

Berger, K. G. (1976). *Food Emulsions*, ed. S. Friberg. Marcel Dekker, New York, USA, chapter 4.

International Association of Ice Cream Manufacturers (1966). *The History of Ice Cream*, Washington, DC.

International Ice Cream Association (1991). *The Latest Scoop*. Washington, DC, USA.

Keeney, P. G. (1972). *Commercial Ice Cream and Other Frozen Desserts*. Circular 533, The Pennsylvania State University, University Park, PA, USA.

Chapter 5

Physical Properties of Dairy Products

M. J. Lewis

Department of Food Science and Technology, University of Reading, UK

The physical properties of milk and milk products are of great importance to the Dairy Technologist, as they will affect most of the unit operations used during their processing. These include fluid flow, mixing and churning, emulsification and homogenisation, as well as heat transfer processes such as pasteurisation, sterilisation, evaporation, dehydration, chilling and freezing. Some of the rheological properties are also used for assessing and monitoring the quality of products, such as yoghurt, cream, butter and cheese.

Raw milk is extremely variable in its composition, and most dairy products can be produced in a variety of ways from this milk. Therefore, it is not surprising to find significant variations in the literature values for the physical properties of these products.

There are two approaches to obtaining data for physical properties. The first is to use data available in the literature, the second is to determine the values experimentally. Although the emphasis in this review will be placed on published literature values, some attention is paid to the experimental methods available, and to the variations that might occur due to modifications in the processing conditions. Wherever literature values are used in process calculations, it is important to ensure that the physical properties are determined under conditions similar to which they are to be applied. Before discussing the properties in more detail, the system of units and dimensions will be briefly discussed. More detailed information is provided by Lewis (1990).

UNITS AND DIMENSIONS

All physical properties can be measured in terms of the fundamental dimensions mass (M), length (L), time (T) and temperature (θ). Three others, added for completeness, are electric current, luminous intensity and the amount of substance (mole). The systems of units commonly encountered are the SI (Système International), the cgs (centimetre, gram, second) and the Imperial System. Scientists, worldwide have opted for the SI system of units, although in many practical situations and textbooks, other systems are encountered.

Properties such as area, velocity, pressure and specific heat are termed 'derived variables' as they can be expressed in terms of the fundamentals. Table I shows how some of these derived physical properties can be expressed in terms of the fundamental dimensions, together with the conversion factors from Imperial to SI units.

TEMPERATURE

Temperature is defined as the degree of hotness of a body. In a spontaneous change, heat is transferred from a high to a low temperature, until thermal equilibrium is achieved.

The two scales encountered in practical measurement are Celsius (°C) and Fahrenheit (°F) with the conversion being $°C = (°F - 32) \times 5/9$. Temperature driving forces or differential temperatures are measured in °F or °C, where $1°F = 5/9°C$. The rate of heat transfer is proportional to the temperature driving force. The absolute scale of temperature (Kelvin (K)) is based on the work performed by an ideal heat engine working between two temperatures; absolute zero is taken as the lowest temperature attainable, i.e. the temperature at which all molecular motion ceases. Although absolute temperatures are not used in practical measurement, many equations require substitution in terms of the absolute temperature, e.g. the ideal gas equation and the Arrhenius equation. The conversion factor is $K = °C + 273{\cdot}15$. Absolute zero on the Fahrenheit scale is given the value 0° Rankine (°R) and the conversion is $°R = °F + 459{\cdot}7$.

The reference temperature used for comparing the sterilisation efficiency during thermal processing operations is 121·1°C (250°F). More recently, 135°C has been selected for ultra-high temperature (UHT) processes (Kessler, 1981). (See also heat treatment, Chapter 7, Vol. I).

TABLE I
Fundamental and derived units

		SI unit[a]	*Imperial unit*	*Conversion factor*
Mass	M	kilogram (kg)	pound (lb)	1 lb = 0·4536 kg
Length	L	metre (m)	foot (ft)	1 ft = 0·3048 m
Time	T	second (s)	second (s)	
Temperature	θ	Kelvin (K)	°Fahrenheit (F)	
Luminous Intensity	cd	Candela		
Electric Current	I	Ampère (A)		
Amount of substance		Mole (mol)		
Area	L^2	m^2	ft^2	$1\ ft^2 = 9{\cdot}29 \times 10^{-2}\ m^2$
Volume	L^3	m^3	ft^3	$1\ ft^3 = 2{\cdot}832 \times 10^{-2}\ m^3$
Velocity	$L\,T^{-1}$	$m\,s^{-1}$	$ft\,s^{-1}$	$1\ ft\,s^{-1} = 0{\cdot}3048\ ms^{-1}$
Acceleration	$L\,T^{-2}$	$m\,s^{-2}$	$ft\,s^{-2}$	$1\ ft\,s^{-2} = 3{\cdot}048 \times 10^{-1}\ m\,s^{-2}$
Force	$M\,L\,T^{-2}$	Newton (N)	lb (force)	1 lb (f) = 4·448 N
Pressure	$M\,L^{-1}\,T^{-1}$	Pascal (Pa)	$lb\,(f)\,in^{-2}$	$1\ lb\,(f)\,in^{-2} = 6{\cdot}895 \times 10^3\ Pa$
Work, Energy	$M\,L^2\,T^{-2}$	Joule (J)	Btu[b]	$1\ Btu = 1{\cdot}055 \times 10^3\ J$
Power	$M\,L^2\,T^{-3}$	Watt (W)	Horsepower (HP)	1 HP = 745·7 W
Density	$M\,L^{-3}$	$kg\,m^{-3}$	$lb\,ft^{-3}$	$1\ lb\,ft^{-3} = 16{\cdot}02\ kg\,m^{-3}$
Dynamic viscosity	$M\,L^{-1}\,T^{-1}$	$N\,s\,m^{-2}$ (Pa s)	$lb\,ft^{-1}\,s^{-1}$	$1\ lb\,ft^{-1}\,s^{-1} = 1{\cdot}488\ Pa\,s$
Kinematic viscosity	$L^2\,T^{-1}$	$m^2\,s^{-1}$	$ft^2\,s^{-1}$	$1\ ft^2\,s^{-1} = 92{\cdot}9 \times 10^{-3}\ m^2\,s^{-1}$
Surface tension	$M\,T^{-2}$	$N\,m^{-1}$	[c]	$1\ dyne\,cm^{-1} = 10^{-3}\ N\,m^{-1}$
Specific heat	$L^2\,T^{-2}\,\theta^{-1}$	$J\,kg^{-1}\,K^{-1}$	$Btu\,lb^{-1}\,°F^{-1}$	$1\ Btu\,lb^{-1}\,°F^{-1} = 4{\cdot}187\ kJ\,kg^{-1}\,K^{-1}$
Thermal conductivity	$M\,L\,T^{-3}\theta^{-1}$	$W\,m^{-1}\,K^{-1}$	$Btu\,h^{-1}\,ft^{-1}\,°F^{-1}$	$1\ Btu\,h^{-1}\,ft^{-1}\,°F^{-1} = 1{\cdot}731\ W\,m^{-1}\,K^{-1}$
Thermal diffusivity	$L^2\,T^{-1}$	$m^2\,s^{-1}$	$ft^2\,s^{-1}$	$1\ ft^2\,s^{-1} = 92{\cdot}9 \times 10^{-3}\ m^2\,s^{-1}$
Latent heat	$L^2\,T^{-2}$	$J\,kg^{-1}$	$Btu\,lb^{-1}$	$1\ Btu\,lb^{-1} = 2{\cdot}326\ kJ\,kg^{-1}$
Heat film coefficient	$M\,T^{-3}\,\theta^{-1}$	$W\,m^{-2}\,K^{-1}$	$Btu\,h^{-1}\,ft^{-2}\,°F^{-1}$	$1\ Btu\,h^{-1}\,ft^{-2}\,°F^{-1} = 5{\cdot}678\ W\,m^{-2}\,K^{-1}$
Overall heat transfer coefficient	$M\,T^{-3}\,\theta^{-1}$	$W\,m^{-2}\,K^{-1}$	$Btu\,h^{-1}\,ft^{-2}\,°F^{-1}$	$1\ Btu\,h^{-1}\,ft^{-2}\,°F^{-1} = 5{\cdot}678\ W\,m^{-2}\,K^{-1}$

[a] SI units: reciprocal form is used here.
[b] British thermal unit.
[c] More often measured in dyne cm^{-1}; very rarely seen in Imperial Units.

It is important to be able to measure, record and control processing temperatures accurately. Mercury-in-steel thermometers are useful for indication, whereas resistance thermometers and thermocouples are used most for recording and control applications. Regular attention should be payed to the accuracy of the readings, as temperature measurement is a critical control point for thermal processing.

PRESSURE

Absolute pressure is defined as the force divided by the area. SI units are expressed in $N\,m^{-2}$ or Pascals (Pa). This is a small unit of pressure so, for many operations, the bar or MPa is used, where

$$1\ \text{bar} = 10^5\,\text{Pa} = 0{\cdot}1\,\text{MPa}$$

Other pressure units commonly encountered are

$$\frac{\text{lb(f)}}{\text{in}^2}\text{(psi)} \quad \text{and} \quad \frac{\text{kg(f)}}{\text{cm}^2}$$

The absolute pressure is used in pressure, volume, temperature (P, V, T) relationships for gases and vapours.

It is well known that there is a relationship between the absolute pressure and the height or head of fluid, supported by that pressure. This is given by

$$P = pgh$$

where P is absolute pressure ($N\,m^{-2}$), g is acceleration due to gravity ($=9{\cdot}81\ m\,s^{-2}$), p is fluid density ($kg\,m^{-3}$) and h is head of fluid (m).

Differential pressures can also be expressed in similar terms, and are commonly used for pressure losses in pipes and fittings, and for pressures (heads) developed in pumping applications.

One standard atmosphere will support a column of mercury, 0·76 m high. In other units, one standard atmosphere is given by

$$1\ \text{atm} = 14{\cdot}69\ \text{psi} = 1{\cdot}013\ \text{bar} = 1{\cdot}033\ \text{kg(f)cm}^{-2}$$

Most pressure gauges measure pressure, above or below atmospheric; the reading being correspondingly referred to as gauge pressure (Fig. 1). A correction may be necessary if the absolute pressure is required:

$$\text{absolute pressure} = \text{gauge pressure} + \text{atmospheric pressure}$$

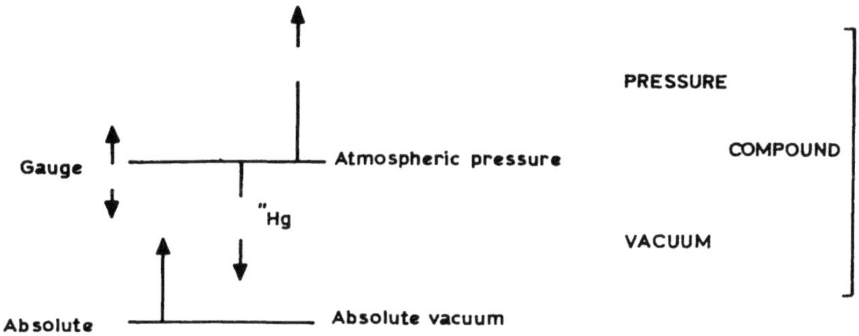

Fig. 1. Relationship between gauge pressure and absolute pressure.

If the pressure is below atmospheric, a vacuum gauge is required. On the Imperial System, vacuum is expressed in terms of inches of mercury ("Hg); this is measured *below* atmospheric pressure. Such units are still commonly used when expressing the vacuum in a can, or during evaporation. Using the metric system, vacuum is expressed as an absolute pressure, measured above zero absolute pressure. The relationship between the two systems is given by

$$\text{absolute pressure (Pa)} = (29{\cdot}92 - X)\ 3385{\cdot}6$$

where X is inches of mercury.

For applications requiring very low pressures, two further units are used, namely the Torr (τ) and micrometre (μm)

$$1\text{mm Hg} = 1\ \text{Torr}(\tau) = 10^3\ \mu\text{m} = 133{\cdot}3\text{Pa}$$

The most common type of pressure gauge is the Bourdon gauge. This is suitable for steam, hot and chilled water and compressed air, but not for foods, as the material enters the Bourdon tube and is difficult to remove. For hygienic operations, involving foods, diaphragm gauges, which are more expensive, should be used.

DIMENSIONAL ANALYSIS

The technique of dimensional analysis can be used to investigate fluid flow, heat transfer and mass transfer problems. The Buckingham pie (π) theory is used to develop relationships between the variables, in terms of dimensionless groups.

For example, a relationship between the heat film coefficient (h) and other physical properties can be derived, in terms of dimensionless groups, for a liquid flowing along a tube, as follows:

$$\frac{hD}{k} = \phi\left(\frac{vD\rho}{\mu}\right)^a\left(\frac{c\mu}{k}\right)^b$$

(Nusselt number) (Reynolds number) (Prandtl number)

This relationship can be investigated experimentally to evaluate the constant (ϕ), and the exponents a and b. Heat film coefficients are given for a wide variety of flow situations by ASHRAE (1985), Kessler (1981) and Loncin and Merson (1979). More information regarding dimensional analysis and dimensionless groups is given by Weast (1988).

DENSITY AND SPECIFIC GRAVITY

Density is defined as the mass of substance divided by the volume occupied; its dimensions are (ML^{-3}) and the SI unit is the kilogram per cubic metre ($kg\,m^{-3}$).

At 4°C water has a density of $1{\cdot}00\ \frac{g}{ml}$ or $\frac{10^{-3}\,kg}{10^{-6}\,m^3} = 10^3\,kg\,m^{-3}$

The addition of most solids, e.g. minerals, sugars, proteins, will increase the density, whereas oils and fats will decrease the density. The density of fluids is usually measured with a hydrometer. The density is temperature dependent, so temperatures should always be recorded.

The density of bovine milk usually falls within the range 1025–1035 $kg\,m^{-3}$. It is generally measured with a special hydrometer, known as a lactometer, and the result can be used to estimate total solids.

The densities of the respective solid constituents are regarded as fat (930), water (1000) and milk solids-not-fat (1614 $kg\,m^{-3}$). BS 734 (1937, 1959) gives information on density hydrometers for use in milk, and tables are supplied for determining the total solids of milk, knowing the density and fat content; temperature correction tables are also presented. Fat contents range between 1 and 10%, and the total solids determination is based on the following equation. Fat is determined separately, usually by the Gerber method.

$$T = 0{\cdot}25D + 1{\cdot}21F + 0{\cdot}66 \quad \text{BS 734 (1937)}$$
$$T = 0{\cdot}25D + 1{\cdot}22F + 0{\cdot}72 \quad \text{BS 734 (1959)}$$

where T is total solids (w/w), D is 1000 (density − 1) (density units are (g ml^{-1})), and F is fat percentage.

Thus, milk at 26°C with a fat content of 3·5% and a density of 1·0320 would be corrected to a value of 1·0322 at 20°C, and have a total solids of 12·95% (1937) or 13·05% (1959). Total solids are normally expressed to the nearest 0·05%. Obviously, the 1959 tables are preferred, but several people are still of the opinion that they tend to overestimate total solids, and that the 1937 formula is more accurate (Egan *et al.*, 1981).

Kessler (1981) presents relationships for the density of whole milk and cream (20% fat), over the temperature range 0–150°C.

$$\text{Whole milk: } p = 1033{\cdot}7 - 0{\cdot}23080\theta - 2{\cdot}46 \times 10^{-3}\,\theta^2$$
$$\text{Cream (20\% fat): } p = 1031{\cdot}8 - 0{\cdot}31790\theta - 1{\cdot}95 \times 10^{-3}\,\theta^2$$

where θ is temperature (°C). Bertsch *et al.* (1982) present information on the density of milk and cream over the temperature range 65–140°C.

Specific gravity is defined as the ratio of the mass of a fixed volume of liquid to the mass of an equal volume of water. Specific gravity is a dimensionless quantity. It is most conveniently measured using a specific gravity bottle, which can be used both for liquids and particulate solids; in the latter case, a solvent is selected which will not dissolve the solid, e.g. toluene.

Specific gravity is less susceptible to changes in temperature, compared to density. The relationship between the specific gravity and density of a material is given by

$$SG_L = (p_L/p_w)$$

Therefore, if the specific gravity of a material is known at a temperature, T°C, its density at T°C will be given by

$$p_L = (SG)_T \times p_w$$

where p_L is density of liquid at T°C, $(SG)_T$ is specific gravity at T°C, p_w is the density of water at T°C (obtained from tables).

Density and specific gravity are useful for monitoring changes occurring during processing, e.g. evaporation, or for checking whether extraneous water has entered the product.

Figure 2 shows how the density of two samples of canned evaporated milk change with temperature. The density increases from 1053 to 1078 over the temperature range 75–25°C. Often total solids is monitored during evaporation by measuring the density and, for the production of evaporated milk, a batch is struck when the desired final gravity is

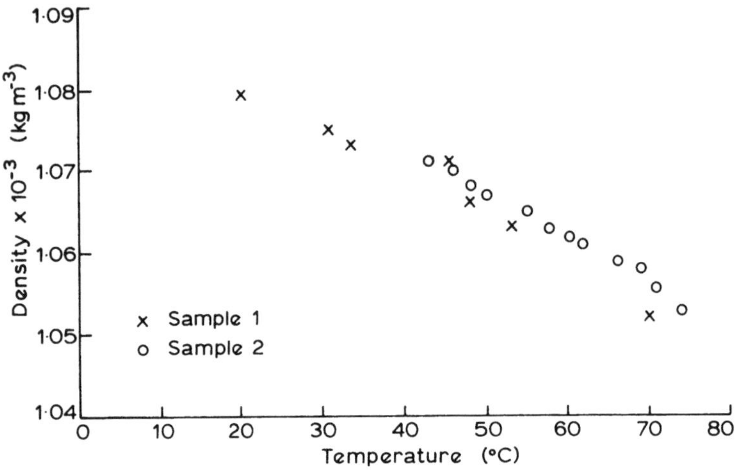

Fig. 2. Graph of density of UK evaporated milk against temperature.

reached. Considerable errors may result if the temperature is not accounted for.

The density of most ice cream mixes falls within the range of 1080–1100. Density measurement has been found to be an extremely simple means of quickly assessing whether an ice cream mix has been watered down during processing; this is important in UHT direct steam injection, where the condensed steam is subsequently removed during the flash cooling process. Dilution of the mix will result in ice cream with an icy texture. Such dilution could also occur during continuous high-temperature short-time (HTST) pasteurisation, and Table II shows how the specific gravity of an ice cream mix, originally at 33·7% total solids, changes as it is diluted with water.

TABLE II
Variation in the specific gravity of ice cream with different total solids

Total solids (%) (w/w)	*Specific gravity*
33·7	1·0968
32·1	1·0918
30·8	1·0878
29·3	1·0826
28·0	1·0796

Author's unpublished data.

When dealing with solids, it is necessary to differentiate between solid density and bulk density, particularly with particulate matter and powders.

The solid or particle density refers to the density of the solid or an individual particle. It is defined, in the normal fashion, as

$$\frac{\text{mass solid}}{\text{volume solid}}$$

and it will take into account the presence of air within the solid. For particulate matter, it can be determined either by flotation using liquids of known specific gravity, or by using a density bottle.

The density of solid constituents have been summarised by Peleg (1983) and Walstra and Jenness (1984) as follows:

	Peleg (1983)	*Walstra and Jenness (1984)*
Sucrose	1590	
Glucose	1560	
Protein	1400	1400
Lactose		1780
Fat	900–950	918
Salt	2160	1850[a]
Water	1000	998·2

[a]Residual milk solids.

Therefore, the density of the food can be estimated from a knowledge of the food composition, using the equation

$$p = \frac{1}{m_1/p_1 + m_2/p_2 + m_n/p_n}$$

where n is the number of components, m_1 to m_n are the mass fraction of components 1 to n, and p_1 to p_n are densities of components 1 to n.

This equation may be useful for estimating the density of solids, such as butter, spreads, cheese, yoghurt and cream, as well as protein concentrates, sweetened condensed milks and ice cream mixes. This approach should be treated with caution for dehydrated dairy products, which may contain air spaces within the particles. This will substantially lower the particle density, and cannot be easily accounted for in this equation because the mass fraction of air is not known. However, an estimate of the volume fraction of air trapped within the particles can be made by calculating the solid density (assuming no air) p_c and measuring the actual solid density p_s. The volume fraction of air trapped within the solid V_a is given by

$$V_a = \frac{p_c - p_s}{p_c}$$

The solid density is important in separation processes, e.g. centrifugation of cheese fines, cyclone operation, and the pneumatic or hydraulic transport of powders and particulate matter.

Bulk density is an important property, particularly for the transportation and storage of bulk particulate material, e.g. fruit, grain, powders. It is defined as the mass of material divided by the total volume occupied. In most cases, it is important to have a high bulk density, and some values for full-cream milk powder are recorded in Table III.

TABLE III
Bulk density, particle density and porosity of some whole milk powders

	Bulk density of powder (g ml^{-1})	*Particle density (g ml^{-1})*	*Porosity (%)*
11% solids: 90°C outlet	0·39	0·802	51
34% solids: 90°C outlet	0·55	1·017	46
47% solids: 90°C outlet	0·66	1·150	42
47% solids: 113°C outlet	0·46	0·913	50
Straight through, instantised	0·63	1·100	43
Re-wet instantised	0·33	1·100	70

All figures obtained by drop packaging 50 g of powder under standardised conditions: Mettler (1980).

The method used for bulk density measurement has a marked effect on the value obtained; these methods are described for milk powders by Lovell (1980). Peleg (1983) discusses the compressibility of powders in more detail.

The bulk density of milk powders is affected by processing conditions, in particular, total solids, with bulk density increasing as the total solids increases. It should also be noted that particle density increases as total solids increase, suggesting that less air is incorporated into the paticles at higher total solids. Injection of air or nitrogen into the product immediately before atomisation may reduce the bulk density, and agglomeration, achieved by a re-wetting process, substantially decreases the bulk density.

In addition, the method of atomisation will affect the bulk density. Early designs of atomiser wheel produced powders of 0·45–0·55 g ml^{-1}, whereas later designs, typified by the vaned wheel, produced bulk densities of 0·55–0·65 g ml^{-1}; later designs have used steam to occlude air from the fluid. Jet nozzles can produce powders with bulk densities as high as 0·83 g ml^{-1}.

The characteristics of sprays produced by pressure nozzles, two-fluid nozzles and centrifugal atomisers are described in more detail by Masters (1991). The porosity of a material is defined as that fraction of the total volume which is occupied by air, between the particles. It does not take into account air within the particle:

$$\text{porosity} = (\rho_s - \rho_B)/\rho_s \quad (\rho_s \text{ is solid density, } \rho_B \text{ is bulk density})$$

Mettler (1980) measured the porosity of 49 samples of full-cream milk powder, and found porosity values ranging from 42·9 to 50·9%. Kjaergaard-Jensen and Neilsen (1982) have reviewed some of the properties of milk powders necessary for recombination technology, and Peleg (1983) has described some other physical properties of powders, such as compaction, flowability and caking.

Overrun

When air is incorporated into a whipped or frozen product, the density decreases. The amount of air incorporated is measured by the overrun, where

$$\text{overrun} = \frac{\text{increase in volume}}{\text{original volume}} \times 100$$

In practical terms, it is most easily measured by comparing the weights of equal volumes of the original liquid and the final aerated product.

$$\text{overrun} = \frac{\text{weight of original liquid} - \text{weight of same volume of aerated product}}{\text{weight of same volume of aerated product}} \times 100$$

The major factors affecting overrun of ice cream and frozen desserts are the total solids content of the mix and the type of freezer used. Overruns between two and three times the total solids content are recommended by some authorities; if the overrun is too low, the product becomes heavy, whereas if it is too high it becomes too light and fluffy. Other factors which affect overrun are reviewed by Arbuckle (1977): they are broken down into those which depress it, and those which enhance it.

For whipping cream, overruns of 100–120% would be expected, although the modern aerosol creams give much higher values. As well as total overrun, it is important to measure the stability of the whipped foam over a period of time (Society of Dairy Technology, 1989).

Specific Volume

An alternative way of quoting the density of materials, particularly gases and vapours is in terms of specific volume (volume/mass), which has units of $m^3\,kg^{-1}$. Gases and vapours are compressible, that is their density is affected by changes in temperature and pressure. The density of an ideal gas at moderate temperatures and pressures can be estimated as follows.

The molecular weight of any gas (expressed in kilograms) occupies 22·4 m^3 at STP. The volume occupied by the same mass at the experimental temperature and pressure can be calculated from the ideal gas equation, and is then used to determine the density of the gas.

The specific volume of wet, saturated and superheated vapours, e.g. steam and refrigerants, are summarised in thermodynamic charts and tables (see 'thermal properties of foods').

RHEOLOGICAL PROPERTIES

Rheology is the study of the deformation of materials, subjected to applied forces. A distinction is usually made between fluids and solids; fluids will flow under the influence of forces, whereas solids will stretch, buckle or break.

An ideal solid is represented by the Hooke solid, and the ideal liquid by the Newtonian liquid (Muller, 1973). Both are structureless (there are no atoms), both are isotropic (they have the same properties in all directions), and both follow their respective laws exactly.

Many materials can exert both types of properties, depending upon the environmental conditions and stresses they are subjected to. For example, butter at 20°C is regarded as a solid, although if the shearing force is sufficiently high, it can be made to flow, or if its temperature is raised to above 50°C, it will melt and behave like a fluid. Some of these aspects of fluid and solid rheology will now be investigated.

Viscosity

The viscosity of a fluid is defined as the internal friction within the fluid. When a fluid is subjected to a shearing force (F) over a surface area (A) it will undergo a deformation, known as flow.

The shear stress is defined as force/area and the rate of deformation, termed the shear rate, is determined by the velocity gradient (see Fig. 3(a)).

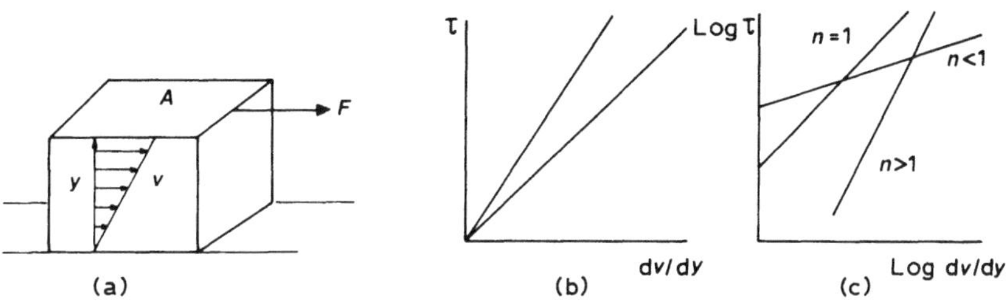

Fig. 3. (a) Deformation of a fluid: (b) and (c) rheograms on ordinary and logarithmic co-ordinates respectively.

For Newtonian fluids, there is a direct relationship between the shear stress (τ) and the rate of shear (dv/dy). The ratio of the shear stress to shear rate is known as the dynamic viscosity or coefficient of viscosity (μ):

$$\mu = \frac{\text{shear stress}}{\text{shear rate}} \quad \frac{\text{N m}^{-2}}{\text{s}^{-1}} \ (\text{N s m}^{-2})$$

Viscosity data are often plotted as shear stress against shear rate, either in ordinary or logarithmic co-ordinates (Fig. 3). Such plots are known as rheograms.

The two main units used for viscosity measurement are the poise (p) (cgs) and the Poiseuille (Pl) or Pascal second (Pa s), (SI):

Shear stress	*Shear rate*	*Dynamic viscosity*
dyne cm^{-2}	s^{-1}	Poise (dyne s cm^{-2})
N m^{-2}	s^{-1}	Poiseuille (N s m^{-2}) (kg m^{-1} s^{-1}) (Pa s)

One Poiseuille is the dynamic viscosity of a fluid which, when subjected to a shear stress of 1 N m^{-2}, gives a shear rate of 1 s^{-1}. The conversion factor is 1 poise $= 10^{-1}$ N s m^{-2}. The viscosity of water at 20°C is $1{\cdot}002 \times 10^{-3}$ N s m^{-2} or 1·002 cp (note that the centipoise is still in common use).

Milk, skim-milk, cheese whey and whey permeate can all be regarded as dilute solutions, and are usually considered to be Newtonian fluids.

The viscosity of all fluids is temperature dependent, the viscosity of liquids, pastes, suspensions and emulsions decreasing with increasing temperaure, between 2 and 10% for each °C; gases increase in viscosity as the temperature rises. Therefore, it is very important to control the temperature accurately when measuring the viscosity, and the temperatures should always be quoted with the results.

Occasionally, it is more appropriate to use the term 'kinematic viscosity,' which is defined as the dynamic viscosity divided by the density. The units of kinematc viscosity are as follows:

	Dynamic viscosity $(ML^{-1}T^{-1})$	*Density* (ML^{-3})	*Kinematic viscosity* (L^2T^{-1})
cgs	poise	g ml^{-1}	$\text{cm}^2\,\text{s}^{-1}$ (Stoke)
SI	N s m^{-2}	kg m^{-3}	$\text{m}^2\,\text{s}^{-1}$

Kinematic viscosity is measured directly by the Ostwald capillary flow type viscometer.

The dynamic viscosity of Newtonian fluids can be measured by capillary flow viscometers (density is also required), falling sphere techniques, or by measuring the flow rate in a horizontal capillary tube under streamline flow conditions. Probably, the most sensitive are the capillary flow viscometers, particularly those with narrow capillaries, giving long efflux times. Methods for measuring viscosity are summarised by Lewis (1990).

Concentration Dependence

The viscosity of solutions increases as the concentration increases, but in a non-linear fashion. At high concentrations, small additional changes in the concentration will lead to rapid changes in the viscosity. This could result in reduced flow rates, higher pressure drops, decreased turbulence and, in heating operations, severe fouling. In concentration processes, such as evaporation, reverse osmosis and ultrafiltration, the extent of concentration may well be limited by viscosity considerations. There is often a transition from Newtonian to non-Newtonian conditions as concentration proceeds.

Non-Newtonian Fluids

As the complexity of fluids increases, there are considerable interactions that result in non-linear relationships between shear stress and shear rate. Such fluids are termed non-Newtonian. Thus in solutions containing macromolecules, either dissolved or in the colloidal form, in suspensions, pastes or emulsions, there may be complex interactions between various components. This effect increases as the concentration increases. Various types of non-Newtonian behaviour are recognisable. To detect non-Newtonian behaviour requires the use of variable speed rotational viscometers. Thus, by altering the speed, it is possible to alter the shear rate; at each shear rate, the corresponding shear stress is measured. Non-Newtonian fluids are characterised by an apparent viscosity (μ_a) where

$$\mu_a = \frac{\text{shear stress}}{\text{shear rate}}$$

Non-Newtonian fluids fall into two major categories, these being time-independent and time dependent. Time-independent fluids show viscosity characteristics that are independent of time, perhaps best illustrated by a pseudoplastic fluid which is the most common type of behaviour. When shear stress is plotted against shear rate, the rheogram of Fig. 4 is obtained. The apparent viscosity at any shear rate is determined from an equation, and can be plotted against the shear rate. Therefore, it can be seen that a pseudoplastic fluid shows a decrease in apparent viscosity as the shear rate increases; this is also called shear-thinning behaviour. If all the readings were repeated whilst reducing the shear rate, they would be identical to those obtained whilst increasing the speed; no hysteresis would be observed. This is another characteristic of a time-independent fluid. Most time-dependent fluids are thixotropic, showing a decrease in apparent viscosity under shear stress, followed by a gradual recovery

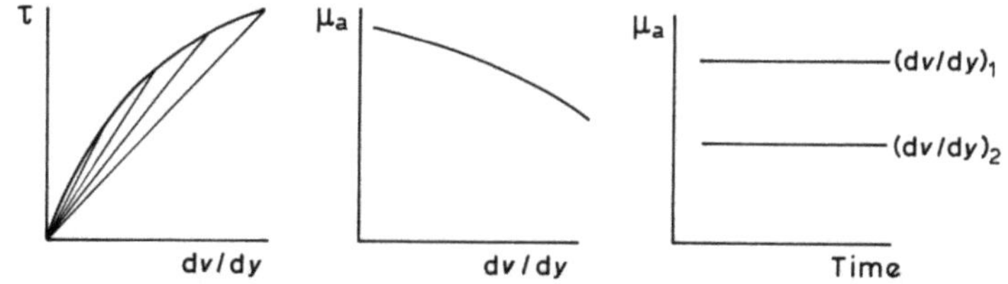

Fig. 4. Rheograms for a pseudoplastic fluid.

when the stress is removed. Time-dependent fluids are usually measured at constant shear stress, and the change in apparent viscosity is measured over a period of time. It is important to ensure that all the experimental conditions are recorded for time-dependent materials. The major types of non-Newtonian behaviour are summarised in Fig. 5 and plastic fluids are covered in the next section. Rotational viscometers are used to characterise non-Newtonian fluids.

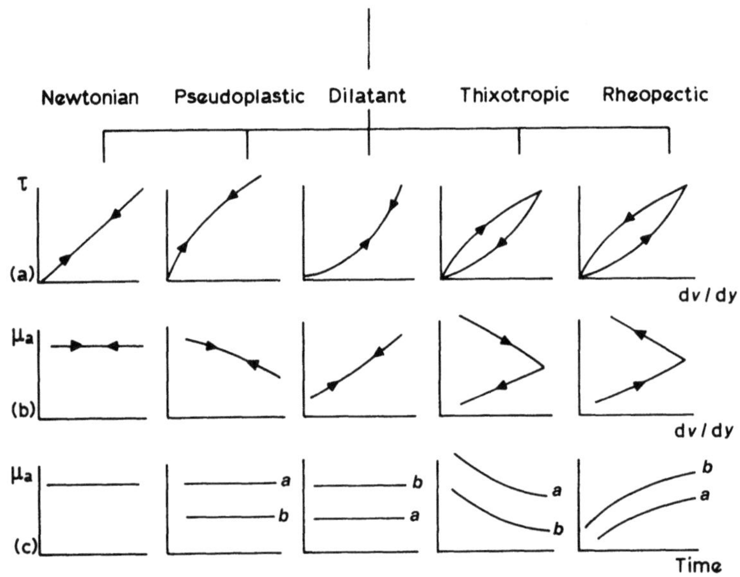

Fig. 5. Rheograms for Newtonian and non-Newtonian behaviour: (a) shear stress against shear rate; (b) apparent viscosity against shear rate; (c) apparent viscosity against time at two different shear rates, *a* and *b*, where $b > a$.

Rotational viscometers rely on a spindle rotating in the test fluid, the force required to overcome the frictional forces being measured. The most common measuring heads are a single spindle, a concentric cylinder viscometer, or a cone and plate viscometer. The Brookfield viscometer is probably the most widely used. In its original design, it consisted of a series of interchangeable spindles which were inserted into the test fluid; a choice of eight speeds were available. The equipment is fairly cheap and robust. Developments include a helipath stand to move the spindles up and down in the liquid, special T-piece adaptors for high-viscosity fluids and gels, a low-viscosity attachment and a special small sample adaptor. A cone and plate viscometer is also available from the same manufacturers. The major disadvantage of using a single spindle is that it is not possible to predict the exact shear rate in the sample. This is ovecome by use of the more expensive concentric cylinder, or cone and plate viscometers. Viscometers are described in more detail by Bourne (1982) and Brennan (1988).

Many time-independent non-Newtonian fluids obey a power law:

$$\tau = k\left(\frac{dv}{dy}\right)^n$$

and a straight line relationship results when log τ is plotted against log (dv/dy).

The consistency index (k) and the power law index (n) are often used to characterise the behaviour of such fluids (Fig. 3(c)).

For all milk and milk products, there is considerable variation in composition and hence viscosity, so that it is recommended that the viscosity be measured wherever possible. The most convenient instrument for measuring low-viscosity fluids is the capillary flow viscometer, which is sensitive enough to be able to detect the small changes which occur when milk is heated or homogenised.

Table IV shows some representative values for the viscosity of whole milk, skim-milk and cheese whey at different temperatures. Most of these fluids exhibit Newtonian behaviour over a moderate range of shear rates.

TABLE IV
Viscosity of milk and whey at different temperatures (average values) (mPa s)

	Temperature (°C)			
	10	*20*	*40*	*80*
Whole milk	2·79	2·12	1·24	0·68
Skim-milk	2·44	1·74	1·03	0·53
Whey	1·71	1·26	0·82	0·68

Interpreted from data in Kessler (1981).

Homogenisation and heat treament both tend to increase the viscosity slightly, with homogenisation giving the milk a creamier mouthfeel. The effects of homogenisation become more pronounced as the fat content increases.

Bertsch and Cerf (1983) used a capillary flow viscometer to measure the viscosity of a variety of milk products in the range 70–135°C; fat contents ranged from 0·03 to 15·5%, and some of the milks were homogenised.

A relationship of the following form was found between dynamic viscosity (μ) and temperature (θ):

$$\ln \mu = a\theta^2 + b\theta + c$$

The viscosity of homogenised milk and cream (up to 15% fat) could be represented by:

$$\begin{aligned}\ln \mu = {} & 3{\cdot}92 \times 10^{-5}\,\theta^2 - 1{\cdot}951 \times 10^{-2}\,\theta + 0{\cdot}666 \\ & + F(-9{\cdot}53 \times 10^{-6}\,\theta^2 + 1{\cdot}674 \times 10^{-3}\,\theta - 4{\cdot}37 \times 10^{-2}) \\ & + F^2(9{\cdot}75 \times 10^{-7}\,\theta^2 - 1{\cdot}739 \times 10^{-4}\,\theta + 9{\cdot}83 \times 10^{-3})\end{aligned}$$

where μ is dynamic viscosity (mPa s), F is fat content (w/w), and θ is temperature (°C).

Data are also presented for non-homogenised milk. High-temperature data are useful for evaluating performance in UHT sterilisers, particularly pressure drops and residence time distribution in the high-temperature sections.

An alternative semi-empirical approach for dispersions is outlined by Walstra and Jenness (1984). It assumes that the increase in viscosity caused by particles results from hydrodynamic interactions only:

$$\eta = \eta_0\left(1 + \left(\frac{1{\cdot}25\,\phi}{1 - \phi/\phi_{max}}\right)\right)^2$$

where η is the viscosity of suspension, η_0 is the viscosity of solvent (water), ϕ is the total hydrodynamic volume, ϕ_{max} is the hypothetical volume fraction giving close packing ($\phi = +\phi_c + \phi_w + \phi_l + \phi_f$), where ϕ_f is the hydrodynamic volume of fat (1·11 ml g^{-1}), ϕ_c is the hydrodynamic volume of casein (3·9), $\phi_w =$ is the hydrodynamic volume of whey proteins (1·5), and ϕ_l is the hydrodynamic volume of lactose (1·0).

This equation is useful for predicting the viscosity of dilute solutions; it breaks down when ϕ increases above 0·6, e.g. for liquids with high fat contents and for whey protein concentrates.

The viscosity of milk products increases as the concentration increases, and some data for concentrated skim-milk are given in Table V, which is compiled from two sources; there is a reasonable agreement between them. Notice the sharp increase in viscosity above 35% total solids.

TABLE V
Dynamic viscosity of concentrated skim-milk at 25°C

Kessler (1981)		*Allen (1980)*	
Total solids	*Dynamic viscosity (mPa s)*	*Total solids*	*Dynamic viscosity (mPa s)*
20	3·8	18·6	4·4
25	5·8	23·8	5·6
30	10·0	29·3	8·8
33	13·0	32·3	16·6
		39·8	59·3
		46·4	1280

Kessler (1984) also records how the viscosity of skim-milk concentrate (25% total solids) changes with temperature.

The viscosity of full-cream evaporated milk will depend upon the degree of forewarming, homogenisation conditions, the type of stabiliser used and the extent of the final in-container heat treatment.

Fernandez-Martin (1984) presents monographs for determining the viscosity of sheep's milk cheese whey over the concentration range 6·9–26·5% total solids and temperature range 5–80°C. The whey was concentrated by reverse osmosis, and most of the concentrates were found to be Newtonian over the shear rates measured. Problems were observed with protein denaturation at the highest temperatures. The equation for viscosity, over this temperaure and total solids range is as follows:

$$\log \mu = 0{\cdot}2214 - 0{\cdot}0131\,\theta + 5{\cdot}5 \times 10^{-5}\,\theta^2 + (0{\cdot}02173 - 1{\cdot}04 \times 10^{-4}\,\theta + 1{\cdot}87 \times 10^{-6}\,\theta^2)\,s$$

μ is dynamic viscosity (mPa s), θ is temperature (°C), and s is solids content (%).

Kjaergaard-Jensen and Neilsen (1982) discuss the factors that might influence the viscosity of recombined sweetened-condensed milks.

Viscosity is one of the main factors which limits the extent of concentration for ultrafiltration and reverse osmosis processes. The protein fraction makes the main contribution to the viscosity. Maubois (1980) examined the possibility of pre-concentrating whey by evaporation prior to ultrafiltration, and he showed how the viscosity changed during ultrafiltration for four different concentrated whey samples of 29, 23, 21·4 and 16% total solids. In each case, viscosity was plotted against the percentage protein (dry weight basis) in the concentrate. If it is assumed that the concentration is not allowed to proceed once the viscosity has reached 20 mPa s, then the protein content attainable decreased as the total solids content of the feed increased. Evaporation appeared to offer no advantage for the production of high-protein powders.

Cream

Freshly separated cream has a fairly low viscosity. The market cream is then standardised to the desired fat content, heat treated, homogenised, cooled and packaged. All these factors can significantly affect the viscosity of the final product, and each cream needs to be treated differently to obtain the best quality product.

For example, single cream is homogenised at fairly high pressures, usually after heat treatment; this increases the viscosity significantly. Filling into cartons using a piston filler will reduce the viscosity, probably due to shear breakdown, but the viscosity then increases during cold storage. Single cream has been shown to observe pseudoplastic or dilatant behaviour, depending upon storage conditions; cream cooled very quickly and stored at a uniform low temperature often shows a dilatant character. When the cream was warmed and then re-cooled, pseudoplastic behaviour was observed. Obviously, the rheological behaviour of cream is very complex, so it is important to record experimental data, shearing history, temperature, storage time and any other relevant information. Further details regarding other types of cream are given by Rothwell (1968) and Society of Dairy Technology (1989).

Kessler (1981) records data for the kinematic viscosity of creams of different fat contents at 20°C (Table VI). The variation of kinematic viscosity with temperature is also shown.

TABLE VI
Kinematic viscosity of cream at 20°C

Fat content (%)	*Kinematic viscosity* ($m^2\,s^{-1}$)
20	$6{\cdot}2 \times 10^{-6}$
35	$14{\cdot}5 \times 10^{-6}$
45	35×10^{-6}

SOLID RHEOLOGY

Ideal solids are considered to be perfectly elastic, that is they revert back to their original length once the stress is removed, provided the elastic limit has not been exceeded. Two common properties used for elastic materials are the modulus of elasticity (E) and shear modulus (G), where

$$\text{modulus of elasticity } (E) = \frac{\text{shear stress}}{\text{strain}} = \frac{\text{force/area}}{\text{extension/original length}}$$

$$\text{shear modulus } (G) = \frac{\text{shear stress}}{\text{angular deformation}}$$

Further information on Young's modulus and shear modulus values for a variety of foods are given by Muller (1973) and Mohsenin (1986). Unfortunately most dairy products do not fall into this category, their behaviour being more complex; these can be described as plastic or viscoelastic.

Some materials will exhibit plastic behaviour. Below a certain limiting shear stress (τ_0), they behave like solids; above τ_0 they behave like fluids and show a plastic viscosity. The two most common forms are the Bingham plastic and the Casson plastic (see Fig. 6).

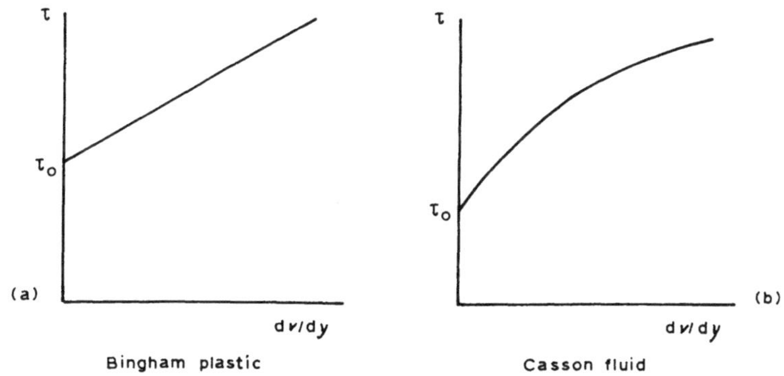

Fig. 6. Rheograms for (a) a Bingham plastic and (b) a Casson fluid.

Products, such as butter, spreads, set yoghurt and cheese, are generally regarded as solids. However, most of these will breakdown and flow, but such breakdown may cause irreversible damage to the product.

The Casson equation is

$$\sqrt{\tau} - \sqrt{\tau_0} = \sqrt{\mu_p}\sqrt{dv/dy}$$

Flow curves for butter and margarine are given by Muller (1983).

VISCOELASTIC BEHAVIOUR

Viscoelastic materials exhibit both elastic and viscous behaviour simultaneously. For some materials, the viscous forces may predominate, whilst for others the elastic forces may do so. Figure 7 shows the strain–time relationship for a typical viscoelastic material, during the application of a constant shearing force and after the force is removed.

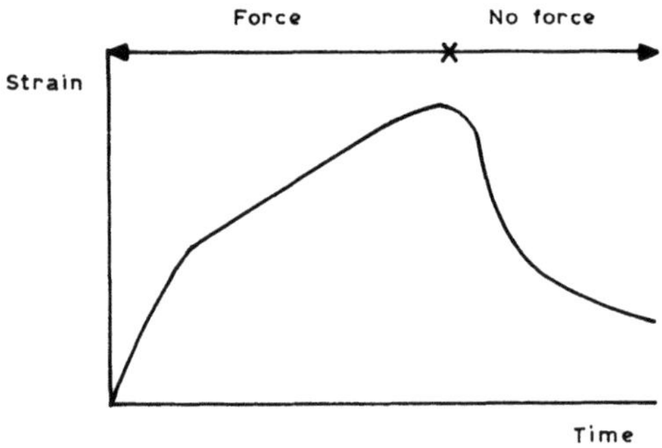

Fig. 7. Strain–time relationship for a viscoelastic fluid, during the application of a constant force, and after removal of the force.

There are several methods for examining viscoelastic behaviour, the main ones being as follows.

(a) To measure the change in strain at constant shear stress, whilst a sample is loaded, followed by removal of that load (Fig. 7). Creep function is the term given to describe the change of strain, at constant stress, and creep compliance is the ratio of the stress divided by the strain. A sample is said to exhibit linear viscoelasticity if there is a straight line relationship between the strain and the applied force, when the samples subjected to different stresses are measured after the same time interval.

(b) To measure the change in stress at constant strain. The relaxation time (t) is defined as the time for the stress to fall to 1/e (36·8%) of its original value. Muller (1973) states that

$$t = \mu / G$$

High relaxation times are associated with materials where the viscous forces predominate, low relaxation times where the elastic forces predominate. If relaxation times are extremely long, some other time period may be selected, e.g. 50 or 70% of the original value.

(c) The use of dynamic tests, whereby the sample is subjected to a harmonic shear strain Y, amplitude Y_0, which gives rise to a harmonic shear stress, within the sample τ, amplitude τ_0 which is out of phase with the shear strain by an angle θ. Cone and plate or concentric cylinder instruments are useful for these tests, the test material being placed in the gap. The harmonic shear strain is set up in one of the elements and the corresponding shear stress is detected by the other element.

If the phase shift $= \theta$, then $\theta = 90°$ for an ideal viscous liquid, $\theta = 0°$ for an ideal elastic material, and $\tan \theta$ is known as the loss factor.

A storage modulus and loss modulus are defined as follows:

storage modulus $(G') = \tau_0 \cos \theta / Y_0$: high for elastic materials
loss modulus $(G'') = \tau_0 \sin \theta / Y_0$: high for viscous materials

Note that $\tan \theta = G''/G'$.

G' and G'' are usually measured as a function of frequency, or temperature. In principle, the frequency plot will give an indication of the time scale of molecular rearrangements taking place within the material in response to the applied strain, whereas the temperature plot will monitor structural changes arising from heating, chilling or freezing (Mitchell, 1987). In general, rheological properties are useful, because as well as describing the flow characteristics of materials, they can be used to monitor certain processes, e.g. curd formation in cheese and yoghurt, as well as in the quality control and assessment of products.

FOOD TEXTURE

Texture is one of the main determinants in the quality of many dairy products. Texture is a psychological property, rather than a physical property. One widely accepted definition is that texture defines the attribute of food material resulting from a combination of physical and chemical properties, perceived by the senses of touch, sight and hearing.

Strictly speaking texture should be evaluated by sensory methods using trained panellists. However, for routine work, sensory methods suffer many disadvantages, and a variety of other methods have been evaluated for a more rapid assessment of food texture. These include rheological methods, chemical tests and, more recently, observations from the miroscope. It should be emphasised that results from such instruments are only valid if they can be shown to correlate with sensory evaluation data. Most of these instruments are designed to subject the food to some form of deformation. A few of them are designed to measure fundamental rheological properties, e.g. μ, G' and G'', whilst others imitate the forces that the food is subjected to during mastication, but the vast majority are empirical in nature, measuring properties that cannot be easily expressed in fundamental terms. The tendency over the past few years has been to move towards more fundamental properties. Some of these instruments, which have been used for dairy products, are listed below:

Efflux viscometers, funnels	—flow properties, yoghurt, creams
Torsion wire viscometers	—custard
Extruders	—butter and other spreads
Penetrometers	—gels, yoghurts, butter

More sophisticated instruments which are widely used are the General Foods Texturometer and the Instron Universal Testing equipment. Texture measurement is comprehensively reviewed by Brennan (1988) and Bourne (1982); the latter gives a good account of the wide range of instruments available. The rheological properties of dairy products have been reviewed by Prentice (1979). Taneya (1979) describes the viscoelastic properties of Cheddar and Gouda cheeses.

Two instruments which have been found to give very useful estimations of the firmness of butter are the extruder (Prentice, 1954) and the cone penetrometer (International Dairy Federation, 1981). A relationship is given between the penetration depth and the yield shear stress (assuming plastic behaviour). Results from both these instruments were found to correlate very well with a sensory assessment of spreadability. On the basis of simplicity, the cone penetrometer has been recommended as a suitable instrument for assessing firmness. A further problem of oiling-off has been identified with some of the softer, more spreadable butters appearing on the market.

The rheology of gels is discussed by Mitchell (1980). Methods for measuring gelation properties, such as gel strength and gelling time, are reviewed by Kinsella (1976). Rheological tests for dairy products involving structural failure have been discussed by Hamann (1983), whilst Prentice (1984) gives a history of some of the rheological work on cheese and butter. The rheological and fracture properties of cheese have been further discussed by IDF (1991).

OPTICAL PROPERTIES

In comparison to rheological properties, the optical properties of milk and dairy products have received less attention. Perhaps the most interesting is the appearance of milk and its products; the presence of absence of visible defects, such as fat separation or coagulation and, in particular, the colour. All these are further important determinants of quality. However, like texture, colour itself is not strictly a true physical property, but a sensory or pyschological characteristic. The physiological basis of vision and colour has been described in greater detail by MacDougal (1986). Nevertheless, a number of instruments have been designed which determine colour on the spectral signal that results from light transmitted through, or reflected by the sample. Other optical properties include refractive index and light scattering. It should be appreciated that the light scattering properties of milk will have a pronounced effect on the appearance of milk, for example, why it appears milky or creamy and why skim-milk has a blue tint.

The colour of milk, and particularly products like cream, butter and cheese, will vary considerably, and will depend on the carotene content of the fat. Generally, additives (colorants) are not permitted in milk and cream, but the colour of milk-based drinks, butter, cheeses and desserts, such as yoghurts, custards and puddings, may be changed by the addition of food colorants, be they natural, nature-identical or synthetic. For certain products there may be a range of permitted colorants; for example, butter may contain α-, β- and γ-carotene, synthetic β-carotene or tumeric (Jukes, 1987). Colours in cheese are also permitted. Regulations should be consulted to ascertan which colorants are permitted in which types of cheese. Obviously, this gives the processor considerable scope for manipulating the final colour of the product.

Instrumental methods are widely used for colour measurement. The most common techniques involve transmission and reflectance techniques. Light is directed at the sample and the signal which is transmitted through, or reflected from, the sample is measured either in a tristimulus colorimeter, or over the visible spectrum, 380 to 750 nm, using a spectrophotometer. In order to decide whether a sample should be measured by reflection or transmission, the Kubella–Munk theory can be used. More detail is given by Francis (1983).

Tristimulus colorimeters use three coloured filters (red, green and blue), each with a transmission curve duplicating the response of the three types of detecting units, known as cones, in the human eye. This is termed the standard observer response. The signal produced from the three filters gives a measure of the amount of each primary colour making up the colour being measured. The signals can be used to evaluate the CIE (Commission Internationale d'Eclairage) *XYZ* tristimulus values; because of the three-dimensional nature of this and other systems, they are often referred to as colour solids. Although not corresponding precisely to the primary colours, *X* is associated with red, *Y* with green, and *Z* with blue. *Y* is also defined as the luminous reflectance or transmittance.

Another type of colour solid widely used is the Hunter *Lab* system, because of its near uniform visual spacing. Data obtained from a tristimulus colorimeter or from a spectrophotometer can be transposed to *Lab* values, where (see Fig. 8)

L represents the lightness, darkness scale (0—black, 100—white);

a represents the red/green scale; positive red; zero grey or neutral; negative green (+100 to −80);

b represents the yellow/blue scale; positive yellow; zero grey or neutral; negative blue (+70 to −80).

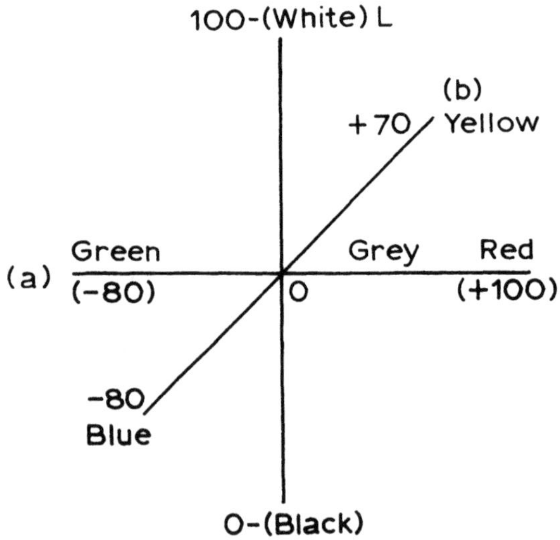

Fig. 8. L ab colour solid.

Much of the reported work on the colour measurement of, and changes in, dairy products uses this system.

For the tristimulus colorimeter, the factors which may affect the values are the nature of the illuminant and the angle of light striking the sample. Such instruments are useful for monitoring changes or differences in colour, but are not so good as the spectrophotometer for the absolute measurement of colour. Spectrophotometers are more sophisticated instruments, which measure the transmission or reflectance spectrum over the entire visible spectrum, 380 to 750 nm, at intervals of typically 10 nm. This information is used in conjunction with the spectrum for the source and the standard observer response, to calculate the *XYZ* or *Lab* coordinates. The *XYZ* and *Lab* systems now seem to be the most widely used, although other systems do exist. Hunt (1987), MacDougal (1988) and Francis (1983) provide more information on some of these, and the latter author also describes some other specialised colourimeters, designed for specific applications. Kent and Smith (1987) describe the findings of a collaborative trial on colour measurement, conducted between 17 laboratories, using a range of colorimeters.

Problems are encountered in colour measurement, because most food materials are not completely opaque or perfectly transmitting. Most reflecting samples absorb some light and most transmitting samples scatter and reflect some light. Sample size and orientation may also influence the colour results, so the experimental method will need to be standardised to obtain meaningful results. Problems in measurement will also be encountered with aerated products and powders; in the latter, particle size will influence the colour measurement.

Sensory techniques involve the panellist matching the colour of the sample against a range of standard colours, under controlled illumination conditions. Two such systems are the Munsell colour system, which makes use of about 1200 different colour reference tiles, and the NCS Colour Atlas, which uses about 1600 different colours. Also used, but to a lesser extent are profiling studies in which the panellists describe the colour of the sample, without reference to any standards, and quantify these using various types of scaling procedures.

Homogenisation will also affect the colour of the milk, more so for whole milk than skim-milk. For skim-milk, there is a slight increase in *L* (lightness), no change in the yellow component and a slight decrease in the green component. For whole milk, there is a considerable increase in lightness and a significant decrease in the yellow component. If the milk is heat treated, this will further complicate the picture.

During heat treatment, the colour of milk changes due to a combination of two effects, which are a whitening caused by an increase in casein micelle size which causes an increase in the reflectance, and a browning effect due to the Maillard reactions; for example, UHT milk may appear to be whiter than the raw milk form which it was produced, whereas sterilised milk appears browner. The kinetics of whitening have been studied by Singh and Patil (1990).

Thus for heat-treated milk and for many other milk products stored at elevated temperatures, the most important chemical reaction is that of Maillard browning (O'Brien and Morrissey, 1989). The factors affecting browning are time, temperature, pH and water activity, with browning taking place more rapidly at intermediate moisture contents. More importantly, browning reactions will also proceed during storage and, as well as affecting the colour and appearance, they will also influence flavour and nutritional value. Such changes will also occur in powders. Such products, when stored a temperatures above 30°C, may become unacceptably brown in a short period of time, i.e. 6–8 weeks, rather than six months under more temperate conditions.

Milk solids form one base material for caramel production. Both the colour and flavour of such caramel is very important. McKenna (1988) has studied the effects of heating at 105, 113 and 121°C for 4, 7 and 10 min on the development of flavour and colour in milk concentrates.

The International Dairy Federation has no publications on the colour of dairy products.

The refractive index (n) of a transparent material is defined as the ratio of the velocity of light in air to that in the medium. The refractive index can easily be measured quickly and accurately to four decimal places. Refractive index values are usually measured using light of a constant wavelength, 589·3 nm, (which corresponds to the sodium D line) and are quoted at a constant temperature, normally 20°C, as the value is affected by both temperature and wavelength.

The refractive index of some materials are as follows: water, 1·3330; milk, 1·3380; and butterfat 1·4620.

The presence of dissolved or suspended solids, such as minerals, sugars, whey proteins and casein micelles, increases the refractive index above that of water and, if the composition of a dairy formulation is known, the refractive index can be estimated (Walstra and Jenness, 1984). However, components which are greater in size than one-quarter wavelength, about 0·1 μm, have no such effect, e.g. fat globules, air bubbles and lactose crystals. Perhaps the most important applications involve using refractive index to estimate the extent of concentration during evaporation or reverse osmosis, as the increase in refractive index is directly proportional to the concentration factor; or as a quick test for checking the purity of butterfat, as fats which may be used for adulteration have significantly different refractive indices.

On the other hand, light scattering is caused by larger particles whose refractive indices are different from that of the surrounding medium.

In milk, the fat globules and casein are primarily involved. Such scattering is a random process, but it is possible to measure the intensity of the light at some angle to the incident beam, as well as the intensity of the light transmitted through the sample. Both transmitted and scattered light (or other forms of radiation such as IR or UV) form the basis of a range of techniques for analysis of the major components in dairy products. Laser scattering techniques now form the basis for particle size measurements, for example, fat globule distributions in emulsions or particle size distributions for powders.

SURFACE PROPERTIES

Milk is a good example of a complex colloidal system. Casein micelles are stabilised by colloidal calcium phosphate, and the dispersed fat phase by the fat globular membrane. The surface area to volume ratios of the dispersed phase components are very large, and Walstra and Jenness (1984) give the surface area to volume ratios and mean diameters of these dispersed phase components as shown in Table VII. The same authors also show how homogenisation pressure will affect the size of the fat globules.

TABLE VII
Surface area to volume ratio and size of dispersed phase components

Component	*Surface area to volume ratio ($cm^2\ ml^{-1}$)*	*Particle diameter*
Fat globules	700	0·1–10 μm
Casein micelles	40 000	10–300 nm
Globular (whey proteins)	50 000	3–6 nm
Lipoprotein particles	100	~10 nm

Adapted from data in Walstra and Jenness (1984).

It is important to maintain colloidal stability during processing, so it is necessary to examine the forces acting at both the interface between milk and air, and on the dispersed and continuous phases.

Surface tension is concerned with the forces acting within the fluid. Molecules at the surface of a fluid will be subject to an imbalance of molecular forces and will be attracted into the bulk of the fluid. Consequently, the surface is said to be under a state of tension.

The surface tension of a liquid can be regarded in two ways, either as the force per unit length acting on a given length of surface, or as the work done in increasing its surface area, under isothermal conditions. The SI units of surface tension are $N\,m^{-1}$ or ($J\,m^{-2}$). The conversion factor is $1\,dyne\,cm^{-1} = 10^{-3}\,N\,m^{-1}$ or $1\,mN\,m^{-1}$. It is the surface tension forces which cause most finely dispersed liquids to form spherical droplets; this is the shape having the minimum surface area to volume ratio.

The methods for determining surface tension are discussed by Levitt (1973). Water has a surface tension value of $72{\cdot}6\,mN\,m^{-1}$ at 20°C, whereas milk has an approximate value of $50\,mN\,m^{-1}$ at 20°C (Jenness *et al.*, 1974). The surface tension is lower than that of water due to the presence of casein, whey proteins and phospholipids, which are all surface active. Bertsch (1983) measured the surface tension of whole milk (4% fat) and skim-milk, in the temperature range 18–135°C, by a drop weight method. The surface tension was found to decrease in an almost linear fashion as temperature increased. There was very little difference between the results for whole and skim-milk, and the combined results could be represented by the equation:

$$Y = 1{\cdot}8 \times 10^{-4}\,\theta^2 - 0{\cdot}163\,\theta + 55{\cdot}6$$

where Y is surface tension ($mN\,m^{-1}$), and θ is temperature (°C).

Jenness *et al.* (1974) concluded that the surface tension of milk decreased slightly as the fat content increased to 4%, thereafter remaining constant. Sweet cheese whey was reported to have a similar value to skim-milk. Homogenisation was found to increase the surface tension slightly. Lipolysis and the liberation of free fatty acids decreased the surface tension, but heat treatment had little effect. The presence of detergent in milk would drastically reduce the surface tension.

There appears to be little published work on the effects of concentration on the surface tension of milk. This will be important in spray-drying operations, as it will affect the drop-size distribution of the spray.

An interfacial tension exists at the boundary of two immiscible liquids, again due to an imbalance of intermolecular forces. The interfacial tension between water and butter oil at 40°C is 19·2 mN m^{-1}. Such a system is unstable and will quickly separate, as an interfacial tension below 10 mN m^{-1} is required to produce a stable emulsion. Surface active components in the milk serum will lower the interfacial tension (Powrie and Tung, 1976), the most effective being α-lactalbumin and interfacial protein. Other emulsifiers and stabilisers may be added and are important in aerated products, such as whipped cream, ice cream and milk concentrates. The interfacial tension of various emulsifiers used in ice cream against palm kernel oil–water and palm kernel oil–casein solutions (0·1%) are given by Berger (1976). The emulsion stability in milk and milk products is discussed in more detail by Graf and Bauer (1976).

The work of adhesion (W_{AB}) is the work required to separate an interface and produce two distinct surfaces:

$$W_{AB} = Y_A + Y_B - Y_{AB}$$

A reduction in interfacial energy will increase the work of adhesion and help to stabilise the emulsion. Emulsions are further stabilised by a high-viscosity continuous phase, and a similar, uniform charge on the surfaces of the particles.

Surface tension forces are important in size reduction and cleaning operations. The Weber number is a dimensionless group, which includes surface tension; its use in connection with high-pressure homogenisers has been discussed by Loncin and Merson (1979). Masters (1991) presents correlations for mean particle diameters for pressure nozzles, two fluid nozzles and centrifugal atomisers. Hygienic design and cleaning of food processing equipment is considered in more detail by Troller (1983) and Jowitt (1980).

THERMAL PROPERTIES OF FOODS

Heat transfer plays an important role in many dairying operations, and in most situations, it is desirable to maximise the rate of heat transfer. This offers economic advantages and generally results in a better quality product. Therefore, drying, heating, chilling and freezing operations are all influenced by the heat transfer processes taking place.

Some of the questions the dairy technologist may be required to answer are as follows.

(a) How much heat is required to process a particular material and what methods are available for conserving energy? In simple heating and cooling operations, this will involve only sensible heat changes, whereas in evaporation and freezing, latent heats will be involved.
(b) What size heat exchangers are required for a particular heating duty? This applies to heaters, evaporators and refrigeration equipment. It involves the solution of steady-state heat transfer equations, and a knowledge of the heat transfer mechanisms and resistances involved.
(c) What are the heating times or freezing times? This involves the solution of unsteady-state heat transfer equations.
(d) What are the requirements for steam, hot water, refrigerants, electricity and compressed air? This involves a knowledge of the thermodynamic properties of these fluids. The thermal properties of dairy products relevant to these processes will now be discussed in more detail.

Sensible and Latent Heats

Sensible heat changes are those which can be detected by a change in temperature, and involve no phase change. The amount of heat (Q) required to bring about a sensible heat change for a batch heating process is given by

$$\underset{(\mathrm{kJ})}{Q} = \underset{(\mathrm{kg})}{\text{mass}} \times \underset{(\mathrm{kJ\,kg^{-1}\,K^{-1}})}{\text{specific heat}} \times \underset{(\mathrm{K})}{\text{temperature change}}$$

For a continuous process, mass is replaced by mass flow rate ($\mathrm{kg\,s^{-1}}$) and the quantity of heat or duty is expressed as $\mathrm{kJ\,s^{-1}}$ (kW). The specific heat of a substance is defined as the amount of heat required to raise a unit mass through a unit temperature rise. The value for water is $4{\cdot}18\ \mathrm{kJ\,kg^{-1}\,K^{-1}}$ (or $1\ \mathrm{Btu\,lb^{-1}\,{}^{\circ}F^{-1}}$). The specific heat of most substances is slightly temperature dependent; this can be overcome by using an average specific heat value for the temperature range being considered. If the variation of specific heat with temperature is known, the heat change can be determined by plotting the specific heat against temperature and evaluating the area under the curve ($m\int C_p\,d\theta$). The different components in foods have different specific heat values, so it should be possible to estimate the specific heat of a food from a knowledge of its composition. Water has the greatest influence on the specific heat.

Lamb (1976) gives a simple equation, considering the food to be a two-component system—water (w) and solids (s)—based on the mass fractions (m) of each

$$c = m_w C_w + m_s C_s$$

Miles *et al.* (1983) make a further distinction, which is quite popular with dairy products, based on water (w), fat (f) and solids-not-fat (snf).

$$c = (0{\cdot}5m_f + 0{\cdot}3m_{snf} + m_w)4{\cdot}18 \ (\text{kJ kg}^{-1}\ \text{K}^{-1})$$

Kessler (1981) has recommended the following equation:

$$\underset{\text{(water)}}{c = 4{\cdot}18m_w} + \underset{\text{(carbohydrate)}}{1{\cdot}4m_c} + \underset{\text{(protein)}}{1{\cdot}6m_p} + \underset{\text{(fat)}}{1{\cdot}7m_f} + \underset{\text{(ash)}}{0{\cdot}8m_A}$$

Therefore, provided the chemical composition is known, specific heats can be estimated reasonably accurately. Values for frozen products can be obtained by substituting the specific heat of ice, in the respective equations. This, however, assumes that all the water is in the frozen form.

Bertsch (1982) describes how the specific heat of milk changes over the temperature range 50–140°C. Fernandez-Martin (1984) describes how the specific heat of sheep's milk whey, concentrated by reverse osmosis, is influenced by temperature and total solids.

$$c = 4{\cdot}216 + (-0{\cdot}0262 + 0{\cdot}092 \times 10^{-3}\,\theta)s$$

where θ is temperature (°C), and s is total solids (%).

The specific heat of milk concentrates has been described by Fernandez-Martin (1972) as

$$c = (m_w + (0{\cdot}328 + 0{\cdot}0027\,\theta)m_s)4{\cdot}18$$

over the temperature range 40–80°C and total solids range (8–30%).

Latent Heat Effects

Latent heat changes are involved when phase changes occur, the major ones being the transition from solid (S) to liquid (L) and from liquid (L) to vapour (V). Under special conditions, the change from solid to vapour can also occur; this is known as sublimation and occurs when the water vapour pressure is maintained below 4·6τ.

The major changes involved with dairy products are: the transition from water to ice (freezing): the removal of water during evaporation and concentration, and the phase changes involved in fat fraction when products are cooled below 50°C (crystallisation).

At atmospheric pressure water boils at 100°C; the latent heat of vaporisation water is the amount of energy required to change 1 kg of water from liquid to vapour, without changing the temperature (2257 kJ kg^{-1}); this provides information on the energy requirements for evaporation. However, as the pressure is reduced and the boiling point decreases, the latent heat value increases. At a pressure of 0·073 bar (absolute), the latent heat value is increased to 2407 kJ kg^{-1}.

Steam is used as the heating medium in evaporation, and the heat lost by the steam causes vaporization of the liquid. In practice, the evaporation process can be regarded as an exchange of latent heat, and in a single-effect evaporator, it takes approximately 1 kg steam to remove 1 kg water vapour.

Many processes are performed under vacuum; this results in a reduction in the evaporation temperature, and an improved heat transfer rate due to the increased driving force between the steam and the evaporating liquid; the amount of energy required increases slightly. Steam economy is effected on evaporation plants by multiple-effect evaporation, and the use of vapour recompression, either by steam ejectors or mechanical recompression.

Steam is a very efficient heat transfer fluid. It has both a very high latent heat value, and a high heat film coefficient. Mixing of air with steam will drastically reduce the heat film coefficient and, in thermal processing operations, may lead to under-processing.

The thermodynamic properties of saturated steam are covered in the steam tables (ASHRAE, 1985). For saturated steam, with no air present, there is a fixed relationship between the temperature and pressure, so pressure gauges are often incorporated into steam lines and retorts as a back-up to temperature readings.

Higher processing temperatures require the use of higher pressure steam. For example, saturated steam for UHT processing (150°C) will be at a pressure of 4·76 bar absolute pressure (a), whereas steam for heating the inlet air to a spray drier (200°C) would be at a pressure of 15·5 bar. If steam is to be injected directly into a product, the steam should be saturated and free from any contaminating particles. Such addition results in a dilution of the milk, and this added water needs to be removed later in the process. Occasionally superheated steam may be required; this is where saturated steam is heated above its saturation temperature. One such use is in the sterilisation of cans, prior to aseptic filling in the Dole process, where the use of superheated steam avoids excessive condensation and wetting of the cans.

The thermodynamic properties of steam and other thermodynamic fluids, such as Freon 12, ammonia, carbon dioxide, cryogenic fluids and immersion fluids are summarised by ASHRAE (1985).

Freezing of Foods

The latent heat of fusion for pure water is 335 kJ kg^{-1}; that is the amount of heat liberated when 1 kg of water is converted to ice at 0°C. Unfortunately the situation for foods is more complex. The presence of solids depresses the freezing point, with most foods starting to freeze at about −1°C. This results in a concentration effect and a further depression of the freezing temperature. Therefore, the food does not freeze at a constant temperature; rather as freezing proceeds, so the temperature falls as more of the ice is converted to water; hence there is a concept of unfrozen water.

Most of the water freezes over the temperature range −1 to −10°C; and by −15°C, more than 90% of the water is frozen. The freezing point of milk is of considerable interest, because it is also used to detect any dilution of the milk. Some of the pitfalls of the method have recently been discussed by Kessler (1984). The amount of frozen water will have a pronounced influence on the texture of ice cream. Freezing point calculations for ice cream and frozen desserts are covered by Arbuckle (1977) and ASHRAE (1982). Some typical figures for an ice cream are given below:

Temperature (°C)	*Percentage frozen water*
−2·75	10
−3·11	20
−3·50	30
−4·22	40
−5·21	50
−6·78	60
−9·45	70
−14·92	80
−30·16	90

Adapted from ASHRAE (1982). Fat = 12·5%. Milk solids = 10·5%. Sugar = 15%.

Smith and Bradley (1983) have described the effects of carbohydrates commonly used in frozen desserts on their freezing points.

Most foods contain substantial quantities of solid, whereas only the water contributes to the latent heat value. On this basis, Lamb (1976) suggested the following equation for determining the latent heat value of a food:

$$L = m_w \times 335 \text{ (kJ kg}^{-1}\text{)}$$

One approach to operations involving sensible and latent heat change is to assume that the material has a melting point of -1°C, and that all the water is converted to ice at that temperature. The operation then comprises the sensible heat change (to the freezing point), a latent heat change, and a further sensible heat change to the storage temperature.

The fact that all the water does not freeze will give an overestimate, but this is balanced by the fact that all the food is considered frozen at -1°C, whereas in fact it is not, which will tend to underestimate the value.

Enthalpy (H) is a thermodynamic function, where $H = U + PV$, where U is internal energy, P is pressure, and V is volume.

It can be shown for operations taking place at constant pressure, that the enthalpy change is equal to the amount of heat (q) absorbed or evolved

$$\Delta H = q = \int C_p \, d\theta$$

Enthalpy data for food are normally presented as the specific enthalpy (kJ kg^{-1}) at different temperatures. Above 50°C, virtually all fats are completely liquid, but as the temperature is reduced, crystallisation takes place and the fat solidifies. The extent of solidification will depend mainly upon the type of fat and the cooling conditions. The relationship between the percentage crystalline solids and temperature is known as the 'melting characteristic'. The melting characteristic and specific enthalpy for butter fat are shown in Table VIII.

Solid–liquid fat ratios can be measured by dilatometry, wide-line nuclear magnetic resonance (NMR) or differential scanning calorimetry (DSC). Reugg *et al.* (1983) describe an improved DSC method for the estimation of the solid content of butter fat, which takes into account the temperature dependence of the heat of melting. Applying this correction gives data which agree more closely with those from NMR techniques. The fractionation of milk fat by cooling and crystallisation, and the composition and properties of the fractions are described by Badings *et al.* (1983*a*,*b*).

TABLE VIII
Enthalpy, percentage crystalline solids and apparent specific heat of butter fat

	Temperature (°C)								
	−40	*−20*	*−10*	*0*	*10*	*20*	*30*	*40*	*50*
Specific enthalpy (kcal kg^{-1})[a]	3	11	17	24	32	45	54	60	65
Crystalline solids (%)	100	98	90	75	56	20	10	0	0
Apparent specific heat (kJ kg^{-1} K^{-1})	1·59	1·84	2·01	3·34	4·39	5·35	3·34	2·09	2·01

Adapted from data in Rha (1975).
[a]1 kcal ≈ 4·2 kJ.

Figure 9 shows some thermodynamic data represented in the form of enthalpy against concentration for whole (full-cream) milk. This chart is extremely useful, as it shows how the heat content changes with temperature and moisture content. In addition, when the temperature is taken below the freezing point, the amount of unfrozen water is also given. Full-cream milk at 80°C has an enthalpy value of 670 kJ kg^{-1}. If this is reduced in temperature to −10°C, the new enthalpy value is 105 kJ kg^{-1}, and the fraction of water that is frozen in approximately 92%. Therefore, the amount of heat removed would equal 565 kJ kg^{-1}. Loncin and Merson (1979) reviewed other sources of data; some other charts are given in Rha (1975), most of which are based on the work of Reidel.

An alternative approach is to use an apparent specific heat value, which accounts for sensible and latent heat changes, i.e. when considerable melting or crystallisation is taken place the apparent specific heat is high; when none is taking place, it is much lower.

Table VIII also shows how the apparent specific heat of butter fat changes with temperature. The apparent specific heat of whey, whole milk and cream containing 20, 40 and 60% fat has been plotted against temperature over the range 0–50°C (Batty and Folkman, 1983).

Some specific heat values for milk products are given in Table IX.

Thermal Conductivity

This is a measure of the rate of heat transfer through a material when conduction is the controlling mechanism. It is applicable to solids, but it can also be measured for fluids when convection is eliminated.

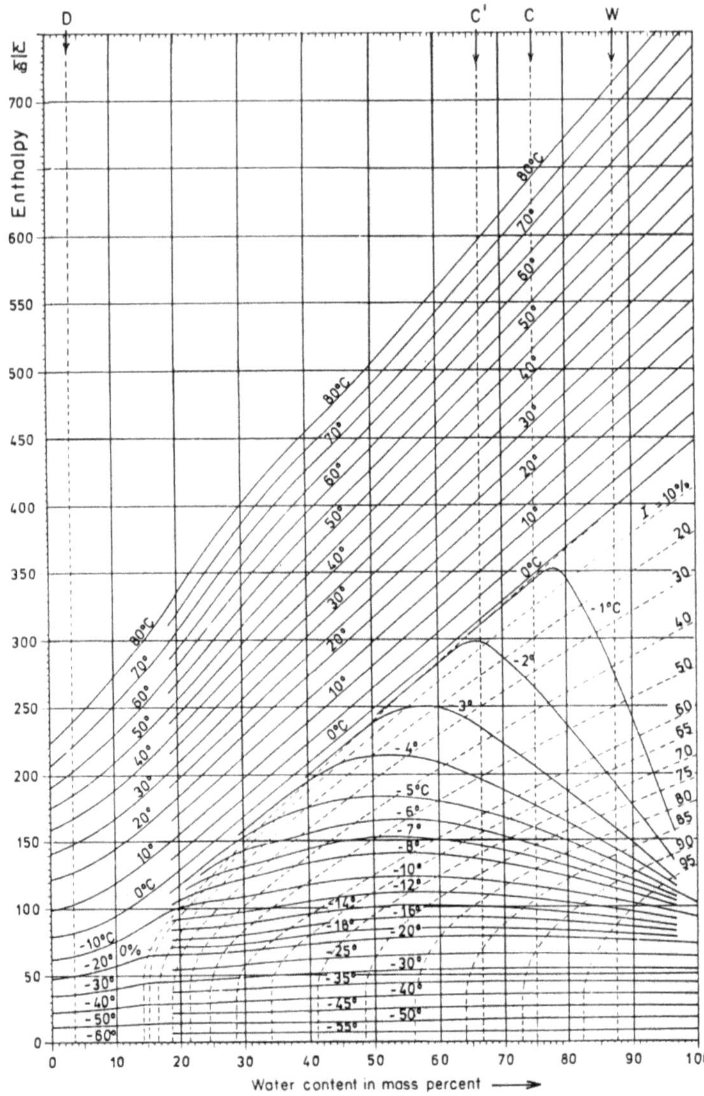

Fig. 9. Enthalpy–concentration diagram for the system dry whole milk/water. The dry solids contain 30% fat. The value of the enthalpy is 0 at the reference temperature −60°C. *I* is the percentage of the water which is frozen; *D* represents the composition of air dried, whole milk powder; *C* and *C′* are condensed milk with 7·5 and 10% fat, respectively; *W* is whole milk. (From Riedel (1976), reproduced by permission of Verlag Hans Carl.)

TABLE IX
Specific heat and latent heat values for some dairy products

	Specific heat (kJ kg⁻¹ K⁻¹)		*Latent heat*
	Above freezing point	*Below freezing point*	*(kJ kg⁻¹)*
Cheese (37–38% moisture)	2·09	1·30	125·6
Roquefort	2·72	1·34	183·8
Cheese low-fat	2·68	1·47	
Cream 15% fat	3·85		
40% fat	3·56	1·68	209·3
Ice cream (58–66% moisture)	3·27	1·88	223·3
Milk	3·85		
Skim-milk	3·98	2·51	305·0
Butter	2·05		53·5

Compiled from data in Polley *et al.* (1980) and ASHRAE (1985).

Thermal conductivity is defined as the steady-state rate of heat transfer through an area 1 m^2 when a temperature driving force of 1 K is maintained over a distance of 1 m.

It can be measured for materials under steady-state or unsteady-state conditions. Methods for measuring thermal conductivity have been described by Mohsenin (1980) and Jowitt *et al.* (1983).

Lamb (1976) gives the following equation for evaluating the thermal conductivity of a food from its moisture content:

$$k = 0{\cdot}0841 + 0{\cdot}568\, m_w$$

It was claimed that there were some large discrepancies below 50% moisture content, and that there was no simple relationship between the themal conductivity of frozen foods and moisture content.

Water is the principal component affecting the thermal conductivity, since it has a higher value than the other chemical components. Ice has a value approximately four times higher than water. It is not so straightforward to predict the thermal conductivity from the composition of complex materials, as it depends on whether the components are considered to be in parallel or series (Miles *et al.*, 1983).

Figure 10 shows how thermal conductivity changes with temperature, for a variety of products. In all cases, the thermal conductivity increases as the temperature increases; again it should be noted that as milk products become more concentrated, their thermal conductivity decreases.

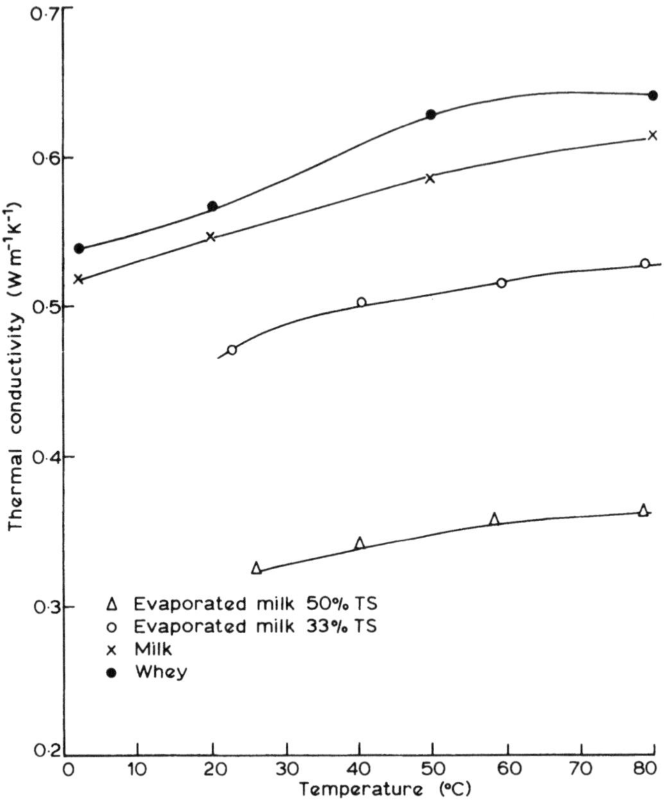

Fig. 10. The relationship between the thermal conductivity and temperature, for a variety of dairy products. (Adapted from data in ASHRAE (1985).)

MacCarthy (1984) has measured the effective thermal conductivity of skim-milk powder using a guarded hot-plate technique. Values ranged from 0·036 to 0·109 W m^{-1} K^{-1} in the temperature range 11·8–49·7°C for bulk densities between 292 and 724 kg m^{-3}. The effective thermal conductivity increased with temperature and with bulk density.

Thermal Diffusivity

Thermal diffusivity (a) is a composite property defined as (k/pc) with units of m^2s^{-1}. It is an extremely useful property in unsteady-state heat transfer problems, because it is a measure of how quickly temperatures change with time, during heating and cooling processes. It is extensively used in unsteady-state heat transfer problems in a dimensionless form known as the Fourier number (*Fo*), where :

$$F_0 = \frac{at}{r^2}$$

where t is heating time, and r is the characteristic dimension of food.

The importance of thermal diffusivity has been discussed by Singh (1982). Unsteady-state heat transfer methods for heating, chilling and freezing are discussed by Jackson and Lamb (1981), Lewis (1990) and Cleland and Earle (1982). Jowitt *et al.* (1983) have reviewed the COST 90 project on the thermal properties of foods. Patel *et al.* (1990) have determined the thermal properties of whey protein concentrates. The thermal and physical properties of these foods play an important role in heat transfer processes, by affecting heat transfer by conduction and convection. Correlations for heat film coefficients are given by Kessler (1981) and ASHRAE (1985).

Sorption Isotherms

A sorption isotherm is a plot of the equilibrium moisture content of a food against relative humidity at a constant temperature. It can most easily be determined by equilibrating the food with atmospheres of different relative humidity and determining the moisture content, once the food reaches a constant weight. If the test material is initially dry, the isotherm is referred to as an adsorption isotherm; if it is wet, it is known as a desorption isotherm. In most cases the two isotherms are similar, but occasionally hysteresis is found. Usually, moisture content is expressed as a percentage on a dry weight basis, i.e. (weight of water/weight of dry solids) × 100.

These isotherms are useful for determining the lowest moisture content attainable in dehydration processes, and for estimating the water activity of a food from its moisture content.

The water activity (a_w) of a food is a measure of the availability of water as a solvent for reactions in food. It is defined as

$$a_w = \frac{\text{water vapour pressure exerted by food}}{\text{saturated water vapour pressure at the same temperature}}$$

It can be seen that if a food is equilibrated in a sealed container (with a relatively small free volume) and the equilibrium relative humidity (RH) of the air equals RH, then ($a_w = \text{RH}/100$).

Therefore, the sorption isotherm also gives the relationship between the water activity and moisture content. Two foods with the same moisture content will not necessarily have the same water activity, and it is the water activity which affects the chemical, microbial and enzymic reaction rates.

Milk has a water activity almost equal to 1·0, as do most other fresh foods; evaporated milk as a water activity of about 0·98, and ice cream mix (40% solids) a value of 0·970. Sweetened condensed milk, containing 55–60% sugar, has a value between 0·85 and 0·89. Most cheese varieties lie between 0·94 and 0·98, whereas dried milk 1·5–4·5% moisture) lies between 0·02 and 0·2 (Troller and Christian, 1978; Walstra and Jenness, 1984). Humectants are compounds which, when added to foods, depress the water vapour pressure and hence the water activity; sugar and salt are two humectants in common use. Esteban and Marcos (1989) found a good correlation between water activity and ash content for processed cheese.

Intermediate moisture foods are regarded as foods whose microbial spoilage is prevented by a low water activity, and which do not require rehydration before consumption. The water activity of such foods lies between 0·2 and 0·85.

There are many published equations relating water activity to moisture content, one of the most useful being the Brunauer–Emmet–Teller (BET) isotherm.

$$\frac{a}{m(1-a)} = \frac{1}{m_1 c} + \frac{c-1}{m_1 c} a$$

where a is water activity, m is water content (% dry weight), c is a constant, and m_1 is a monomolecular layer water content (as above). The factor m_1 is a measure of the monomolecular layer, i.e. the amount of water strongly bound to that material.

Iglesias and Chirife (1982) have compiled sorption isotherms for a wide variety of foods. For each isotherm, they used curve-fitting techniques to select the best two-parameter equation (from a choice of nine) to describe that isotherm. The results for some dairy products are summarised in Table X together with calculated values of the monomolecular layer value. Some dairy products exhibit a broken isotherm, particularly during the adsorption stage. This is found with substances containing lactose, and is attributed to a transition from the amorphous to the crystalline form. No such change is noted for the desorption isotherm. Figure 11 shows examples for defatted milk and cheese whey.

TABLE X
Summary of sorption isotherm data for dairy products

	Temperature (°C)	Type[a]	a_w range	Equation[b]	B_1	B_2	Monomolecular layer per cent (dry weight)
Cheese Emmental	25	A	0·1–0·8	1	1·1889	5·9967	3·3
	25	D	0·1–0·8	1	1·4435	11·9777	3·7
Cheese Edam	25	A	0·1–0·8	1	1·0668	4·9692	3·3
	25	D	0·1–0·8	1	1·2540	8·5716	3·5
Casein	30	—	0·1–0·8	2	2·1510	0.0044	7·6
β-Lactoglobulin[c]	25	A	0·1–0·8	2	1·5211	0·0166	6·6
Non-fat dry milk[c,d]	30	D	0·1–0·8	1	1·9684	72·3080	6·5
	37·8	D	0·1–0·8	1	1·9927	67·8072	6·1
Skim-milk[c,d]	20	A	0·1–0·8	4	−3·0113	1·3983	2·8
	34	D	0·1–0·8	1	2·0544	54·3870	4·7
	34	A	0·1–0·8	1	1·7764	23·8439	4·0
	14	D	0·1–0·8	1	2·6527	290·2579	
Whole milk[c,d]	24·5	D	0·1–0·8	4	−1·4503	3·1356	3·1
	24·5	A	0·1–0·8	1	2·1884	37·9004	3·5
Sweet whey[d]	24·5	D	0·1–0·8	3	3·1279	3·0619	—
Whey protein concentrate[c]	24·0	A	0·12–0·86	1	1·4806	17·6165	4·8
Yoghurt[c]	25·0	A	0·1–0·8	1	1·0529	6·4806	4·1
	45·0	A	0·1–0·8	4	−3·6752	0·2732	3·0

Compiled from data in Iglesias and Chirife (1982).
[a]Type of isotherm: A, adsorption; D, desorption.
[b]The following equations are relevant:
(1) Halsey's equation $a_w = \exp(-B(2)/X^{B(1)})$
(2) Henderson's equation $1 - a_w = \exp(-(B(2)X^{B(1)})$
(3) Iglesias and Chirife's equation $X = B(1)[a_w/(1-a_w)] + B(2)$
(4) Kuhn's equation $X = \dfrac{B(1)}{\ln a_w} + B(2)$ X = moisture content (% dry weight basis)
[c]Isotherms are also given at other temperatures, or alternatives are given.
[d]Broken isotherms.

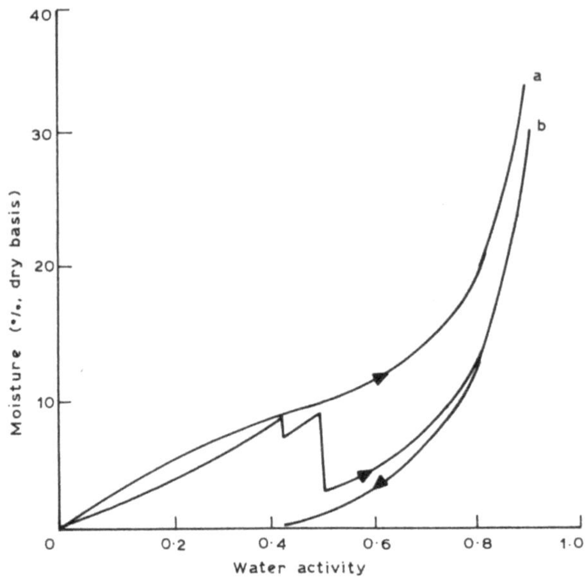

Fig. 11. Sorption isotherms for (a) defatted milk (30°C) and (b) cottage cheese whey (24°C). (Adapted from data in Iglesias and Chirife (1982).)

Watt (1983) discusses the use of a three-parameter GAB equation for fitting sorption isotherm data.

The results of collaborative trials on determining water sorption isotherms are reviewed by Jowitt *et al.* (1983). An extensive bibliography on sorption isotherms and water activity has been produced by Wolf *et al.* (1985).

ELECTRICAL AND DIELECTRIC PROPERTIES

Electrical Conductance

Electrical resistance and conductance provide a measure of the ability of a material to transport an electric current, resistance usually being preferred for solids and conductance for liquids.

If the total electric resistance (R) is measured across a material, length (L) and cross-sectional area (A), then

$$R = \frac{\rho_r L}{A}$$

where ρ_r = resistivity (ohm m).

The specific conductance (K) is the inverse of resistivity

$$K = \frac{1}{\rho_r} = \left(\frac{L}{RA}\right) : (\text{mho m}^{-1})$$

The reciprocal ohm is known as a mho or Siemen (S). The specific conductance is measured by resistance techniques, the cell usually being calibrated with a liquid of known specific conductance (Levitt, 1973).

The cgs unit (mho cm^{-1}) is still commonly encountered, where

$$1 \text{ mho m}^{-1} (\text{Sm}^{-1}) = 10^{-2} \text{ mho cm}^{-1}$$

Most dairy products are poor conductors of electricity. Bovine milk has values ranging from 0·004 to 0·0055 mho cm^{-1} (Jenness *et al.*, 1974). There is little difference between varieties, as the major contribution arises from potassium and chloride ions. It might be expected that increasing the concentration of milk solids would increase the specific conductance, but the relationship is not so straightforward. The conductance of concentrated skim-milk was shown to increase to a maximum value of about 0·0078 mho cm^{-1} at 28% total solids, after which it decreased; this was explained by the extremely complex salt-balance between the colloidal and soluble phases.

The presence of fat tends to decrease the specific conductance, the specific conductance of milk fat being less than 10^{-16} mho cm^{-1}.

It has been suggested that conductivity measurements may be useful for monitoring processes where such changes occur. For example, in the decationisation of sweet whey using Amberlite IR 120 in its hydrogen form, the conductance rose from 5·2 to 8·15 mmho cm^{-1}. The demineralisation level was 75% and the pH fell from 6·22 to 1·7. Increasing the pH to 3·0 using Amberlite IRA 47 in the OH^{-1} ion form, decreased the conductivity to 0·165 mmho cm^{-1}, and gave a demineralisation level of 94% (Kanekanian, 1983).

The development of acidity occurring during many fermentations has also been observed to increase the conductivity. Mastitic milk also has an increased conductivity due to its raised content of sodium and chloride ions.

The conductivity of dairy products will also be important in ohmic heating processes. Electrical conductivity increases as temperature increases, according to the following equation:

$$K_T = K_{25}(1 + M(T - 25))$$

where K_T is the electrical conductivity at temperature T°C (Sm^{-1}), M is the proportionality constant (°C^{-1}) (usually about 0·02), and T is temperature (°C).

Dielectric Properties

The dielectric properties of foods are currently receiving more attention, due mainly to the advent of dielectric and microwave heating processes, which involve selected frequencies in the range 13·56–40·68 MHz (dielectric) and 896–22 125 MHz (microwave).

The two properties of interest are the dielectric constant (ε') and the dielectric loss factor (ε'').

The dielectric constant (ε') is a measure of the amount of energy that can be stored when the material is subjected to an alternating electric field. It is the ratio of the capacitance of the material being studied to that of a vacuum (or air) under the same conditions.

In an AC circuit containing a capacitor, the current leads the voltage by 90°. When a dielectric is introduced, this angle may be reduced. The loss angle (θ) is a measure of this reduction and is usually recorded as the loss tangent ($\tan\theta$). Energy dissipation within the dielectric increases as the loss tangent increases. A new factor, known as the dielectric loss factor (ε''), is introduced, which is a measure of the energy dissipated within the sample, where $\varepsilon''=\varepsilon'$ tan.

During microwave and dielectric heating, the power (P_0) dissipated within the sample is given by

$$P_0=55{\cdot}61\times10^{-14}FE^2\varepsilon''$$

where P_0 is adsorbed power ($W\,cm^{-3}$), f is frequency (Hz), and E is electric field strength ($V\,cm^{-1}$). Materials which absorb microwave energy well are known as 'lossy' materials.

Values for the dielectric constant and the dielectric loss factors of a wide variety of foods are given by Mohsenin (1984) and Mudgett (1982). Both these properties are affected by the moisture content and temperature of the sample, and the frequency of the electric field. Dielectric properties of materials are measured over a wide frequency range, using a variety of instrumental methods.

Mudgett *et al.* (1974) measured the dielectric properties of milk. At 1000 MHz, the dielectric loss factor increased from 22 to 32 over the temperature range 22–55°C, whereas at 3000 MHz, the values were lower, decreasing linearly from 18 to 15 over the same temperature range. Loss factors at 300 MHz were much higher.

Kent (1987) has compiled values for butter, milk concentrates, milk powders and liquid skim-milk. Results are not directly comparable as different frequency ranges and temperatures were used. A brief summary is provided below:

Product	*Frequency range*	ε'	ε''
Butter	10^5–10^7 Hz	4–6	0·01–0·2
Milk powders	10^5–10^7 Hz	2·0–3·9	0·025–0·118
Milk concentrates	10 Hz	NA	0·55–2·35
Skim-milk	3000 MHz	55·0–74·5	12·1–24·4

See Kent (1987) for further details.
NA— not available.

Since dielectric properties are dependent on moisture content, they are also being used as a means of estimating the moisture content of materials. For powders this is not so straightforward as these properties are also affected by the particle density. Kress-Rogers and Kent (1987) have determined ε' and ε'' for milk powder of different particle densities at 4% moisture content, and have used this information to develop a combined parameter which is almost independent of particle size and density differences.

REFERENCES

Allen, C. R. (1980). *Milk and Whey Powders*, Society of Dairy Technology, Wembley, London, UK.

Arbuckle, W. S. (1977). *Ice Cream* (3rd edn). AVI, Westport, CT, USA.

ASHRAE (1982). *Applications Handbook*. ASHRAE Inc., Atlanta, GA, USA.

ASHRAE (1985). *ASHRAE Handbook, Fundamentals*, ASHRAE Inc., Atlanta, GA, USA.

Badings, H. T., Schaap, J. E, Jong, C. De and Hagedoorn, H. G. (1983*a*). *Milchwissenschaft*, **38** (2) 95.

Badings, H. T., Schaap, J. E., Jong, C. De and Hagedoorn, H. G. (1983*b*). *Milchwissenschaft*, **38** (3) 150.

Batty, J. C. and Folkman, S. L. (1983). *Food Engineering Fundamentals*. John Wiley, New York, USA.

Berger, K. G. (1976). *Food Emulsions*, ed. S. Friberg. Marcel Dekker, New York, USA.

Bertsch, A. J. (1982). *Lait*, **62**, 265.
Bertsch, A. J. (1983). *J. Dairy Res.*, **50**, 259.
Bertsch, A. J. and Cerf, O. (1983). *J. Dairy Res.*, **50**, 193.
Bertsch, A. J., Bimbenet, J. J. and Cerf, O. (1982). *Lait*, **62**, 250.
Bourne, M. C. (1982). *Food Texture and Viscosity*. Academic Press, New York, USA.
Brennan, J. G. (1988). *Sensory Analysis of Foods*, ed. J. R. Piggott. Elsevier Applied Science Publishers. London, UK.
British Standard Number 734(1937).
British Standard Number 734(1959).
Cleland, A. C. and Earle, R. L. (1982). *Int. J. Refrig.*, **5** (3), 134.
Egan, H., Kirk, R. S. and Sawyer, R. (1981). *Pearson Chemical Analysis of Foods*. Churchill Livingstone, Edinburgh, UK.
Esteban, M. A. and Marcos, A. (1989). *J. Dairy Res.*, **56**, 665.
Fernandez-Martin, F. (1972). *J. Dairy Res.*, **39**, 65.
Fernandez-Martin, F. (1984). *J. Dairy Res.*, **51**, 445.
Francis, F. J. (1983). *Physical Properties of Foods*, ed. M. Peleg and E. B. Bagley. AVI, Westport, CT, USA.
Graf, E. and Bauer, H. (1976). *Food Emulsions*, Ed. S. Friberg. Marcel Dekker, New York, USA.
Hamann, D. D. (1983). *Physical Properties of Foods*, ed. M. Peleg and E. B. Bagley. AVI, Westport, CT, USA.
Hunt, R. W. G. (1987). *Measuring Colour*. Ellis Horwood, Chichester, UK.
Iglesias, H. A. and Chirife, J. (1982). *Handbook of Food Isotherms*. Academic Press, New York, USA.
International Dairy Federation (1981). Bulletin No. 135, Evaluation of the Firmness of Butter, Brussels.
International Dairy Federation, (1991). Bulletin No. 268, Rheological and Fracture Properties of Cheese, Brussels.
Jackson, A. T. and Lamb, J. (1981). *Calculations in Food and Chemical Engineering*. Macmillan Press, London, UK.
Jenness, R., Shipe Jr, W. F. and Sherbon, J. W. (1974). *Fundamentals of Dairy Chemistry*, ed. B. H. Webb, A. H. Johnson and J. A. Alford. AVI Westport, CT, USA.
Jowitt, R. (1980). *Hygienic Design and Operation of Food Plant*, Ellis Horwood, Chichester, UK.
Jowitt, R., Escher, F., Hallstrom, B., Meffert, H. F. Th., Spiess, W. and Vos, G. (1983). *Physical Properties of Foods*. Applied Science Publishers, London, UK.
Jowitt, R., Escher, F., Kent, M., McKenna, B. and Roques, M. (1987), *Physical Properties of Foods—2*, Elsevier Applied Science, London, UK.
Jukes, D. J. (1987), *Food Legislation in the UK—A Concise Guide*, Butterworths, London, UK.
Kanekanian, A. D. A (1983) PhD thesis, Reading University, Reading, Berkshire, UK.
Kent, M. (1987), *Electrical and Dielectric Properties of Food Materals*. Science and Technology Publishers, London, UK.
Kent, M. and Smith, G. L. (1987) *Physical Properties of Foods—2*, ed. R. Jowitt *et al.* Elsevier Applied Science Publishers, London, UK.

Kessler, H. G. (1981). *Food Engineering and Dairy Technology*. Verlag A, Freising, Germany.
Kessler, H. G. (1984). *Milchwissenschaft*, **39**(6) 339.
Kinsella, J. E. (1976). *Critical Reviews in Food Science and Nutrition*, **7**, 219.
Kjaergaard-Jensen, G. and Neilsen, P. (1982). *J. Dairy Res.*, **49**, 515.
Kress-Rogers, E. and Kent, M. (1987), *J. Food Engng.*, **6**, 345.
Lamb, J. (1976). *Chem. and Indust.*, **24**, 1046.
Levitt, B. P. (1973). *Findlays Practical Physical Chemistry*, Longmans, Harlow, UK.
Lewis, M. J. (1990). *Physical Properties of Foods and Food Processing Systems*. Ellis Horwood, Chichester, UK.
Loncin, M. and Merson, R. L. (1979). *Food Engineering Principles and Selected Applications*. Academic Press, New York, USA.
Lovell, H. R. (1980). *Milk and Whey Powders*. Society of Dairy Technology, Wembley, London, UK.
MacCarthy, D. (1984). *Engineering and Food* (Vol. 1), ed. B. M. McKenna. Applied Science Publishers, London, UK.
MacDougal, D. B. (1986). *Food Chem.*, **21**, 283.
MacDougal, D. B. (1988). In: *Sensory Analysis of Foods*, ed. J. R. Piggott. Elsevier Applied Science Publishers, London, UK.
Masters, K. (1991). *Spray Drying Handbook*. Longman Scientific and Technical, Harlow, Essex, UK.
Maubois, J. L. (1980). *J. Soc. Dairy Technol.*, **33**(2), 55.
McKenna, A. B. (1988). *NZJ. Dairy Sci. Technol.*, **23**(4), Supplement, 363.
Mettler, A. E. (1980). *Milk and Whey Powders*. Society of Dairy Technology, Wembley, London, UK.
Miles, C. A., Van Beek, G. and Veerkamp, C. H. (1983). *Physical Properties of Foods*, ed. R. Jowitt, *et al.* Applied Science Publishers, London, UK.
Mitchell, J. R. (1980). *J. Texture Studies*, **11**, 315.
Mitchell, J. R. (1987). *Food Technology International Europe*, *249*, Sterling Publications, London, UK.
Mohsenin, N. N. (1980). *Thermal Properties of Foods and Agricultural Materials*. Gordon and Breach, London, UK.
Mohsenin, N. N. (1984). *Electromagnetic Radiation Properties of Foods and Agricultural Products*. Gordon and Breach, New York, USA.
Mohsenin, N. N. (1986). *Structure, Physical Characteristics and Mechanical Properties of Plant and Animal Materials*. Gordon and Breach, London, UK.
Mudgett, R. E. (1982). *Food Technol.*, **36**(2), 109.
Mudgett, R. E., Smith, A. C., Wang, D. I. C. and Goldblith, S. A. (1974). *J. Ed. Sci.*, **39**, 52.
Muller, H. G. (1973). *An Introduction to Food Rheology*. Heinemann, London, UK.
O'Brien, J. M. and Morrissey, P. A. (1989). IDF Bulletin No. 238, Brussels.
Patel, M. T., Kilara, A., Huffman, L. M., Hewitt, S. A. and Houlihan, A. V., (1990). *J. Dairy Sci.*, **73**(6), 1439.
Peleg, M. (1983). *Physical Properties of Foods*, ed. M. Peleg and E. B. Bagley. AVI, Westport, CT, USA.
Polley, S. L., Snyder, O. P. and Kotnour, P. (1980). *Food Technol.*, **11**, 76.

Powrie, W. D. and Tung, M. A. (1976). *Principles of Food Science Part I, Food Chemistry*, ed. O. R. Fennema. Marcel Dekker, New York, USA.
Prentice, J. H. (1954). *Lab. Practice*, **3**, 186.
Prentice, J. H. (1979). *Food Texture and Rheology*, ed. P. Sherman. Academic Press, London, UK.
Prentice, J. H. (1984). *Measurements in The Rheology of Foodstuffs*. Elsevier Applied Science Publishers, London, UK.
Rha, C. (1975). *Theory Determination and Control of Physical Properties of Food Materials*, ed. C. Rha. D. Reidel, Dordrecht, The Netherlands.
Riedel, L. (1976). *Chem. Mikrobiol. Tech. Lebensm.*, **4**, 177.
Rothwell, J. (1968). *Proc. Biochem*, **3**(1), 19.
Ruegg, M., Moor, U. and Blanc, B. (1983). *Milchwissenschaft*, **38**, 10.
Singh, R. P. (1982). *Food Technol.*, **36**(2), 87.
Singh, R. R. B. and Patil, G. R. (1990). *Milchwissenschaft*, **45**, 367.
Smith, K. E. and Bradley Jr, R. L. (1983). *Dairy Sci.*, **66**(12), 2464.
Society of Dairy Technology (1989). *Cream Processing Manual*. Society of Dairy Technology, Wembley, London, UK.
Taneya, S. (1979). *Food Texture and Rheology*, ed. P. Sherman. Academic Press, London, UK.
Troller, J. A. (1983). *Sanitation in Food Processing*. Academic Press, New York, USA.
Troller, J. A. and Christian, J. H. B. (1978). *Water Activity and Food*. Academic Press, New York, USA.
Walstra, P. and Jenness, R. (1984). *Dairy Chemistry and Physics*. John Wiley, New York, USA.
Watt, I. C. (1983). *Physical Properties of Foods*, ed. R. Jowitt, *et al.* Applied Science Publishers, London, UK.
Weast, R. C. (1988). *Handbook of Chemistry and Physics* (69th edn). CRC Press, Cleveland, OH, USA.
Wolf, W., Speiss, W. E. L. and Jung, G. (1985). *Sorption Isotherms and Water Activity of Food Materials*. Science and Technology Publishers, London, UK.

Chapter 6

Modern Laboratory Practice—1: Chemical Analyses

T. Andersen, N. Brems, M. M. Børglum, S. Kold-Christensen, E. Hansen, J. H. Jørgensen, and L. Nygaard

A/S N. Foss Electric, Hillerød, Denmark

Since milk is used as human food, there is a natural interest in its composition. This interest may serve several purposes:

— Nutritive value of dairy products.
— Feeding of cows.
— Payment schemes.
— Processing, yield and product control.
— Breeding of cows.
— Control of hygiene and health status of herds.

A multitude of chemical analytical methods—gravimetric, volumetric, titrimetric—have been developed. A few have been established as reference methods, but others have been found suitable for routine analysis. A general feature of the chemical methods is, however, that analysis of large numbers of samples is inconvenient, instantaneous results are impossible, and cost per analysis is relatively high. Automation of these chemical methods has to some extent remedied the disadvantages, but the biggest leap forward has been made with the appearance of instruments based on the measurement of physico-chemical properties of the milk constituents, with no or very little sample preparation and handling. Typical examples are the infra-red (IR) and near-infra-red (NIR) instruments. These modern analytical methods perform quick, accurate and cheap analyses, making possible extensive control of cows, milk, processing and products.

CHEMICAL ANALYSIS

First a short review of traditional chemical methods will be presented. It should be noted that well-established procedures, like the Röse-Gottlieb method for fat determination and the Kjeldahl method for quantitative determination of total nitrogen and estimation of protein, are still the ultimate methods to be consulted when evaluating new instrumental methods. Calibration and performance checks of modern instruments also require the use of chemical reference methods. A selection of FIL-IDF and AOAC approved methods is listed in Table I.

Fat Analysis

Reference methods

The Röse–Gottlieb principle, a gravimetric method, is accepted as the reference method for fat, and all other methods should be checked against it. The procedure is, in short, as follows:

The sample is weighed precisely into a test tube. Ammonia solution and ethanol are added, and the milk fat is extracted two or three times with a mixture of diethylether and light petroleum. The combined extracts are then evaporated, and the weight of the oven-dried residue—which must be soluble in light petroleum—is determined.

The Mojonnier and Schmid–Bondzynski–Ratzlaff (SBR) methods are similar to Röse–Gottlieb. SBR is preferred when increased amounts of free fatty acids from decomposed milk fat are present, due to processing or to aging of the milk. In this method, hydrochloric acid replaces the ammonia solution.

The above methods are rather slow and time-consuming, and special attention must be paid to the highly flammable solvents used. Results are available after two days, although using a centrifuge to separate the water from the solvent reduces the delay.

Routine methods

The Gerber method, a volumetric method, is often used for routine analysis, using graduated and calibrated butyrometers.

Sulphuric acid and amyl alcohol are pipetted, along with the milk sample, into a butyrometer, which is then shaken and centrifuged. The volume of the separated upper layer of milk fat is read from the butyrometer scale at an elevated temperature.

TABLE I
A selection of (a) FIL-IDF standards and (b) AOAC methods

(a) FIL-IDF standards

Component	*FIL-IDF*	*Principle*	*Status*
Fat	1B:1983	Röse–Gottlieb	Reference
	22A:1983	Röse–Gottlieb	Reference
	16A:1971	Röse–Gottlieb	Reference
	9A:1969	Röse–Gottlieb	Reference
	13A:1969	Röse–Gottlieb	Reference
	116 :1983	Röse–Gottlieb	Reference
	5A:1969	SBR	
Protein	20:1962	Kjeldahl	Reference
	25:1964	Kjeldahl	Reference
	92:1978	Kjeldahl	Reference
	98:1980	Dye-binding	Routine
Lactose	28A:1974	Chloramine-T	
	79:1977	Enzymic	
	43:1967	Cuprous oxide	
Water	21A:1982	Drying-oven	Reference
	4A:1982	Drying-oven	Reference
	15A:1982	Drying-oven	Reference
Somatic cells	Doc. 168:1984	Microscopic count	Reference
	Doc. 168:1984	Coulter Counter	Routine
	Doc. 168:1984	Fossomatic	Routine
	Doc. 168:1984	Auto-Analyzer	Routine

(b) AOAC methods

Components	*AOAC*	*Principle*	*Status*
Fat, protein, lactose, water	16·078–·087	Infra-red absorption	IRMA Milko-Scan Multispec
Fat	16·063–·068 (+1st suppl.)	Light scatter	Milko-Tester Anritsu Milk Checker
Protein	cfr.		
	7·021–7·024	Kjeldahl	Kjel-Foss
	16·042–16·045	Dye-binding	Pro-Milk
Somatic cells	46·086–46·104	Optical	Auto Analyzer
	46·105–46·109	Fluorescence	Fossomatic

FIL-IDF: Federation Internationale de Laiterie—International Dairy Federation, Bruxelles, Belgium. Documents and standards, currently revised.
AOAC: Official Methods of Analysis of the Association of Official Analytical Chemists, 13th edn, 1980, Washington, DC, USA.

The Babcock method, which does not involve amyl alcohol, is a similar, widely used method. These procedures provide quick results, and are reliable if carefully performed; deviations from the reference method are well documented. A disadvantage is the use of sulphuric acid, and the necessity of cleaning the butyrometers.

Protein Analysis

Reference method

The work of Johan Kjeldahl, described in 1883, is used as the basis for determining total nitrogen. One variant of this procedure, adopted by the IDF as a reference method, is as follows:

A precisely weighed quantity of milk is placed in a Kjeldahl flask containing mercury oxide (catalyst) and potassium sulphate (increasing boiling point). Concentrated sulphuric acid is added, and the mixture is heated and boiled until the sample is completely digested yielding NH_4^+. The flask is cooled, and the solidified contents are then redissolved in water. An excess of sodium hydroxide solution (including a sulphide to precipitate the mercury) is added, and the released ammonia is distilled via a condenser into a boric acid solution containing an acid-base indicator. The collected ammonia is titrated with standardised hydrochloric acid. The amount of ammonia present, and thus the amount of nitrogen, can then be calculated. For milk, the empirical factor of 6·38 is used to estimate the percentage of protein in the sample.

A manual Kjeldahl analysis takes several hours, and care must be taken to avoid accidents with the corrosive chemicals involved.

Routine method

The dye-binding method is suited for routine analysis, and dyestuff used is Amido Black 10B, 'milk testing quality'. Raw, whole or homogenised milk or skim-milk with 2–5% protein can be analysed directly. The principle is as follows:

Amido Black in excess forms an insoluble dye-protein complex by reaction with basic amino-acid residues (e.g. lysine with an ε-amino group) in the proteins at a buffered pH of 2·4. The light absorption of the buffered dye solution is recorded.

Milk and the dye solution are mixed, and the insoluble complex is filtered off (or, alternatively, centrifuged). The light absorption of the filtrate containing an excess of dye is measured. The percentage of protein is obtained from a calibration curve, which is prepared by determining protein content by the Kjeldahl method and corresponding read out of light absorption.

Although the Kjeldahl method determines total nitrogen and the dye-binding method quantitates basic amino-acid residues in protein (and not non-protein nitrogen), good correlation can be obtained between the two methods.

Lactose Determination

Many methods are available for the estimation of lactose. Most of them presuppose that no other interfering sugars are present, and some commonly used methods illustrate this.

Titrimetric method with chloramine-T

This method is based on the reducing property of lactose. Fat and milk proteins are precipitated with a tungstic acid reagent. To part of the lactose-containing filtrate, solutions of potassium iodide and a known quantity of chloramine-T are added, and the mixture is left to react Hydrochloric acid is added to stop the reaction, and the unreacted chloramine-T is titrated with standardised sodium thiosulphate solution, using soluble starch to determine the end-point.

This method is applicable to normal fresh milk, and requires the minimum of laboratory equipment.

Copper reduction method (Munson–Walker method)

This method is also based on the reducing property of lactose. Ground cheese is dissolved in water, and the fat and protein are precipitated with Carrez reagents (zinc sulphate and potassium ferrocyanide solutions). Part of the lactose-containing filtrate is added to Fehlings solution (cupric sulphate/alkaline tartaric acid solution). The mixture is then heated and boiled under strictly controlled conditions to precipitate cuprous oxide. The cuprous oxide is collected on a filter crucible and dried. By weighing the amount of cuprous oxide, the content of lactose can be calculated using an empirical factor from a table.

This method is applicable to other milk products that contain no reducing sugars other than lactose.

Polarimetric method

This method is based on the ability of lactose to rotate polarised light. The fat and milk proteins are precipitated (acidic mercuric salt solution), and the optical rotation of the filtered solution, which is proportional to the lactose concentration, is measured. The lactose content can be calculated using the specific rotation. The method can be used for milk to which no other optically active components (sugars) have been added.

Enzymic method

The lactose in the sample is hydrolysed to glucose and β-galactose in the presence of a β-galactosidase (Gal) enzyme. The β-galactose which is formed is oxidised by nicotinamide adenine dinucleotide (NAD^+) in the presence of β-galactose dehydrogenase (Gal-DH). The amount of the reduced form of NAD^+, NADH—as measured by an increase in extinction at 340 nm—is proportional to the amount of β-galactose.

$$\text{(1) lactose} + H_2O \xrightarrow{\text{(Gal), pH 6·6}} \text{glucose} + \beta\text{-galactose}$$

and

$$\text{(2) } \beta\text{-galactose} + NAD^+ \xrightarrow{\text{(Gal-DH), pH 8·6}} \text{galactonic acid}$$
$$+ NADH + H^+$$

The method is applicable to products to which sugars other than β-galactose have been added. Added β-galactose can be determined by reaction (2), but L-arabinose interferes.

Products such as milk, condensed milk, cheese and ice cream can be analysed by dissolving the product, and precipitating the fat and protein with perchloric acid, trichloroacetic acid, or Carrez reagents.

Another enzymatic scheme can also be used:

$$\text{(1) lactose} \longrightarrow \text{glucose} + \beta\text{-galactose} \quad (\beta\text{-galactosidase})$$
$$\text{(2) glucose} \longrightarrow \text{glucose-6-phosphate} \quad \text{(hexokinase, ATP)}$$
$$\text{(3) glucose-6-phosphate} + NADP^+ \longrightarrow \text{6-phosphogluconate} + NADPH \quad \text{(glucose-6-phosphate dehydrogenase)}$$

NADPH-extinction is then measured; added glucose can be determined by using reactions (2) and (3).

Total Solids and Water

The sample is weighed into a flat-bottomed dish and dried in an oven at 102°C for several hours until constant weight. The content of water or of total solids is then calculated by simple subtractions.

Drying time is dependent on the product. Dilution with water, use of sand for even spreading, or preheating of the sample on a steam bath may be necessary (cheese and sweetened condensed milk).

Somatic Cell Count

A direct microscopic count of stained somatic cells (leucocytes and epithelial cells) is recommended by IDF as a reference method. The procedure is, in short:

10 μl of the milk sample is spread in two thin films (of 1 cm^2 each) on a microscopic slide and the films are dried. The fat globules are removed, and the cells are fixed and stained with methylene blue, all in one process. After being rinsed in water and dried, the prepared slide is placed under a microscope with high magnification. By counting the number of stained cells (nuclei) in a large number of microscopic fields, the number of somatic cells per ml milk sample can be calculated. A coefficient of variation of 5% or less at a level of 500 000 cells per ml can be obtained by trained personnel for reference purposes.

MODERN ANALYTICAL METHODS

Today, most traditional, chemical procedures for analysis of the main milk components can be replaced by quick, easy and reliable instrmental methods. Attention will be focussed here on fat and protein analyses, and the count of somatic cells. The choice of instrumentation will be based on considerations such as: price of instrument, number of samples per hour, price per sample, auxiliary equipment required, method of presenting the sample to the instrument, output of data, staff required, and ease of repair and service. An additional factor is the extent to which the instrument must be regularly checked to ensure:

Accuracy: Agreement with a reference method must be acceptable. Recalibration may be necessary, for instance due to seasonal variations in the milk constituents.

Precision, as expressed by repeatability (repeated measurements on the same sample) and reproducibility (between laboratory agreement) must be acceptable (Grappin, 1984).

Some of the options are given in Table II, but before going into detail about the various instruments, a few basic concepts should be reviewed,

TABLE II
Some instruments available for the routine measurement of the components of milk and milk products

Instrument	*Principle*	*Components*	*Manufacturer*
Milko-Tester	Light scatter	Fat	1
Milk Checker			2
Kjel-Foss	Kjeldahl	Protein	1
Pro-Milk	Dye-binding	Protein	1
Milko-Scan	Infra-red	Fat, protein,	1
Multispec		lactose and	4
IRMA		water	3
Instalab	Near infra-red	Fat, protein,	6
Infra Analyzer	Reflectance	lactose and	5
Neotec		water	7
Fossomatic	Fluorescence	Somatic cells	1
Coulter Counter	Conductivity		8

1. A/S N. Foss Electric, Hillerød, Denmark.
2. Anritsu Electric Co., Ltd., Kanagawa, Japan.
3. Sir Howard Grubb Parsons & Co. Ltd, Newcastle upon Tyne, England.
4. Shields Instruments Ltd., York, England.
5. Tecnicon Industrial Systems, Tarrytown, NY, USA.
6. Dickey-john Corp., Auburn, IL, USA.
7. Neotec Corp., Silver Springs, MD, USA.
8. Coulter Electronics Ltd., Luton, England.

N.B. This list is not exhaustive.

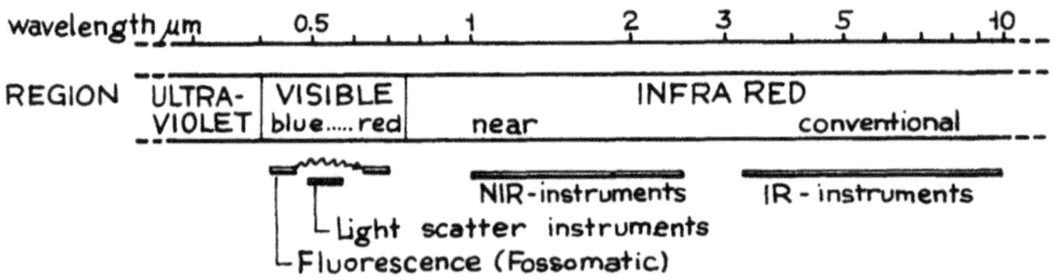

Fig. 1. Part of the electromagnetic spectrum: logarithmic scale.

namely the electromagnetic spectrum (Fig. 1), and the relationships expressed by the Lambert-Beer Law (Figs 2 and 3):

$I = I_0 \times 10^{-\varepsilon \cdot l \cdot c}$ or $A = -\log_{10} I/I_0 = \varepsilon \cdot l \cdot c$, $T = I/I_0$
I_0 = intensity of incident light (radiation)
I = intensity of transmitted light (radiation)
(I_s = scattered light, I_r = reflected light)

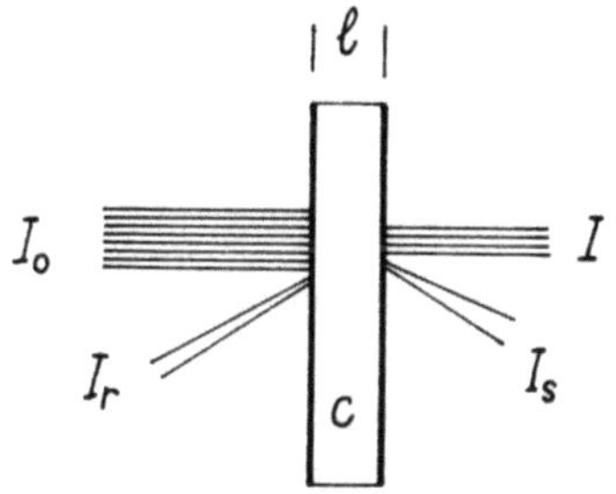

Fig. 2. Curvette with milk, illustrating light reflection (I_r), scatter (I_s) and transmission (I).

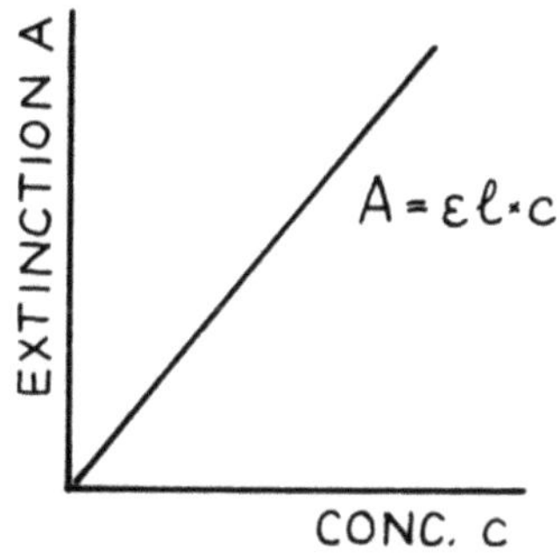

Fig. 3. Graphic representation of the Lambert–Beer Law.

ε = (molar) extinction coefficient, absorption coefficient (varies with wavelength, solvent, temperature)
c = concentration of the solute
l = path (cell) length
A = extinction, absorbance, absorption
T = transmission, transmittance, transparency

Instruments for Fat Analysis

A widely used—and AOAC-approved—method, based on measuring the light scatter caused by milk fat globules in a homogenised milk sample, is represented by the Milko-Tester.

Light scatter photometry for fat determination

When visible light passes through a layer of milk, transmission is highly affected. This fact has motivated researchers to isolate the effect of fat content from the effects of other factors, in order to obtain a photometric test for fat in milk to replace the traditional Röse–Gottlieb and Gerber methods. Butterfat absorbs a negligible portion of energy from visible light, and the effect of fat, as measured by a visible-light photometer, is essentially due to refraction and reflection.

Refraction occurs when light passes from one medium into another with a different refractive index (RI). For butterfat the RI is about 1·454, while the surrounding medium with lactose, proteins and salt dissolved in water has an RI if about 1·33. This makes each fat globule behave like a condenser lens (Fig. 4).

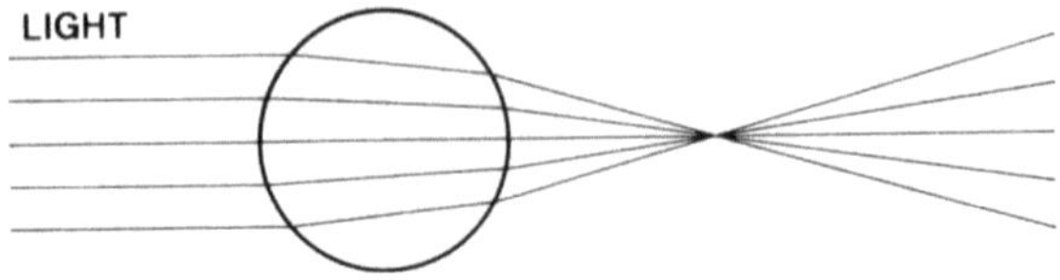

Fig. 4. Light refraction in a fat globule (idealised).

When a thin layer of milk is placed in the light path in a photometer, the transmitted light forms a cone-shaped bundle of beams. The photocell receives a certain central part of these, and the amount is determined by the angle of acceptance of the photocell seen from the cuvette; the outer part of the bundle is lost. This produces a photometer response that reflects the presence of the fat globules (Fig. 5). Casein micelles also affect light scatter, but interference from casein micelles can be eliminated by solubilisation with a chelating agent (EDTA) at pH 9·5–10. This is uncomplicated, and a test on skim-milk confirms that this procedure works well. Even for a condensed skim-milk, casein scatter is completely suppressed.

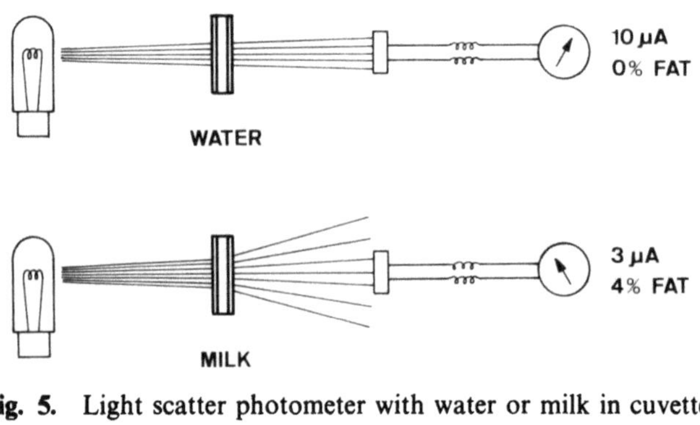

Fig. 5. Light scatter photometer with water or milk in cuvette.

The main problem in converting light scatter effects into a reliable photometric method for fat determination has been to understand and control the relationship between actual fat content, light scatter, fat globule size distribution, wavelength of the light, and the RI's for fat and milk serum. Within this complex problem, the effects of globule size distribution have been the subject of intensive research, theoretical as well as practical. Globule size distribution in raw milk varies significantly between breeds and between individual animals, with fat globule sizes of up to 10 μm. When light of, say 0·6 μm wavelength travels through globules of diameters of several wavelengths, it is a consequence of the RI difference between fat and serum that the light shows a phase shift depending on the distance it travels. Goulden (1958) has illustrated this by a graph of the scattering coefficient for monodisperse emulsions (Fig. 6).

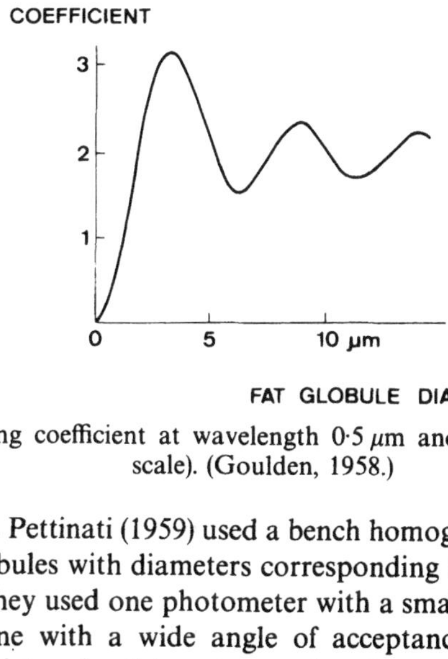

Fig. 6. Scattering coefficient at wavelength 0·5 μm and RI = 1·33/1·45 (not to scale). (Goulden, 1958.)

Haugård and Pettinati (1959) used a bench homogeniser for raw milk to produce fat globules with diameters corresponding to the first linear part of this curve. They used one photometer with a small angle of acceptance and another one with a wide angle of acceptance; a nomogram was constructed to determine fat content and an homogenisation index.

A photometer combining a wide angle of acceptance and a motor operated high-pressure homogeniser adjusted for optimal suppression of the globule size effect has been developed, and this formed a solid basis for the family of 'Milko-Testers', of which the first model was introduced in 1964.

The light scatter fat tester

A few years later improved models were marketed: the Milko-Tester Automatic for payment analysis of herd and cows' milk (180 samples per hour), the Milko-Tester Mk III for raw milk and dairy products, and the Milko-Tester Control for on-line standardisation in the dairy plant.

These instruments illustrate how basic photometric principles can be realised in practice.

(1) A 1·5 ml milk sample is diluted tenfold with a 5% EDTA solution at pH 9·5, and this perfectly suppresses caesin scatter, which otherwise would interfere with fat determination. The dilution also makes it possible to use a wide light path (0·4 mm) in the curvette, which can then be used for butterfat contents of up to 10% without compromising the validity of the Lambert-Beer Law. The combination of a tenfold dilution by a basic solution and the 0·4 mm space of the cuvette eliminates the risk of the cuvette being blocked by acidified milk samples.

(2) High pressure homogenisation is carried out in four consecutive steps of 40 kp each at a stabilised temperature of 50 or 60°C. The Milko-Tester Minor has a simplified one-stage, hand-operated homogeniser which gives slightly reduced peformance.

The homogenisation changes the globule size distribution:

Parameter	*Raw milk*	*Homogenised milk*
Average size, d_{vs}, μm	3–4	0·3–0·4
Standard deviation, μm	1–2	0·2–0·25

This serves two purposes: variations in globule size distributions between individual animals and between breeds are suppressed, and all globules are reduced to a size where they contribute to light scatter with the same weight as they contribute to fat content, which is the meaning of the linear part of the curve shown in Fig. 6. This agreement holds for all kinds of raw milk samples, but for homogenised products, some Milko-Testers provide a calibration that is different from its raw milk calibration by about −0·3% fat. This calibration is essentially unaffected when products from different homogenisers are tested, even if they have been homogenised at different pressures.

(3) Light scatter is determined by a photometer with a wide angle of acceptance, empirically designed for minimising the effect of globule size variations. Monochromatic light is not necessary, but a blue filter is used in some models to keep near-infra-red radiation from disturbing the stability of the photocell.

(4) Electronic data processing is used to produce the log function of the Lambert-Beer Law, to provide adjustments of linearity and sensitivity, and for analog-to-digital conversion.

Performance and limitations

A repeatability of 0·01–0·02% fat is typical, and accuracy, compared to the Röse–Gottlieb reference method, is typically 0·06% fat for single cow samples, and 0·02% for bulk milk samples. For high-fat products, standard deviations of 1.0–1·5% of the actual fat content can be expected. The Milko-Tester Minor has slightly higher standard deviations for precision and accuracy.

Sample preparation may be necessary: acidified products must be neutralised, condensed milk and cream must be diluted, powder must be dissolved, and cheese must be grated, dissolved and homogenised. In some cases, specific calibrations may be needed. This requirement applies first of all not only to homogenised products, but also to products containing additives that directly influence the optical density, such as chocolate, coffee and fruit flavours. Vegetable emulsions can be tested if a specific calibration is applied, but mixes of butterfat and vegetable fat can only be tested if the ratio between them is known and fixed. Air bubbles in the prepared samples must be avoided.

The accuracy and stability of the calibration are affected by the refractive index (RI) of the butterfat. The major part of the standard deviation of 0·06% for the accuracy of single cow samples is due to RI variations, and this explains why accuracy is better for herd milk, and much better for bulk milk. The practical aspect of this is that seasonal RI variations are reflected as calibration problems for the fat tester. A typical winter RI can be as low as 1·4530, while the summer RI may be 1·4560. For a typical raw milk with 4 per cent fat, this means a difference of nearly 0·2% far between summer and winter calibrations, making it necessary to re-calibrate the instrument each spring and autumn. This is just as important for bulk milk tests as for herd and single cow milk tests.

A minor limitation of the light scatter fat tester is that it measures globular fat only, but for good milk samples it is no problem to calibrate the fat tester using Gerber as a reference method. However, for samples that have been stored for a long time with inadequate preservation, the appreciable amounts of free fatty acids that can form are sufficient to explain the dramatic deviations that may be observed between fat tester and Gerber analyses. The light scatter fat tester is excellent for skim-milk tests in the dairy plant to check separator performance.

Instruments for Protein Analysis

One type of instrument is an automated or semi-automated version of the Kjeldahl procedure which measures total nitrogen.

Another type of instrument is based on the dye-binding procedure: Amido Black forms an insoluble complex with protein. Although an indisputable, versatile analysis method, the Kjeldahl analysis in dairy chemistry is mainly used to obtain reference values for protein. For routine analysis, quick and reliable alternative methods designed especially for milk analysis are available. Dye-binding is one alternative, and another, infra-red absorption, will be described later.

Automated Kjeldahl

Operation

In the Kjel-Foss Automatic (Fig. 7), six Kjeldahl flasks are placed in a carousel in one part of the instrument. A titration unit, containers with the necessary solutions of chemicals, and the electronics for operation

Fig. 7. Kjel-Foss Automatic for Kjeldahl analysis. (Courtesy of A/S N. Foss Electric, Denmark.)

occupy the other part. The Kjeldahl flasks rotate among the six positions with a holding time of three minutes at each position:

Position 1. Add three tablets containing catalyst and potassium sulphate to flask. Press dispenser arm for addition of approx. 10 ml hydrogen peroxide and 10–16 ml concentrated sulphuric acid (the volume depends on the fat content of the sample). Add precisely weighed sample: milk is added from a syringe; solid dairy products are wrapped in nitrogen-poor paper and dropped in. Close flask with lid.

Positions 2 and 3. Sample is digested at about 410°C. Heart is supplied by a gas burner in each position and after 2 × 3 min, the sample is fully digested. A fan at position 4 cools the flask and its contents. At the end of the 3 min period, the lid is opened and 140 ml of water is added for dilution.

Position 5. A solution of sodium hydroxide and sodium thiosulphate (50 ml) is dispensed into the flask. Steam is led to the flask and the ammonia is distilled through a condenser into a flask containing 50 ml of acid-base indicator solution. The indicator, which is originally red, gradually turns green as ammonia is absorbed. The original red colour is recorded by a photocell. When a colour change is registered, 1 per cent sulphuric acid will be added until the original red colour is regained. The amount of dispensed sulphuric acid is registered, and from this figure, the total nitrogen is calculated and displayed or printed.

A choice is possible between '0·5 g' or '1·0 g' as the sample size, and the results can be expressed in percentage nitrogen or percentage protein. Three built-in factors for calculating protein are available (one for dairy products, one for food products and grains, and one especially for wheat); a fourth may be defined by the user for special purposes.

The Kjeldahl flask is emptied in position 6, ready for another sample in position 1. The total time for analysis of one sample is 12 min. With the Kjel-Foss running continuously, a result will be displayed every 3 min corresponding to 100 samples every 5 h. Start-up and calibration last about 30 min.

The Kjel-Foss Automatic is suitable for the analysis of a great number of samples, but a semi-automated system may suffice if smaller numbers of samples are involved. Several units are available for facilitating the Kjeldahl analysis, including digestion units, distillation and titration units. A very long list of products ranging from earth and blood to food, grain and animal feeds can be analysed, but here we are mainly interested in milk and dairy products. Although protein, and thus nitrogen, is rather low in milk compared to other products, precision is good (CV less than

1 per cent), and accuracy and correlation to the manual, classical Kjeldahl method is completely satisfactory.

Most improvements of the Kjel-Foss Automatic have focussed on reducing safety risks, and on the growing concern about pollution by mercury. An alternative catalyst based on antimony (although not a completely harmless replacement) has been developed, and its properties are comparable to those of the mercury catalyst.

Semi-automated dye-binding method

Principle

This is the dye-binding method, using Amido Black, described in the 'Chemical Analysis' section.

Operation

The equipment, a Pro-Milk, is shown in Fig. 8. The procedure is, in short: turn the mixing assembly into the neutral position: place a piece of filter paper on the support and mount the transparent tube on top.

Fig. 8. Pro-Milk equipment for semi-automated determination of the protein by dye-binding. From left to right: read-out unit, percentage protein displayed; measuring unit with photometer; mixing and filtering assembly with tube and filter assembly and rubber ball; syringe for 1 ml milk sample: dispenser and container for dye solution. (Courtesy of A/S N. Foss Electric, Denmark.)

A milk sample (1 ml) is drawn-up into the syringe and dispensed into the tube. Dispense dye solution (20 g) into the tube, close the tube with a stopper, and tilt the mixing assembly two or three times for thorough mixing. The mixed content is filtered by applying air pressure to the tube with the rubber ball. The pressure is then released, and the protein content—as calculated by the reduction in blue colour—is read by a photometer in the measuring unit. The result, in percentage protein, is displayed on the read-out unit.

After removal (and cleaning of the tube and filter paper), the assembly is ready for a new sample.

The Pro-Milk must be regularly checked against a standard dye solution and the actual working solution. Its measuring range is 2·0–5·5% protein.

Products and performance

Directly applicable products are raw milk, homogenised whole milk and skim-milk. Other milk products like whey protein, ice cream mix and yoghurt can be examined following appropriate sample preparation such as dilution, precipitation or neutralisation (A/S N. Foss Electric). Investigations of the possibilities for applications have been carried out by the National Dairy Reseach Centre, Ireland (O'Connell and McGann, 1972; McGann and O'Connell, 1972; McGann *et al.*, 1972a; McGann *et al.*, 1972b) and the Institut für Chemie der Bundesanstalt für Milchforschung, Kiel, Germany (Thomasow *et al.*, 1971).

Repeatability, expressed as a coefficient of variation, is better than 0·5% on milk, and the standard deviation for accuracy against Kjeldahl is 0·045%. Fifty to sixty milk samples can be analysed per hour by trained personnel. Fully automated instruments based on the dye-binding principle were formerly produced by A/S N. Foss Electric, but development of IR instruments for measuring fat and protein (and lactose) have made these superfluous.

Multi-component Analysis

Two types of instrument are available for multi-component analysis of dairy products.

One type, the infra-red (IR) instrument, makes use of the characteristic bands of absorption of the main milk components in the 3–10 μm wavelength range of the infra-red spectrum. Another type, the near-infra-red reflectance (NIR) instrument, is based upon diffuse reflectance in the near-infra-red range (more precisely 1–2·5 μm), of the spectrum, IR instruments are used for liquid or liquefied product applications, mainly raw milk, while NIR instruments may be the method of choice for solid materials, such as milk powders. The most versatile models of both types of instruments can determine both fat, protein, lactose and water/solids.

Infra-red instruments

The development of infra-red (IR) instruments for the determination of milk composition on a large scale started in 1961, when Goulden applied for a United Kingdom patent for the quantitative analysis of milk by infra-red absorption (Goulden, 1961). Goulden's work provided the basis for the first commercial instrument, the IRMA (Infra-red Milk Analyzer) developed by the firm Sir Howard Grubb Parsons & Co. Ltd. The method employed in the IRMA was officially adopted by the Association of Official Analytical Chemists (AOAC) in 1972 (Biggs, 1972), and since then, a number of IR instruments have been developed and marketed. Infra-red instruments are now used all over the world, both for centralised payment and dairy herd improvement tests, and for the analysis of milk and milk products at dairy plants.

Infra-red absorption

Vibrational energy changes in molecules occur in the infra-red region of the electromagnetic spectrum. The most useful vibrations, from the point of view of the organic chemist, occur in the narrow range of 2·5–16 μm (1 μm $= 10^{-6}$ m). Functional groups have characteristic vibration frequencies within well-defined regions of this range. All infra-red milk analysers are basically spectrophotometers which measure the amount of infra-red energy transmitted through a sample. When the molecule is exposed to infra-red radiation at a frequency similar to that of the vibration within the atomic group, infra-red energy will be absorbed as a result of this oscillation.

The main interference effects of the IR method are due to the fact that water has a relatively high extinction coefficient in the IR region of the spectrum. Water is probably the strongest known infra-red absorbing compound, and even thin films show intense absorption throughout most of the 2·5–16 μm region.

Figure 9 shows three spectra recorded in a classical double beam infra-red spectrophotometer. Figure 9(A) is the spectrum obtained from water using a narrow path length (39 μm) cell in the sample beam and atmospheric air in the reference beam. Figure 9(B) shows the spectrum

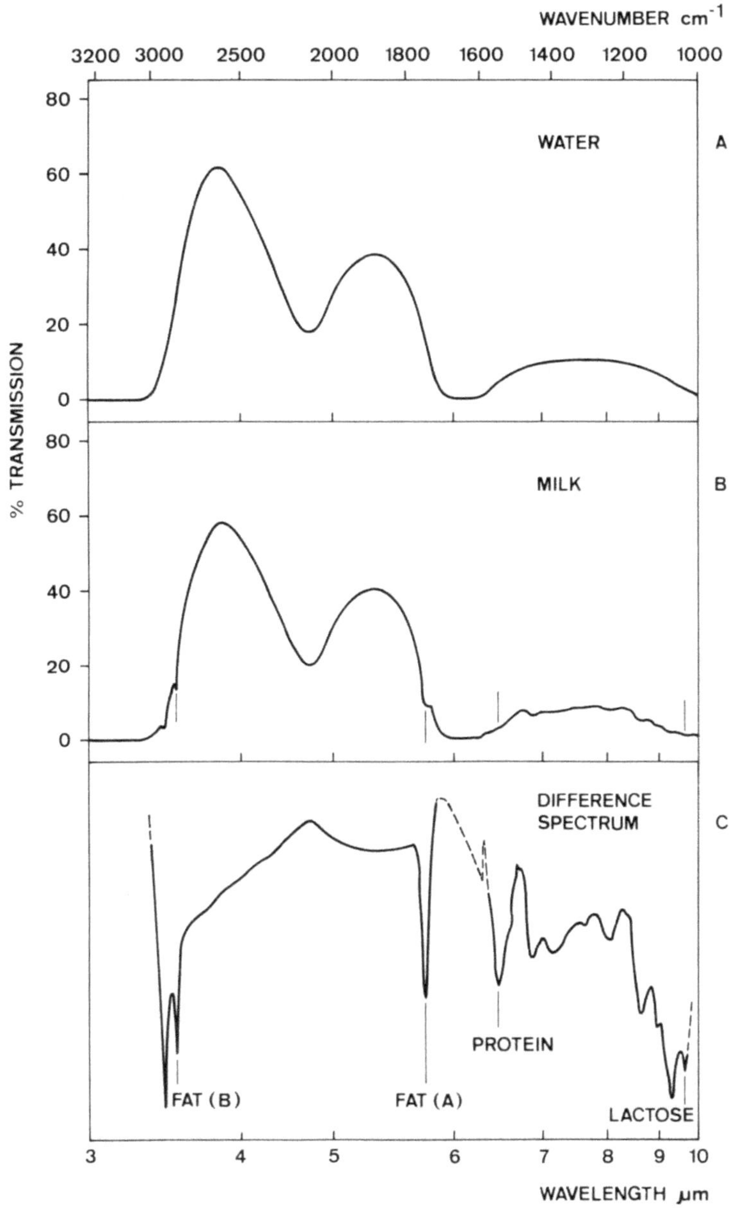

Fig. 9. IR spectra of water versus air, milk versus air, and milk versus water (see text).

obtained from milk under the same conditions, and as can be seen, the spectrum of milk is almost identical to that of water. This is due to the fact that 85% of milk is water. However, small differences can be seen around 3·5, 5·7, 6·5 and 9·6 μm. These differences are magnified in the last spectrum. Figure 9(C), which is a difference spectrum for milk versus water. This spectrum is obtained by placing a milk sample in one beam of the spectrophotometer and a water sample in a matching cell in the other beam. The instrument automatically subtracts the water absorption from the milk spectrum. The resulting spectrum shows the bands due to absorption by fat, protein and lactose at 3·5 and 5·7, 6·5 and 9·6 μm respectively.

The alternative wavelength at 3·5 μm for fat determination has been introduced recently (Nexø *et al.*, 1981).

Interfering components and water displacement

The main interfering effect of the IR method is explained by the high extinction coefficient of water in the whole IR region of the spectrum. Changes in the level of water concentration affect the readings obtained from all components, but since water concentration is dependent on the amount of other components present, equations which define the concentration of one component as a function of the readings obtained from all three components have been successful in correcting for these interference effects. For example, take the equation $F = bF_i + cP_i + dL_i$, where F is the fat content of the sample, and F_i, P_i and L_i are the instrument signals for fat, protein and lactose. As it reflects the absorption of energy by fat, b is large, whereas c and d are small, because they correct only for the interference effects of protein and lactose respectively.

Water absorption is very strong in the spectral region of interest. So the net result at any selected component wavelength is usually that the other two components, because they displace a more strongly absorbing medium, impart a negative displacement or absorptivity error to the result. There are two exceptions to this rule. One is protein absorption at the lactose wavelength, which is slightly stronger than the water it displaces (Shields, 1975). The other exception is at 3·5 μm, where the absorption of water is not so strong and the secondary components (protein and lactose) increase the total absorbancy additively (Sjaunja, 1982). Mineral components also interfere, partly because they displace water, but also because some of them can change the absorptivity of water. However, minerals are present in low and almost constant concentration in normal cow's milk, and can be accounted for when the instrument is calibrated.

Light scatter

In addition to attenuation by absorption, a beam of radiation passing through milk is attenuated by scattering from both fat globules and protein micelles (Goulden, 1964). Light scatter is proportional to $(d/\lambda)^4$, where d is the particle diameter and λ is the wavelength (Jøndrup, 1980). This means that the light scatter increases with increasing particle size and decreasing wavelength (Shields, 1975). To avoid light scatter, the particle diameter must as a rule be less than 1/3 of the wavelength, which for milk means a particle diameter less than 1·5 μm (Jøndrup, 1980). Small and homogeneous particle sizes are obtained by homogenising the milk samples before the absorption is measured. The most effective homogenisation is achieved when the milk fat is melted before the sample is introduced to the instrument. This is one of the reasons why samples must be pre-heated to 40°C before they are analysed. Another reason for pre-heating is that it can be difficult to make representative subsamples of cold raw milk where the milk fat often gathers at the top of the sample.

Instrumentation

There are different ways to compensate for the strong water absorption in the milk. Here three principles will be mentioned, and the types of instrument to which they are applied are briefly described. The different principles are shown schematically in Fig. 10.

Single-wavelength double-beam double-cell principle

The first infra-red milk analyser, the IRMA, was based on a conventional double-beam spectrophotometer modified to automatically select, in turn, characteristic absorption wavelengths for the components in the milk to be analysed. In this way, the absorption of a milk sample at a specific wavelength was compared to that of water at the same wavelength.

In the IRMA, the energy from the infra-red source is divided optically into two beams: one, the sample beam, passes through a cell containing milk; the other, the reference beam, passes through a matching cell containing distilled water. The two beams are then recombined at the entrance slit of a monochromator by a reciprocating mirror which transmits the energy of the alternate beams at a frequency of 10 Hz. The monochromator, which includes a diffraction grating and a KBr prism, filters the energy into narrowly selected ranges of wavelengths, which are focussed onto a thermocouple that serves as a detector. Because of the difference in absorption by milk and water, there is an alternating energy level in the radiation reaching the detector. The resultant tiny voltage is

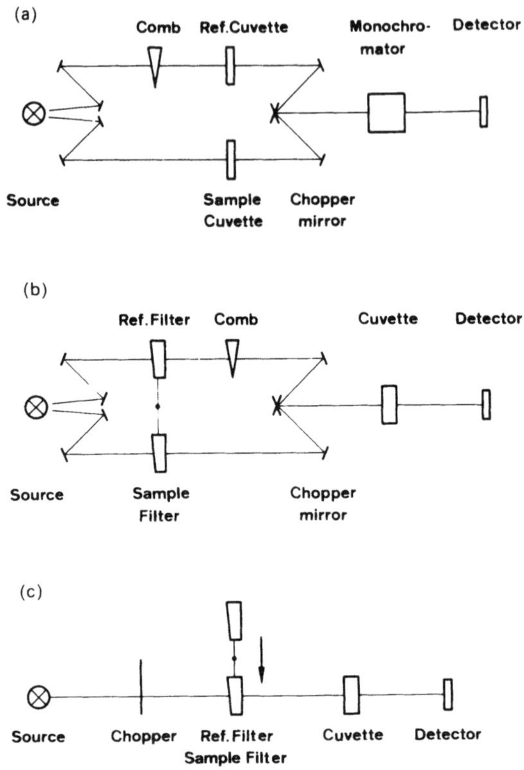

Fig. 10. Optics in various IR instruments. (a) Single-wavelength double-beam double-cell; (b) dual-wavelength double-beam single-cell; (c) dual-wavelength single-beam single-cell.

amplified, rectified and used to drive a servo motor which moves an attenuating comb into or out of the reference beam until equal energy is detected in the two beams. The distance moved by the comb reflects the concentration of the milk component in the sample cell. A potentiometer fixed to the shaft of the attenuator translates the required beam attenuation into a corresponding DC voltage, which can be used to produce a read-out.

Dual-wavelength double-beam single-cell principle

In contrast to the principle mentioned above, the dual-wavelength system does not use a water cell for reference. Instead, a wavelength is chosen close to the sample wavelength where neither fat, protein nor lactose contributes to the signal.

In 1975, the first single-cell dual-wavelength infra-red milk analysers were introduced (Milko-Scan 203 and 300, Fig. 11). The same principle is utilised in the Multispec analysers (Shields, 1982). In these types of instruments, the radiation from the source is also divided into a sample beam and a reference beam. The beams are passed through interference filters—a sample filter and reference filter, respectively—which select narrow bands of wavelengths. The sample wavelength is characteristic for the actual component being measured, and the reference wavelength is, preferably, close to this, but not coincident with either fat, protein or lactose absorption.

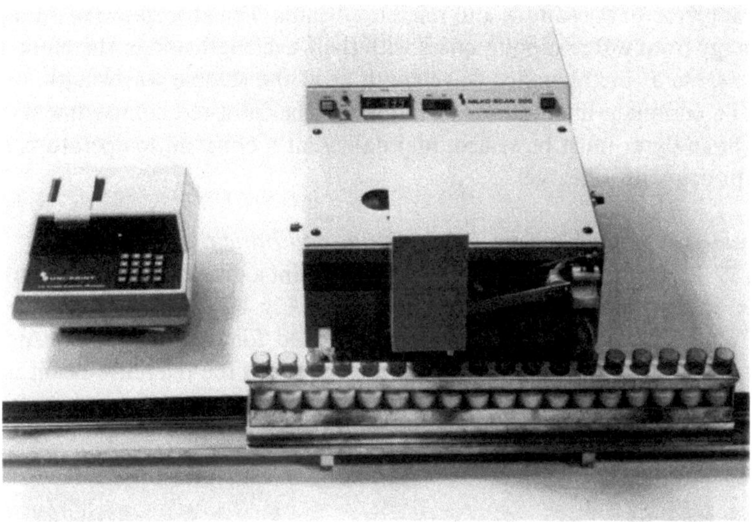

Fig. 11. 1975 version of IR instrument (Milko-Scan 300) for automatic fat and protein analysis. Measures 300 samples per hour. (Courtesy of A/S N. Foss Electric, Denmark.)

The two beams are recombined using a rotating chopper mirror, and then passed through a narrow-path length cuvette containing the homogenised milk sample. The transmitted energy is collected by a mirror and focussed onto the detector. A servo system and an attenuating comb are used to balance the energy of the two beams, just as in the IRMA, and the signal generated is proportional to the movement of the comb. Following a logarithmic-to-linear conversion, the signals from each component are stored and used in pre-determined equations to electronically estimate component concentrations. There are several advantages of this system compared to the IRMA. One is that there is a water displacement effect in both beams, which to a great extent balances out variations in the water or solvent content of the sample. Another advantage of the dual-wavelength system is that the measurements are less sensitive to the degree of homogenisation, because light is scattered in both beams. However, the refractive index of butterfat varies in the region of an absorption band, and as light scatter depends on the refractive index, a difference arises between the sample and the reference beam, depending on the fat globule size. Finally, having only one cell means that any dirt in the cuvette windows will attenuate the sample and reference beam equally. A disadvantage of the dual-wavelength system is an incomplete compensation for water vapour absorption in the sample and reference beams. The absorption of infra-red energy from water vapour varies with the wavelength and is, therefore, not the same at the reference wavelength as at the sample wavelength.

To eliminate fluctuations in humidity, the infra-red compartment of a milk analyser must be sealed, maintained at a constant temperature, and kept dry with silica gel.

Dual-wavelength single-beam single-cell principle

In 1979, a principle was introduced that meant a dramatic simplification of the optical system (Milko-Scan 100 series, Fig. 12).

Where previous instruments compensated for the background 'noise' (water absorption) by attenuating the energy in the reference beam until

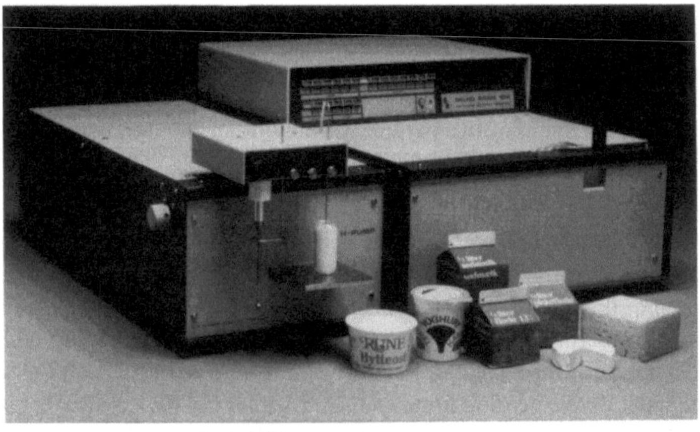

Fig. 12. 1983 version of IR instrument (Milko-Scan 100) for three- or four-component analysis. Measures 90 samples per hour (fat, protein, lactose), manually presented to instrument. (Courtesy of A/S N. Foss Electric, Denmark.)

null-balance with a comb was obtained, the read-outs of the single-beam instrument are calculated electronically from the ratio between the sample wavelength energy and the reference wavelength energy. The interference filters are arranged in a filter wheel which rotates to bring each filter in turn—two filters per component—into the path of the infra-red beam (Nexø *et al.*, 1980). In a single-beam instrument, the number of mirrors is reduced to two, which in turn reduces the travelling distance of the infra-red radiation to about 27 cm. This reduced light path, and the fact that the sample and reference beams follow exactly the same path, make this type of instrument less sensitive to water vapour than the earlier double-beam versions. The high stability and accuracy of the single-beam instruments is due to the reduced number of mirrors, the fact that the mechanically moving servo comb is no longer necessary, and to improved interference filters and electronics.

Reliability of data

Fat determination. As mentioned before, two different wavelengths, 5·7 μm and 3·5 μm, can be used to determine fat in milk. The absorption at 5·7 μm is due to stretching vibrations in the C=O bonds of the carbonyl group in fat, and this measurement 'counts' the number of fat molecules regardless of the length and weight of the individual fatty acids (Fig. 13).

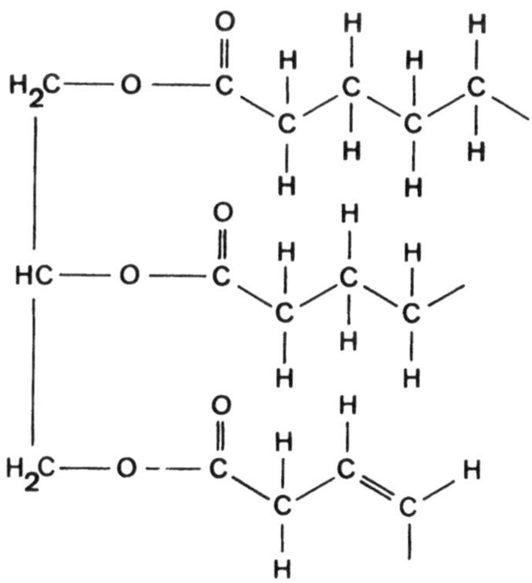

Fig. 13. Structure of butterfat molecule.

If the average chain length (mean molecular weight) of the fatty acids is changed, the number of triglyceride molecules per unit weight will change too, and an error will occur in the results unless the change is compensated for by recalibrating the instrument. The composition of butterfat varies with season, region, breed, cow and stage of lactation (Grappin and Jeunet, 1972), and this means that an instrument using the 5·7 μm filter must be recalibrated when, for instance, going from winter to summer.

The absorption at 3·5 μm is due to stretching vibrations in the saturated C-H bonds of the fatty acid chains. This measurement is, therefore, related to both the size and the number of fat molecules in the sample, as the number of carbon-hydrogen bonds increases substantially in proportion to the molecular size. Both $-CH_3$ and $-CH_2$ groups absorb infra-red energy at 3·5 μm, but the C-H stretching is markedly reduced by the presence of double bonds adjacent to these groups. The absorption decreases as a function of the degree of unsaturation (Mills and Van de Voort, 1982), but even so, the 3·5 μm determination is less sensitive to variations in refractive index than the 5·7 μm determination is, because it reflects the variation in chain lengths (Fig. 13).

Another advantage of the 3·5 μm wavelength is that the measurement includes free fatty acids that may have formed during storage; these cannot be measured at 5·7 μm. Protein and lactose contribute to the absorption at 3·5 μm, but their interference is removed by means of suitable correction constants in the equation that calculates the fat content from the instrument signals. The difference in result between the two wavelengths is, in many cases, negligible, but analyses of individual cow's milk should be performed using the 3·5 μm filter for optimal performance.

Protein measurement. The wavelength for protein determination is 6·5 μm, and it is the nitrogen–hydrogen bonds within the peptide bonds that are responsible for the IR absorption (Fig. 14). Thus, the measurement represents the number of amino acids rather than their weight, but as the composition of protein in milk is fairly constant, this causes no problems.

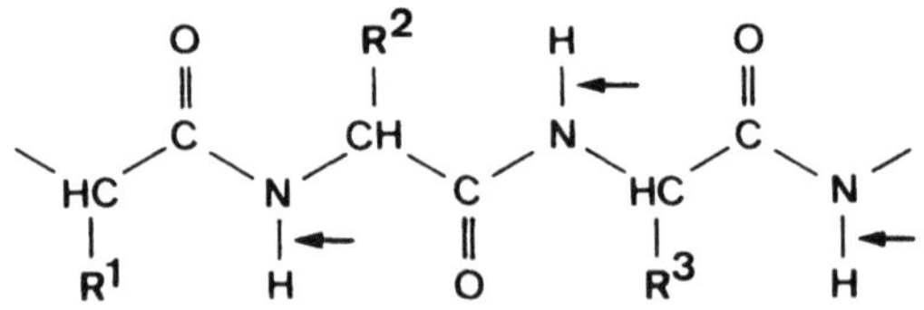

Fig. 14. Primary structure of protein.

In contrast to the Kjeldahl analysis, which is the reference method, the infra-red measurement does not include non-protein-nitrogen.

Lactose measurement. Lactose is measured at 9·5 μm, and the absorption is mainly due to bending vibrations in the C–OH bonds (see Fig. 15) (Goulden, 1956).

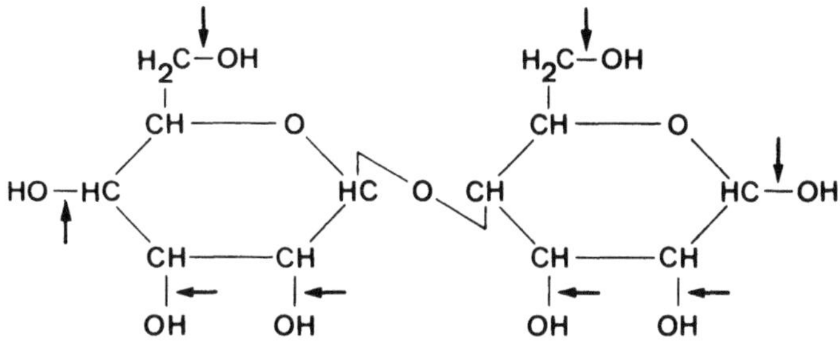

Fig. 15. Lactose.

Accuracy of results

Infra-red milk analysers have been tested and approved in several countries, and all of them have obtained AOAC approval (Biggs, 1972, 1978, 1979a, b; Van de Voort, 1980). Biggs has proposed a set of performance specifications based on manufacturers' claims (Biggs, 1979), and instruments with performances corresponding to these specifications automatically comply with the requirements of the AOAC-approved method. The specifications are based upon analysis of eight different milk samples, and are as follows:

> Repeatability (standard deviation of duplicate instrument estimates): fat, protein and lactose: SD ⩽ 0·02%, conc. 2–6% and total solids: SD ⩽ 0·04%.
> Accuracy (standard deviation of differences between instrument estimates and values found by reference methods):
> fat, protein and lactose: SD ⩽ 0·06%, conc. 2–6% and total solids: SD ⩽ 0·12%.

The best accuracy is obtained with instruments that have been calibrated for the type of milk to be analysed (herd, individual cow, homogenised, unhomogenised, consumers' milk, etc.).

Operation of infra-red instruments
The operation of infra-red instruments is simple. In central laboratories where thousands of analyses are performed each day, fully automated milk analysers (Fig. 16) are used, whereas dairy plants with a limited number of samples prefer semi-automatic instruments. In both types of instruments, the zero-point and calibration must be checked regularly to ensure that no instrument drift takes place.

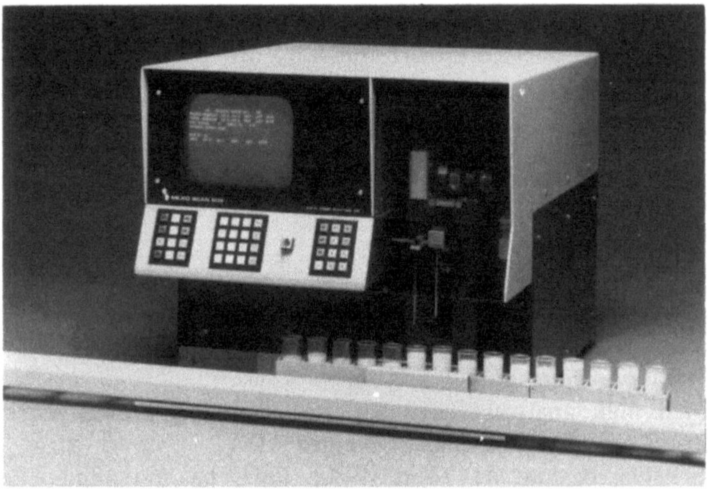

Fig. 16. 1984 version of programmable IR instrument (Milko-Scan 605) for automatic analysis of up to five components. Measures 600 samples (fat only) or 360 samples (fat, protein, lactose) per hour.

The zero-point is checked with demineralised water, and the calibration check is performed with pilot samples that have a known content of fat, protein and lactose.

Samples must be preheated to 40°C in a waterbath, and gently shaken or stirred to ensure an even distribution of the butter fat before they are presented to the instrument. Pushbuttons on the control panel of the instrument allow different analysis programs to be selected, depending on which parameters the user wants to measure. Most infra-red instruments can electronically calculate total solids (TS) or solids-non-fat (SNF) from the measured contents of fat, protein and lactose, plus an added constant for mineral content. Certain instruments are also provided with a pair of filters for the direct determination of water.

Analysis of dairy products other than fluid milk

Any dairy product which is liquid, or can be converted into a liquid form, can be analysed in an infra-red instrument, provided that the TS content is low enough for the pump to work, and that the concentration of each single component does not exceed about 15%. Soured milk products must be neutralised to redissolve the precipitated proteins; whipping cream must be diluted to bring the fat content down to within the instrument's measuring range; cheese must be grated, neutralised, diluted and homogenised; and milk powders must be dissolved in water. Because of the sample preparation (dilution, neutralisation) and/or the presence of lactic acid and various additives, the read-outs for most dairy products other than milk must be corrected. Lactic acid, for instance, has absorption bands near the wavelengths for protein and fat (5·7 μm).

Infra-red milk analysers have gradually replaced turbidity tests for fat (e.g. Milko-Testers) due to the increased interest in protein content, and they have fostered new applications of fat, protein and lactose determinations, where tedious and time-consuming standard analyses previously limited the availability of these data. The newest micro-processor-controlled, infra-red instrument analyses 360 samples per hour for fat, protein, lactose and solids, and needs as little as 3·6 ml of milk per analysis. If the only results needed are for fat, the measuring speed of this instrument is 600 samples per hour (Fig. 16).

Somatic Cell Count

Mastitis

Mastitis is a complex disease of various degrees and consequences. There are many steps from the normal and healthy cow to the cow affected by serious mastitis. Two main diagnoses are:

1. Clinical mastitis: visible changes in the milk (acute or subacute).
2. Subclinical mastitis: no visible changes in the milk.

It is rather easy to detect the first group. The farmer will often have the cow treated with antibiotics, leading to a cure. Subclinical mastitis, on the other hand, is difficult to detect, and many investigations have shown that about 40% of the average herd is subclinically infected.

Mastitis is due to the penetration of bacteria into the udder, and the host reacts to this by trying to combat the infection. This is done simultaneously in several ways. The main way is by secretion of white blood cells (leucocytes) from the udder into the milk, where an elimination of bacteria takes place. Consequently, by counting the number of these white blood cells, one can obtain a measure of how serious the infection is.

IDF classification of quarter milk samples is given in Table III.

TABLE III

Classification	*Somatic cell count*	*Pathogen present*
Normal secretion	< 500 000 per ml	no
Non-specific mast.	> 500 000 per ml	no
Latent infection	< 500 000 per ml	yes
Mastitis	> 500 000 per ml	yes

Economic aspects

It is very well known that both clinical and subclinical mastitis induce a remarkable decrease in milk yield, and that this decrease is, in most cases, irreversible. It has also been shown that there is a very close connection between the number of somatic cells and yield. Not only is yield related to somatic cell count (SCC), but the milk constituents will also be influenced; the per-volume content of fat, casein and lactose decreases as the SCC increases (Fig. 17).

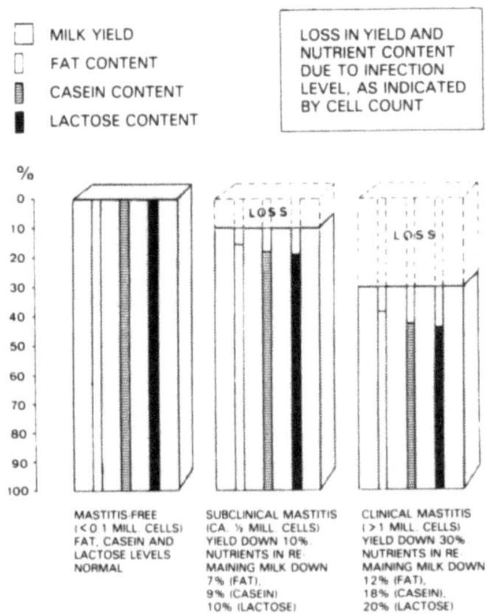

Fig. 17. (Courtesy of A/S N. Foss Electric, Denmark.)

Fossomatic

Principle of electronic somatic cell counting

A/S N. Foss Electric, Denmark, has developed a fast and automated method to measure the number of somatic cells in milk. The method is called fluoro-opto electronic cell counting. A milk sample is warmed to 40°C, stirred, and automatically pipetted, with a hot buffer solution, to a chamber in a revolving rack carousel. A dye solution (containing ethidium bromide) is added (Fig. 18). In the next carousel position, the mixture is stirred, and by means of air pressure, part of the solution is passed into a microsyringe and transferred to the edge of a vertical rotating disc. The tiny film of diluted milk, 0·5 mm wide and 10 μm thick, is then passed under a microscope. By means of a lens system and dichroic mirror, blue light (400–500 nm) is passed to the milk film. The dyed cells in the milk then fluoresce, emitting red light which is transmitted through the microscope to a photo-multiplier. The impulse from each cell is counted, and the total count is displayed, calculated as the number of somatic cells per ml milk. After each sample, the tubing is rinsed, and the mixing chamber in the carousel is washed and dried. The rotating disc is constantly rinsed and dried; the rinsing liquid is heated before use.

Two versions of the Fossomatic instrument are available: a fully automated version that can count cells in 180 or 215 samples per hour, and a semi-automated version (with a flow system different from that shown in Fig. 18) that can handle about 90 samples per hour; both are able to count up to 10 000 000 cells per ml. Repeatability is good, with the coefficient of variation as good as 4–5%. The standard deviation, compared to direct microscopic somatic cell count (DMSCC), is less than 10%; the correlation with DMSCC is 0·960. The precision and accuracy must be checked daily using pilot milk samples.

Coulter Milk Cell Counter

Principle

The Coulter principle is based on the difference in electrical conductivity between milk particles and added diluents, in that while the milk particles act as insulators, the diluent acts as a good conductor. By means of two immersed electrodes, an electrical current is established over a small aperture, and the milk particles suspended in an electroltye are forced through the aperture. As each particle displaces the electrolyte in the aperture, a pulse proportional to the particle volume is produced. After

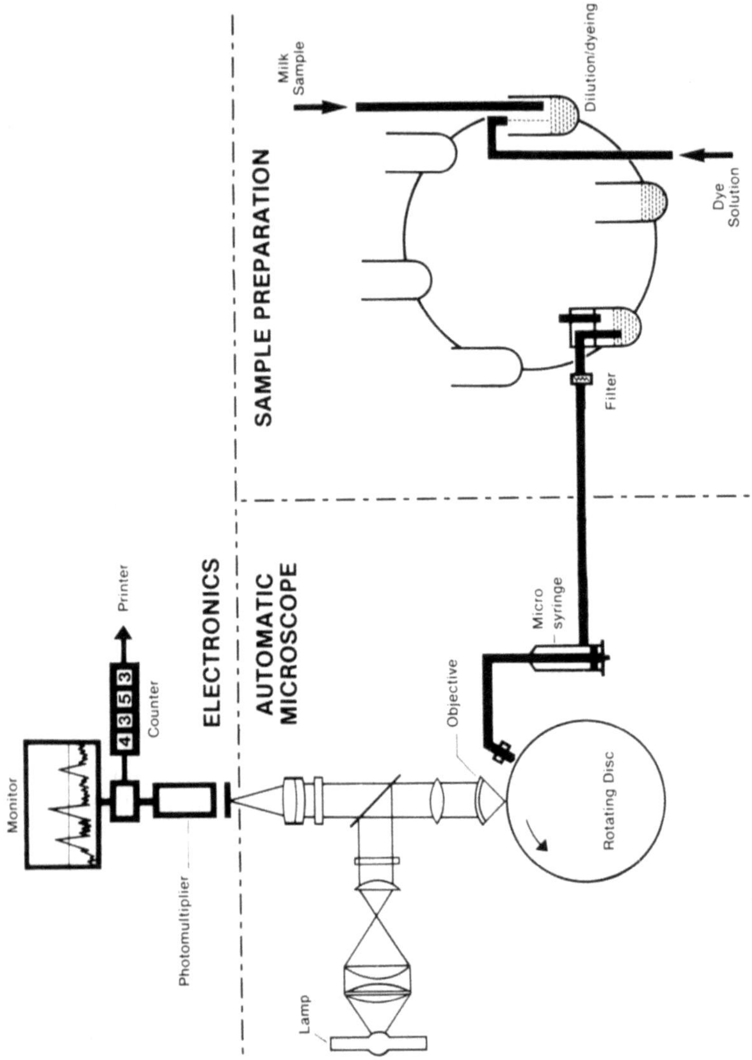

Fig. 18. Operation of a Fossomatic: schematic. (See text for further details.)

initial fixation of each milk sample according to procedures recommended by the manufacturer, the Coulter Counter dilutes and incubates the sample, and counts the number of particles present in it.

The Coulter Milk Cell Counter consists of the following sub-units:

1. *The pneumatic power supply*, which powers the circuits in the diluter section.
2. *The main unit*, composed of:
   (a) an electronic power supply that powers the electronic circuits in the Analyzer Section and Teleprinter;
   (b) a diluter section to accept milk samples which have been pre-fixed;
   (c) an analyser section;
   (d) a trolley holding containers for chemicals.
3. *A teleprinter* for printing results.

Operation

1. Samples are prepared by manually fixing the somatic cells, prior to incubation, in order to preserve their size: add Somafix® or formaldehyde solution to the milk sample. Heat sample for a short time, or leave at room temperature for 18–24 h before counting.
2. Sample vials (1–50) are placed in a carousel, and Somaton® blanks are added to two holes. After stirring, 2 ml of fixed milk is drawn through the sample valve, and approx. 45 μm of milk is transferred to a reaction tube with a precise volume of Somaton® to give a dilution of 1:100. (The sample valve is purged and rinsed before the next sample is taken.) The reaction tubes are incubated in a glycol bath maintained at 80°C for 10 min. During this reaction, fat globules are eliminated so that they do not interfere with the cell count.
3. On reaching the counting station, the diluted milk sample is gently stirred, and the cells are then counted using a 140 μm aperture tube. Suspended, fixed somatic cells pass through the aperture, and the resulting, amplified pulses are displayed on an oscilloscope monitor. A threshold value, linear with cell volume, is set to discriminate against pulses caused by fat, debris, air bubbles and contaminating substances, so that only particles the size of fixed somatic cells are counted. A digital register records the number of cells present in 0·3 ml of cell suspension, and the count is corrected and printed. After counting, the tube is rinsed. The capacity is 210 samples per hour.
4. Precision and accuracy should be checked daily.

FUTURE DEVELOPMENTS

A need for chemical analysis is indicated by:

- analysis for payment purposes and herd improvement;
- process control during manufacturing of dairy products;
- official demands and requirements.

Payment Purposes

Methods of settling accounts with the milk supplier have developed parallel to the appearance of analysis instruments with large capacities and direct output to EDP systems for processing of data and calculation of payments. The main component in milk payment was originally—and still is—butterfat, but protein content is gradually being introduced as a milk pricing parameter. There is a tendency towards greater control of raw milk quality, depending on the kind of product to be made.

In cheese production, not only as a whole but also with reference to the production of different specific types of cheese, there is a need to ensure that milk quality (especially protein content) is optimal for processing. Similar requirements as to the quality of raw materials can be made in the manufacturing of other dairy products, such as dried milk powder, condensed milk, etc.

Besides analysis of raw materials, there is an increasing demand for control of the hygienic quality of milk. Somatic cell counts and bacterial counts have been introduced in payment systems in a rapidly increasing number of countries. As more attention is paid to hygiene, accepted threshold limit values will be reduced, and this again increases the demand for instruments with better sensitivity.

Process Control

In the manufacture of dairy products, it is highly desirable to have ways of monitoring and controlling the production processes. Quick and reliable analytical instruments provide a basis for optimal processing.

Standardisation of milk according to fat content is controlled by analysing samples of the finished product from the pipeline in the dairy. Based upon the results, the flow of cream into the skim-milk can be increased or decreased to obtain the desired fat content. Increased instrument accuracy means that more narrow limits can be established for fat content in the finished product. A reduction of fat by 0·01% may result in savings of more than $25 000 per year in some dairies.

In various processes, standardisation of protein content can be expected—for example, in the production of milk powder. Research and development in milk processing will add to our knowledge of the influence of individual milk components on the finished products, which may increase applications and even result in a need for measuring novel parameters.

Official Demands

A growing interest among consumers in the constituents, quality and age of food products has resulted in intensified control and legislation in many countries. Regulations that require declaration of contents and shelf-life are examples. To meet these demands, suitable analytical methods must be available. The introduction of new standards can, typically, come about in two ways:

1. In some caes, a producer of instruments develops and introduces to the market a new method of analysis which slowly gains greater use. The attention of the authorities is drawn to the method, and legislation makes it officially accepted.
2. In other cases, parameters are regulated by legislation which is to come into force within a specified time limit. This leaves time for potential producers of instruments to develop and market suitable apparatus to meet the demands made by the legislation.

There is, thus, a close relationship between arising needs for measurement and instruments offered on the market. Technological developments in physics, chemistry and electronics, including computer technology, offer many possibilities to the producers of analysis instruments. The technological possibilities within manufacturing of instruments are great. Not only single instruments, but also for total solutions, including sample handling, sampling, measurement, and recording and processing of data in integrated systems will, in the future, be offered. Great investments will be made in research and development to the benefit of milk suppliers, the dairy industry, and consumers.

REFERENCES

A/SN. Foss Electric. Pro-Milk Mk II Applications. DK-3400 Hillerød, Denmark.
Biggs, D. A. (1972). *J. Assoc. Off. Anal. Chem.*, **55**, 488.
Biggs, D. A. (1978). *J. Assoc. Off. Anal. Chem.*, **61**, 1015.
Biggs, D. A. (1979a). *J. Assoc. Off. Anal. Chem.*, **62**, 1202.
Biggs, D. A. (1979b). *J. Assoc. Off. Anal. Chem.*, **62**, 1211.
Goulden, J. D. S. (1956). *J. Sci. Food Agric.*, **7**, 609.
Goulden, J. D. S. (1958). *Trans. Faraday Soc.*, **54**, 941.
Goulden, J. D. S. (1961). *Nature*, **191**, 905.
Goulden, J. D. S. (1964). *J. Dairy Res.*, **31**, 273.
Grappin, R. and Jeunet, R. (1972). *Le Lait*, **52**, 324.
Grappin, R. (1984). Challenges to contemporary dairy analytical techniques. Special publication no. 49, pp. 77–90. The Royal Society of Chemistry, London.
Haugård, G. and Pettinati, J. D. (1959). *J. Dairy Sci.*, **42**, 1255.
Jøndrup, P. (1980). *Elektronik*, **8**, 10.
McGann, T. C. A. and O'Connell, J. A. (1972). *Laboratory Practice*, **21**, 489.
McGann, T. C. A. *et al.* (1972a). *Laboratory Practice*, **21**, 628, 650.
McGann, T. C. A. *et al.* (1972b). *Laboratory Practice*, **21**, 865.
Mills, B. L. and Van de Voort, F. R. (1982). *J. Assoc. Off. Anal. Chem.*, **65**, 1357.
Nexø, S. A. *et al.* (1980). US Pat. 4,236,075.
Nexø, S. A. *et al.* (1981). US Pat. 4,247,773.
O'Connell, J. A. (1970). *Laboratory Practice*, **19**, 1119.
O'Connell, J. A. and McGann, T. C. A. (1972). *Laboratory Practice*, **21**, 552.
Shields, J. (1975). B.Ph. Thesis, University of York, England.
Shields, J. (1982). US Pat. 4,310,763.
Sjaunja, L. -0. (1982). Studies on milk analysis of individual cow milk samples. Report no. 56, Swedish University of Agricultural Science, S-75007 Uppsala, Sweden.
Thomasow, J. *et al.* (1971). *Milchwissenschaft*, **26**, 474–81.
Tolle, A. (1971). *International Dairy Federation Annual Bulletin*, part 2, p. 3.
Van de Voort, F. R. (1980). *J. Assoc. Off. Anal. Chem.*, **63**, 973.

Chapter 7

Modern Laboratory Practice—2: Microbiological Analyses

G. L. Pettipher

Reading Scientific Services Ltd, The Lord Zuckerman Research Centre, Reading, UK

Milk and milk products are highly perishable foodstuffs which are of considerable economic and nutritional importance, especially in the Western hemisphere. In its original state, milk is an excellent foodstuff for man if it is obtained from healthy cows and produced under hygienic conditions. Although milk production methods have generally improved over the last few decades, the microbiological quality of some raw milk supplies still causes concern. Udder disease remains widespread, and consumers of raw milk still risk food poisoning. Refrigeration of farm milk can mask the effects of unhygienic practices, such as defectively cleaned milking equipment. Inadequate processing, poor packaging techniques and insufficient refrigeration can all substantially reduce the shelf-life of milk products and cause considerable economic loss.

Microbiological methods have an important role to play in the Dairy Industry. They are used to protect the Public Health, and can reduce economic losses by the early detection of inadequate processing, packaging or refrigeration. This can be achieved by monitoring the microbiological quality of raw milk supplies, bulk milk and finished milk products immediately after production and during storage.

THE NEED FOR MICROBIOLOGICAL TESTING

Public Health Aspects

Raw milk was once regarded as the most dangerous item in our diet, giving rise in England and Wales between 1912 and 1937 to about 65 000 deaths. The eradication of tuberculosis and brucellosis in cattle, the hygienic production and the heat treatment of milk have greatly reduced the incidence of food poisoning. However, there is recent evidence that the incidence of milk-borne disease has begun to increase (Galbraith *et al.*, 1982), mainly due to unpasteurised, defectively pasteurised or recontamination of the milk after pasteurisation. These outbreaks are mainly due to *Salmonella* or *Campylobacter* species. Liquid market milk and milk for further processing which has been correctly pasteurised will be quite safe provided that recontamination is prevented.

The major sources of bacteria in raw milk are the udder (interior and exterior) and the milking equipment (Cousins and Bramley, 1981). In 1979, a survey of some 500 British herds showed that about one-third of dairy cattle were infected by sub-clinical mastitis. These infections may be caused by *Staphylococcus* sp., *Streptococcus* sp., *Escherichia coli* and *Campylobacter* sp., some of which are pathogenic for man. Between milkings the cows' teats may become soiled with dung, mud and bedding litter. If not removed before milking, this dirt, which may contain $>1\times10^{10}$ bacteria g^{-1} (of which $>1\times10^{9}$ g^{-1} may be psychrotrophic and $>1\times10^{6}$ g^{-1} may be spore-formers), contaminates the milk during milking. When milking equipment is solely responsible for high counts in the milk, then the cleaning and disinfection must be seriously defective, and the equipment would have a count of around 1×10^{9} bacteria m^{-2}.

Clearly the microbiological monitoring of raw milk supplies is necessary to protect the Public Health, especially where the consumption of raw milk is concerned.

Economic Aspects

The numbers and types of microorganisms in milk immediately after production directly reflects the microbial contamination during production. High numbers of certain types of bacteria in the milk can reduce the value of the raw supply to the processor by reducing the time for which the raw material can be stored, reducing the shelf-life of the product or, importantly, causing taints in the finished product.

There is a considerable variation in the incidence of thermoduric and psychrotrophic organisms in fresh raw milk, which probably reflects the different conditions on the individual farms. *Microbacterium lacticum* and bacterial spores show nearly 100% survival after pasteurisation. Fortunately, only a small proportion of the thermoduric organisms are also psychrotrophic (McKinnon and Pettipher, 1983), otherwise the shelf-life of refrigerated pasteurised products would be considerably reduced. Psychrotrophic organisms in raw milk supplies are of considerable importance to the Dairy Industry, as they reduce the time that the raw material can be stored before processing. Milk with an initial psychrotroph count of 1×10^4 ml^{-1} even at 5°C may contain $> 1 \times 10^6$ ml^{-1} after 3 days storage (Cousins *et al.*, 1977). Pasteurisation will destroy these bacteria, but heat-resistant enzymes may persist and could be detrimental to the quality of the milk products (Law, 1979).

Poor processing, for example, inadequate pasteurisation, recontamination after pasteurisation or recontamination during packaging, may result in a finished product with a much reduced shelf-life. Clearly, it is necessary to use microbiological methods to detect these mistakes as early as possible, in order that they can be speedily rectified and economic losses kept to a minimum.

THE MICROBIOLOGICAL TESTING OF MILK AND MILK PRODUCTS

On the basis of Public Health and economic considerations, it is necessary to test the microbiological quality of milk at a number of points along the chain from producer to consumer. These include the following:

(a) raw milk supplies;
(b) bulk milk arriving at the processor's premises;
(c) monitoring bulk milk during storage prior to processing; and
(d) monitoring the processing and quality control of finished products.

In many European countries in which the Dairy Industry is highly developed, the microbiological testing of farm supplies is carried out in central testing laboratories. It is the responsibility of the processor to monitor the microbiological quality of incoming bulk supplies, milk during storage and finished products.

Central Testing

In England and Wales there are five central testing laboratories, the largest is responsible for testing 7500 supplies and the smallest 5000 supplies. Each laboratory is fitted with automated equipment and operates every day of the year. Test results are transmitted to the Milk Marketing Board's Headquarters (Fig. 1).

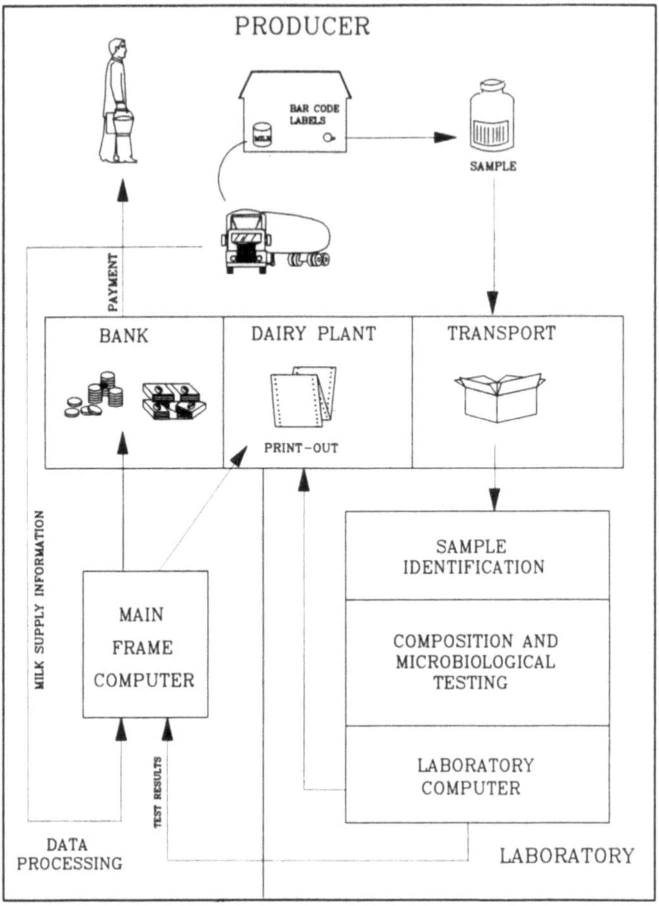

Fig. 1. Schematic diagram of the system of quality payment testing of farm milk.

Milk samples are collected from each farm by the tanker driver once a week on unspecified days. After the bulk milk has been mixed, the samples are taken using a disposable sterile dipper and collected in disposable, sterile, plastic pots. Samples are identified with bar code labels which are unique for each producer. The bar code is read by an IR light pen at the point of testing to avoid misidentification of the samples. The samples are transported in insulated containers fitted with ice packs to hold the temperature below 4°C for 48 h. Samples should reach the central laboratory within 24 h of collection.

Milk samples are tested for bacteriological quality by a plate colony count. A small volume of milk is removed from the sample and mixed with agar in a Petri dish. The Petri dish is then incubated for 3 days at 30°C, and the bacterial colonies formed counted automatically by a colony counter. Three bands of milk are used for payment purposes, these are

band A—2×10^4 bacteria ml^{-1} or fewer
band B—$>2 \times 10^4 - 1 \times 10$ bacteria ml^{-1}
band C—$>1 \times 10^5$ bacteria ml^{-1}

Band B supplies receive the standard payment and band A supplies receive a premium. Supplies classified as band C are subjected to a penalty. Certain rules are applied to safeguard producers from the effects of a single high count in a month.

Antibiotic failures and unsatisfactory hygiene or compositional quality results are made known to the producers and buyers within 24 h of the test result being available. The producer's monthly payment from the Board is calculated from the volume of milk produced and its value, as determined from the results of the various tests.

Testing in Dairies and Creameries

Dairies and creameries still perform quality-assurance tests on milk, but these are mainly on whole tanker loads and doubtful farm supplies. Incoming tanker milks are assessed for bacterial content either by the use of dye reduction methods or by a plate colony count. Both of these methods have major disadvantages; dye reduction methods are insensitive and unreliable, and cultural methods take 2–3 days to give a result. If the plate colony count method is used, the milk will probably have been processed before the bacterial content is known.

Increasing pressure is now being brought to bear on processors by their major customers regarding the bacteriological specifications laid down for both raw milk supplies used for processing and the finished products. Ideally, dairies and creameries need to use rapid methods for testing raw milk prior to acceptance, and methods which will successfully predict the eventual keeping quality of their finished products.

MICROBIOLOGICAL METHODS FOR THE TESTING OF RAW MILK

There are many different methods available to the microbiologist for testing the microbiological quality of milk. The choice of test will vary with the application. For example, it is preferable if tests used in a dairy or creamery are rapid so that any necessary remedial action can be taken quickly. If only small numbers of samples are involved, then increased labour requirements or running costs may be acceptable to achieve a rapid result. In central testing laboratories, rapidity is not as important as the tests on farm supplies are retrospective, the sample generally takes 6–24 h to reach the laboratory. In this case, automation, low running costs and high accuracy and precision are given greater consideration than rapidity in selecting the best method to use.

Most of the microbiological methods available can be used to assess the microbiological quality of raw milk. Frequently, the total bacterial count is measured, and if this is low, it is assumed that there is low risk to Public Health and that the quality of the milk is satisfactory. However, the initial total viable count is of little value in predicting the count after refrigerated storage. Many of the general microbiological methods can, with certain modifications, be used for specific applications for example, the detection of post-pasteurisation contamination, sterility testing and the detection of spoilage organisms such as yeasts. In order to simplify the discussion, the most widely used microbiological methods will be described for the application to raw milk, and then specific applications will be considered separately, as a number of microbiological methods can be used for a single specific application.

Methods for estimating bacterial numbers in raw milk can be divided into two groups, direct and indirect. Direct methods count cells directly, either by the ability of viable cells to grow and form colonies, or microscopically. Indirect methods measure either a chemical constituent, enzyme, metabolite or changes produced by bacteria during growth. This measurement is converted into bacterial numbers by reference to a calibration curve. The 'true' numbers of bacteria for the calibration curve are usually assessed by a direct method. Generally, direct methods are more sensitive and accurate than indirect methods, and cultural methods take longer to give a result than microscopic or indirect methods.

The adjective 'rapid' has been ambiguously used when describing microbiological methods. In terms of speed, 'rapid' has meant a few minutes or a number of hours. Miniaturised, modified and automated methods have also been described as 'rapid' since the processing time is shorter than the standard manual technique, even though the result may not be obtained for 1–3 days. The term 'rapid' should be reserved for tests which give a result in 1 h or less.

One of the first rapid methods in the food industry was probably the 'sniff and taste' test used at the creamery to determine the acceptability of raw can milk. Milks which smelled off and tasted sour, and therefore probably contained greater than 1×10^7 bacteria ml^{-1}, could be immediately rejected by this method. The test became inappropriate with the introduction of refrigeration as spoilage was no longer due to milk-souring organisms. Spoilage by psychrotrophic bacteria is less easily detectable by sensory assessment. Non-quantitative methods have no place in a modern Dairy Industry.

Direct

Viable

Plate colony count

The plate colony count isolates bacteria in a quantitative manner. A range of dilutions, usually 10-fold, of the sample is prepared in a sterile diluent, and 1 ml of each dilution is mixed with melted agar cooled to 45°C and then allowed to solidify (American Public Health Association, 1972). The organisms present in the sample are fixed within the agar gel. The pour plates are then incubated for 2–3 days at 30°C, during which time viable organisms replicate and form visible colonies. Individual colonies on a plate containing 30–300 colonies are counted, and then the count per ml calculated by multiplying the colony count by the appropriate dilution factor.

The plate colony count suffers from a number of disadvantages. Two or more adjacent cells may give rise to only a single colony. Some organisms may be unable to grow in the medium or at the incubation temperature selected, and some may be damaged by exposure to the warm agar and unable to replicate.

One of the major disadvantages of the plate colony count is that it takes 2–3 days to give a result. The plate colony count is very sensitive because, in principle, any viable cell when placed in an appropriate medium and incubated at a suitable temperature will give rise to a colony. No other microbiological method can detect with certainty the presence of a single, viable bacterium in a large volume of sample.

The plate colony count is generally accepted as the reference method for the microbiological analysis of milk. Rightly or wrongly, new microbiological methods are compared against the plate colony count, and must show good agreement before they are accepted as suitable alternative methods. It is important to note that the plate colony count itself is also subject to errors, and provides only an estimate of the 'true' bacterial numbers. These errors, which are mainly due to sampling and decrease with increasing bacterial numbers, are often in the range ±10–40%.

Automation of the plate colony count enables as operator to perform analyses more rapidly than by the manual method. Using machines like the Colworth 2000, decimal dilutions of food suspensions can be made automatically and plated to a maximum of 2000 plates per day per technician with obvious savings in operator time (Sharpe *et al.*, 1972). However, for plate colony count methods, the lengthy time taken to give the result is mostly due to the incubation period required for colony formation. Automation of the preparation of the plates does little to decrease the overall time required for the technique.

Plate loop method

Savings in time and materials can be made by removing the need for the dilution series in the standard plate colony count. Various methods based on the Thompson plate loop method have been described for milk (Bradshaw *et al.*, 1973; Fleming and O'Connor, 1975). In the technique, two loops which retain 0·01 and 0·001 ml dip into the sample. A Petri dish is positioned under each loop and the loops are flushed with diluent, agar is then added and the contents of the Petri dish mixed. These machines can plate 10^{-2} and 10^{-3} dilutions from 300 samples h^{-1}. The technique is suitable for enumerating bacteria in the range 3000–300 000 ml^{-1}. Of the various factors influencing the volume transferred by the loop, the speed of the loop as it emerges from the milk is the most important (King and Mabbitt, 1984). Increasing the speed two-fold results in a 23% increase in the volume transferred. The depth of the dip and changes in the temperature of the sample also cause a significant change in the volume transferred. These authors suggested that to obtain the transfer of 1 μl at the speeds commonly used in the plate loop method, the internal diameter of the loop should be 1·30 mm and not the 1·45 mm in use at that time.

One of the most commercially successful automated plate loop machines is the Petrifoss (A/S N. Foss Electric, Denmark) which is widely used in the central testing laboratories of Europe. The instrument can 'plate' 300 samples h^{-1}, and is, therefore, ideally suited to laboratories which analyse hundreds or thousands of samples each day. The manufacturer claims that the relationship between the Petrifoss result and that of a reference method has a correlation coefficient of 0·99.

Spiral plate

The Spiral Plate method for enumerating bacteria also avoids the use of a dilution series. By using a varying rate of sample application, it needs only one plate to obtain counts over a range which would require 2 or 3 plates in the standard plate colony count. The Spiral Plater deposits a known volume of sample on a rotating plate in an ever decreasing amount in the form of an Archimedes spiral (Gilchrist *et al.*, 1973). After incubation different colony densities are apparent, closely packed or confluent in the centre to well isolated at the outside (Fig. 2). A counting grid which relates the area of the plate examined to the volume of the sample is used to convert the count in a given area of the plate to the number of bacteria per ml of sample. This can be done automatically in a few seconds by a laser colony counter or image analyser. The count per ml is calculated from the area of the plate scanned to reach a pre-determined number of colonies. Using the Spiral Plater and laser colony counter, milks with bacteria in the range 500–300 000 ml^{-1} can be analysed in about one-third the preparation time of the standard plate colony count. Contradictory results of comparison of the Spiral Plate with the plate colony count have been reported. Some workers found no difference between the two methods (Peeler *et al.*, 1977), whereas others found the overall geometric mean of the Spiral Plate count 33% lower than the plate colony count (O'Connor and Fleming, 1979).

Droplet technique

Miniaturisation of the plate colony count also reduces processing times. In the Colworth Droplet technique, which can be used to enumerate bacteria in foods, the plates are 0·1 ml droplets of agar (Sharpe and Kilsby, 1971). The agar is used as diluent thereby saving materials. Five replicate droplets of each of three dilutions can be prepared in about 45 s, enabling operators to process about three times as many samples as by the standard plate colony count (Fig. 3). For milk, the droplet technique after 48 h incubation gives similar results to the standard plate colony count (Fondén and Strömberg, 1978).

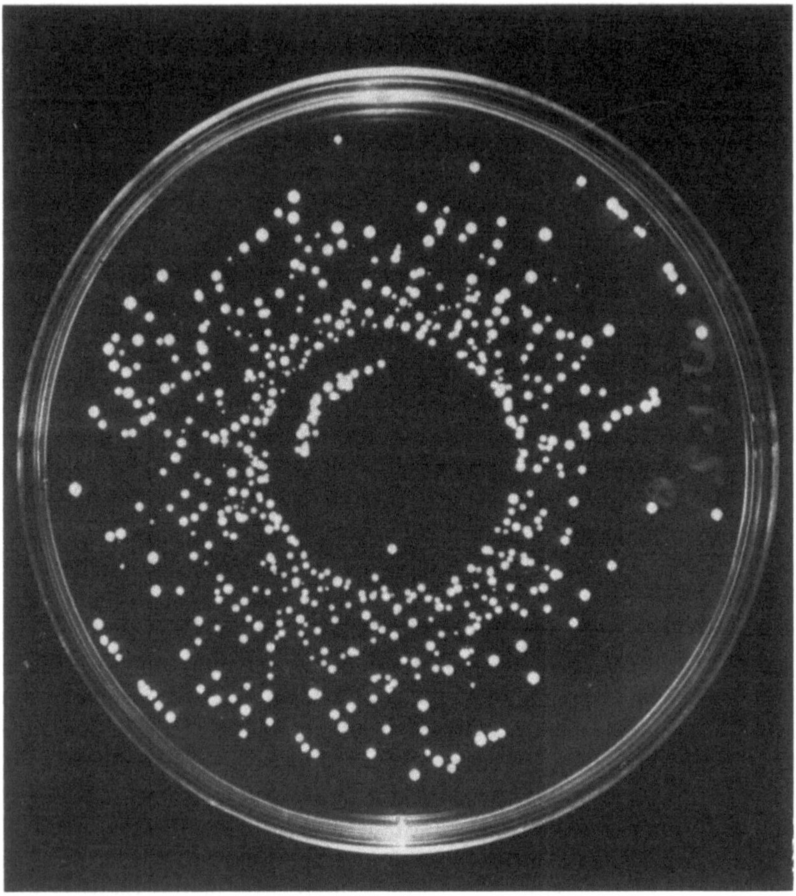

Fig. 2. *Escherichia coli* colonies on a Spiral Plate.

Hydrophobic grid membrane filter

One direct method for counting viable bacteria retained on membrance filters, following filtration of the sample, involves placing the filter on a pad soaked in nutrient mdium and counting the colonies formed after 1–3 days incubation. This method, which cannot be considered as rapid, has been used to enumerate coliform bacteria in raw and pasteurised milk (Fifield *et al.*, 1957; Claydon, 1975).

A variation on this method is a Hydrophobic Grid Membrane Filter (HGMF) described by Sharpe and Michaud (1974). The techique reduces the need for dilutions of the sample before enumeration, and hence

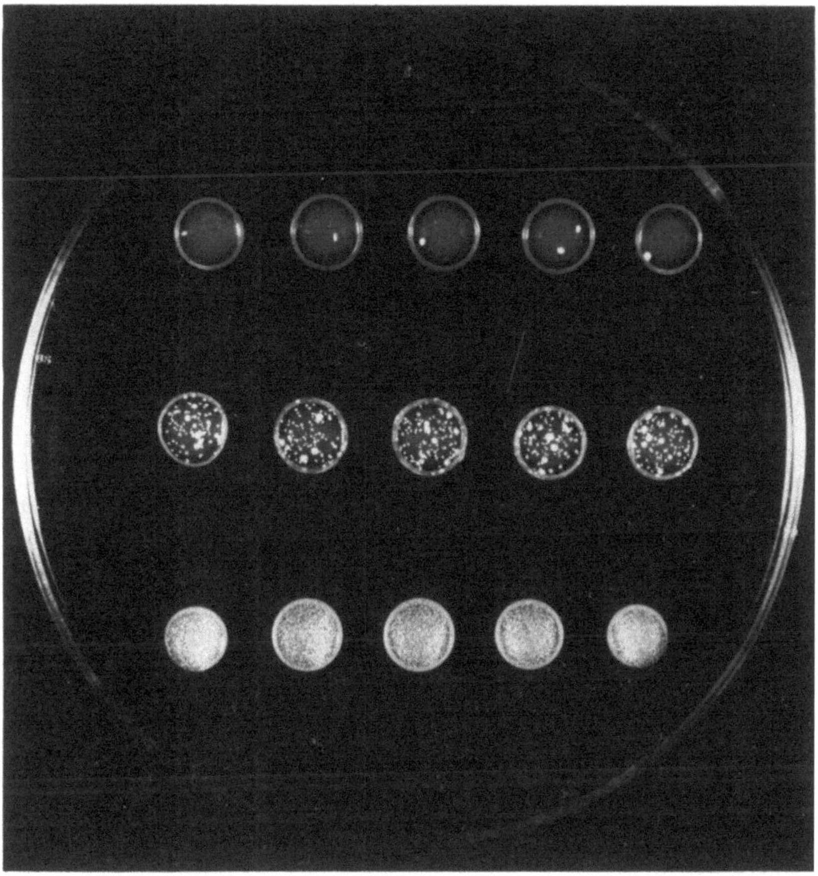

Fig. 3. *Escherichia coli* colonies in agar droplets.

reduces processing time, and gives better recovery than conventional filters, especially when high numbers of bacteria are involved (Sharpe and Michaud, 1975). The HGMF has a square grid pattern printed in hydrophobic material, such as wax, on a conventional membrane filter. This divides the filter into a number of compartments, usually 2000–4000 depending on the size of the grids. The device functions as a most probable number technique, eliminating size variations in colonies and preventing lateral spread (Fig. 4). This greatly facilitates the automated counting of the colonies. Growth in a grid cell does not necessarily equal one colony in the late colony count, since frequently a grid is inoculated with more than one bacterial cell. Coinicident inoculation is allowed for

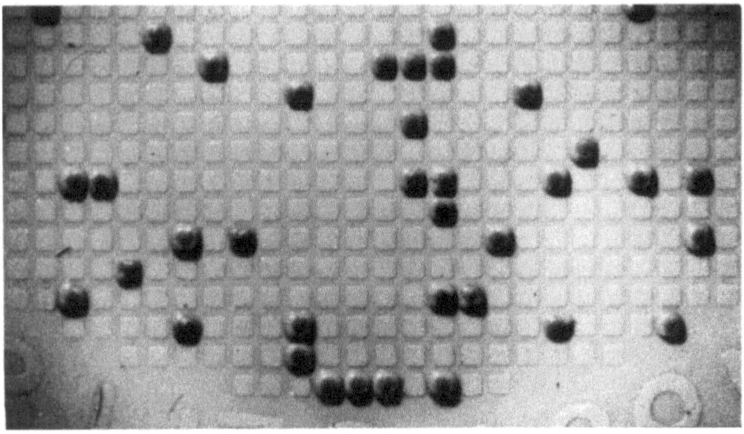

Fig. 4. *Escherichia coli* colonies on a Hydrophobic Grid Membrane Filter.

in calculating the count per ml, e.g. growth in 3649 grid cells gives a most probable number of 30 000. Hydrophobic Grid Membrane Filters have been successfully used to enumerate coliforms in a variety of foods (Sharpe *et al.*, 1979). Pre-treatment of milk and milk products with either a proteolytic enzyme, a detergent, or both, improves the filtration sufficiently to enable 3–5 g to be filtered (Peterkin and Sharpe, 1980). No bacteriological results for these products were reported. This method of assessing bacterial numbers in milk may have the advantage over the plate colony count in reducing operator time but, like all methods relying on bacterial growth, it requires a lengthy incubation period.

Microscopic

Breed smear

One of the first methods to use microscopy for counting bacteria in milk involved the preparation of milk films, staining with methylene blue, and then microscopic examination (Breed and Brew, 1916). The method, which is still in use today, has the advantages of taking less than 1 h to complete, stained milk films can be stored as a relatively permanent record, and tentative identification of the types of bacteria can be made during examination. The disadvantages of the method include the use of a very small sample volume (0·01 ml) which leads to increased errors, a large microscope factor of about 500 000 which limits the sensitivity, failure of bacteria to stain, irregular distribution of bacteria in the films, and operator fatigue resulting from prolonged use of the microscope. This direct miroscopic method is particularly popular in the US, where it is used to differentiate raw milk supplies into three classes (American Public Health Association, 1972).

Direct epifluorescent filter technique

The Direct Epifluorescent Filter Technique (DEFT) was originally developed as a rapid method for counting bacteria in raw milk (Pettipher *et al.*, 1980; Pettipher, 1983). The DEFT uses membrane filtration, fluorescent staining and epifluorescence microscopy. Pre-treatment of the milk with an enzyme and surfactant disperses somatic cells and fat sufficiently to enable at least 2 ml to filter through a 0·6 μm polycarbonate membrane filter; bacterial cells remain intact and are concentrated on the membrane. After staining with acridine orange, the bacteria fluoresce orange–red under blue light and can easily be counted using an epifluorescence microscope. The DEFT count is rapid, taking 25 min to complete, is inexpensive, correlates well with the plate colony count ($r > 0{\cdot}9$), and is suitable for milks containing 6×10^3–10^8 bacteria ml^{-1}. The DEFT count showed >90% agreement with the plate colony count in the classification of milk samples into groups of more or less than 10^4, 10^5, and 10^6 bacteria ml^{-1}. The method is both sufficiently rapid for monitoring tanker and silo milk, and sensitive enough for grading farm milk on the basis of bacteriological quality. No other method can give an accurate, sensitive count of bacteria in milk in as short a time as the DEFT.

One of the major disadvantages of the manual DEFT is operator fatigue associated with prolonged use of the microscope. This can be eliminated by the use of a semi-automated counting system based on an image analyser which enables operators to count up to 50 DEFT slides h^{-1} (Fig. 5). The semi-automated count of bacteria on DEFT slides agrees well with the corresponding visual DEFT count, $r = 0{\cdot}94$ (Pettipher and Rodrigues, 1982).

The DEFT is now operating satisfactorily in more than 50 laboratories and in more than 11 countries. The Jersey Milk Marketing Board was quick to see the advantages of the method and has been using the semi-automated DEFT count to assess the bacteriological quality of farm milk since April 1982. The DEFT has been included in the revised version of British Standard 1984: 4285 '*Methods of Microbiological Analysis for Dairy Purposes*'.

Recently, the French company Biocom has marketed an instrument (Cobra) which automates fully the sample filtration, staining, rinsing, drying and counting stages of DEFT (Fig. 6). A recent evaluation has

Fig. 5. Image analyser for the semi-automated counting of bacteria on DEFT slides. (Reproduced by courtesy of Perceptive Instruments, Halstead, UK.)

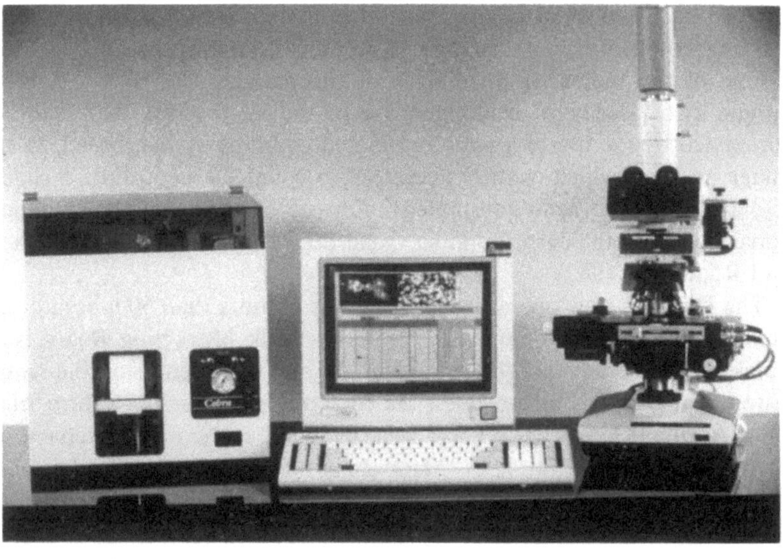

Fig. 6. The COBRA, an automated DEFT system, showing the sample preparation and counting units. (Reproduced by courtesy of Biocom, France.)

shown that the Cobra instrument can be used successfully to grade raw farm and tanker milks at the 20 000, 50 000 and 100 000 bacteria ml^{-1} levels (Table I). The Cobra instrument can be used to analyse batches of 24 samples simultaneously with more convenience and rapidity than the standard DEFT. One operator can probably process at least 96 samples h^{-1} and results are available in *c.* 30 min. Compared with the standard DEFT, miniaturisation has led to substantial cost savings (approximately 80%) on disposables. The Cobra instrument is widely used in laboratories of French dairies.

TABLE I
Grading of raw milk samples by the Cobra compared to plate counts[a]

Grading level	*Percentage correctly classified*			*Percentage incorrectly classified*[b]	
	Below limit	*Above limit*	*Total*	*High Cobra or low plate count*	*Low Cobra or high plate count*
20 000/ml	65	20	85	9	6
50 000/ml	79	13	92	6	2
100 000/ml	86	9	95	3	2

[a]From Pettipher *et al.* (1992).
[b]Results obtained for 109 samples.

Bactoscan

The Bactoscan instrument (A/S N. Foss Electric, Denmark) uses fluorescence microscopy for the automated counting of bacteria in milk. The Bactoscan works on the same principle as the Fossomatic, an instrument developed for automated somatic cell counts in milk. The basic difference between the two systems is that in the Fossomatic system, the somatic cells are stained directly in the diluted milk and subsequently counted, whereas in the Bactoscan, somatic cells, sediment and fat globules are first removed to prevent incorrect counting (Fig. 7). Somatic cells and casein micelles are dissolved chemically, and then the bacteria are separated by continuous centrifugation in gradients formed with solutions of dextran and sucrose. The bacteria recovered from the gradient are incubated with a protease to remove residual protein, and then stained with acridine orange and counted. For counting, the treated sample is applied in a thin film on the surface of a rotary disc and passed under a microscope objective. The fluorescent impulses in the miroscope

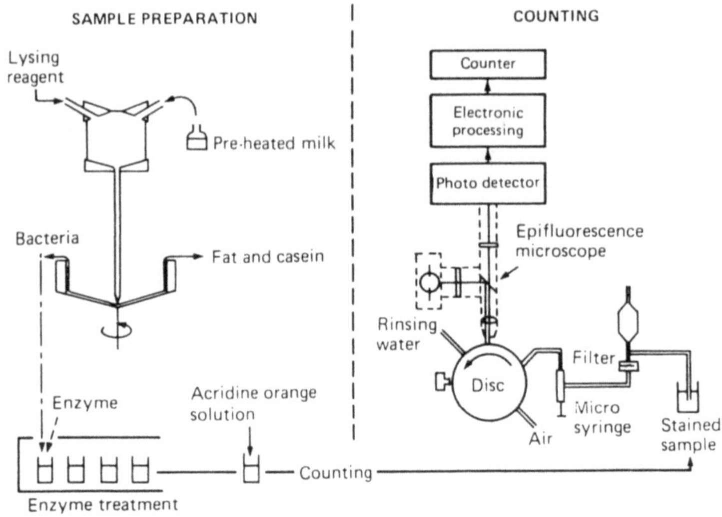

Fig. 7. Diagrammatic representation of sample preparation and counting by the Bactoscan. (From Keilwein (1982), reproduced by permission of Verlag Th. Mann, Germany.)

image are converted into electrical impulses and recorded. The instrument can analyse 80 samples h^{-1} with an analysis time of 7 min. It would seem to be suitable for use in central testing laboratories.

For milk, the relationship between the impulses registered by the Bactoscan and the plate colony count correlates reasonably well, $r = 0{\cdot}88$ (Fig. 8). The manufacturer claims a correlation between the Bactoscan

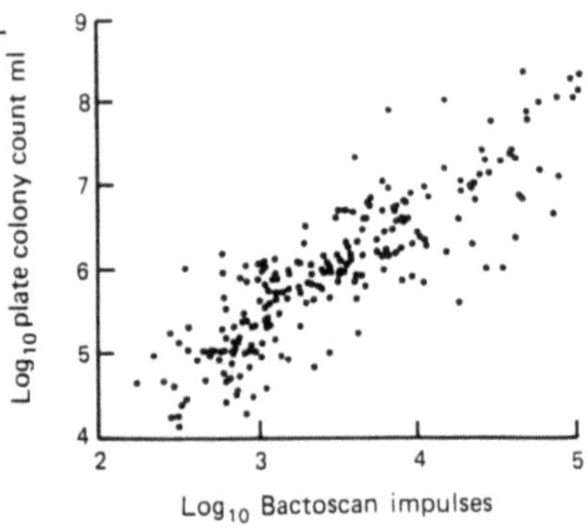

Fig. 8. Relationship between $\log_{10}$ plate colony count per ml and $\log_{10}$ impulses registered by the Bactoscan instrument for samples of refrigerated milk. (From Keilwein (1982), reproduced by permission of Verlag Th. Mann, Germany.)

count and the plate colony count of 0·8–0·9, and this was bettered during the evaluation reported by Saarinen (1984). The Bactoscan can be used to rapidly detect milks of poor bacteriological quality, and could be used to grade milks at the 100 000 bacteria ml^{-1} level, the first penalty band in a number of countries. During recent years, improvements have been made to the instrument and its lower sensitivity for raw milk is *c.* 30 000 bacteria ml^{-1} (Suhren *et al.*, 1991).

Flow cytometry

Although it has been used principally for medical applications (Grogan and Collins, 1990), flow cytometry is an emerging technology with applications in rapid microbiological testing. Flow cytometry is an optically based technique which for liquids can provide cell-by-cell analysis, counting and classifying cells according to size, intensity of staining, amount of light scatter etc. A French-based company, Chemunex S. A. has developed and marketed an instrument which, together with the reagents, is called the ChemFlow system (Fig. 9). In this system, target cells are first fluorescently labelled and are then counted by the ChemFlow analyser using the principles of flow cytometry. Viable

Fig. 9. The ChemFlow system of reagents, automatic sample injector and flow cytometer. (Reproduced by courtesy of Chemunex SA, France.)

cells contain enzymes which hydrolyse the fluorochrome precursor (substrate) and release a fluorochrome which concentrates intracellularly. In 'clean' systems, such as lemonade, the ChemFlow can detect as few as 50 yeast ml^{-1} with results available in less than 1 h (Pettipher, 1991).

As with most techniques, there are associated problems with flow cytometry. Solid samples require some form of homogenisation and separation of the microorganisms prior to staining and analysis. This pre-treatment may affect the metabolic state of the cells and hence their subsequent staining reactions. The sensitivity of the technique varies with sample type and is determined largely by product interference (noise). Due to product interference, flow cytometry is unlikely to be of use in counting bacteria in raw milk, but it is useful for counting somatic cells and certain specific microbiological applications.

Indirect

Dye reduction

Dye reduction tests are based on the ability of bacterial enzymes, such as dehydrogenases, to transfer hydrogen from a substrate to a redox dye which then undergoes a change in colour. The rate of reduction depends on the enzyme activity, and this has been used as an index of the number of bacteria present in the milk. In general, the reduction time is inversely related to the bacterial content of the sample when incubation with the dye commences.

Dye reduction tests were developed during the 1930s. The most commonly used redox dyes are methylene blue and resazurin. Resazurin has the advantage over methylene blue in that reduction from purple through pink to colourless can be quantified using a comparator. A resazurin test with a 10 min incubation time is used in England and Wales to grade milk on arrival at the creamery. The 10 min resazurin test is rapid and requires small amounts of equipment, but only milks containing large numbers of actively growing bacteria ($>10^6$ ml^{-1}) will fail. Methylene blue reduction tests provide a reasonable estimate of the number of bacteria in non-refrigerated, bulked raw milk. The relationship between $\log_{10}$ plate colony count and methylene blue reduction time had a correlation coefficient (r) of 0·9 (Lück, 1982). Poorer agreement between the two methods was observed for refrigerated, bulked raw milk, $r=0{\cdot}62$ (Fig. 10).

There are a number of reasons why dye reduction times and numerical estimates of bacterial populations fail to correlate; the reducing activities

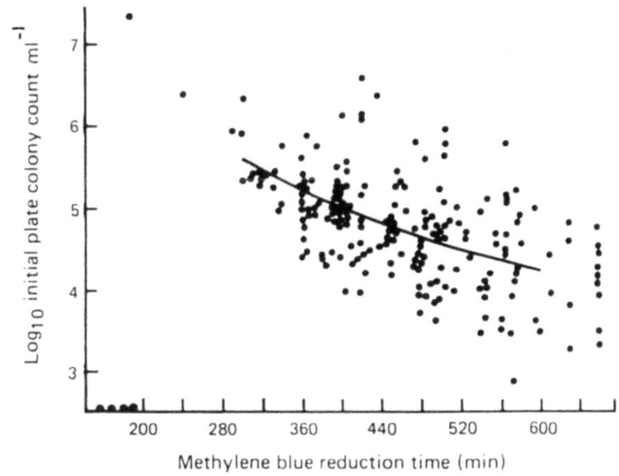

Fig. 10. Relationship between $\log_{10}$ initial plate colony count and methylene blue reduction time for samples of refrigerated raw milk. Line represents fitted regression curve. (From Lück (1982), reproduced by permission of Verlag Th. Mann, Germany.)

of the different bacterial species vary, the reducing activity of bacteria is not lowered by clumping whereas the plate colony count and microscopic clump count are lowered, and substances such as antibiotics may inhibit bacterial growth. In addition, somatic cells at levels of about $1 \times 10^6\,m^{-1}$ reduce resazurin at a rate not dissimilar to that resulting from the same number of bacteria. Dye reduction tests are not suitable for classifying milk with low bacterial counts of $< 10^5\,ml^{-1}$. Pre-incubation of the samples before testing improves the sensitivity of the method but prolongs the time needed, making it no longer a rapid method.

Electrical methods

The growth of microorganisms results in changes in the composition of the culture medium as nutrients are converted into metabolic end products. Complex uncharged molecules, such as carbohydrates or lipids, are catabolised to smaller charged molecules such as lactic acid and acetic acid. Charged molecules such as proteins and polypeptides, are converted via amino acids into ammonia and bicarbonate. As growth proceeds, these processes lead to a decrease in the overall impedance of the medium as conductance and capacitance increase. Electrical changes in microbial cultures provide a means of detecting microorganisms and their metabolic effects.

Impedance is the resistance to flow of an alternating current through a conducting material. It is a complex quantity which is dependent upon the frequency of the alternating current employed, and can be described in terms of resistors, capacitors and inductors. The model of a resistor and capacitor in series seems to explain the electrical circuit formed when a pair of metal electrodes is placed into a microbiological medium (Hadley and Senyk, 1975). The relationship for this model is as follows:

$$Z^2 = R^2 + \frac{1}{(2\pi f C)^2}$$

where Z is impedance, R is resistance, C is capacitance, f is frequency of the alternating current. Impedance changes can be detected by passing a small alternating current through the medium and comparing the impedance of the inoculated medium with that of the uninoculated medium.

Electrical changes are a function of bacterial growth and replication. The threshold for detection depends upon the organism(s) and the media. In general, most media give detection thresholds at 10^6–10^7 organisms ml^{-1}. If the sample inoculated into the medium contains fewer organisms than this, there will be no detectable signal until the organism(s) have replicated sufficiently to reach the threshold number (Fig. 11). The detection time depends upon the initial number of viable organisms present in the sample and their specific growth kinetics. Variations in growth rates of different organisms may give rise to errors in the estimation of bacterial numbers. Similar numbers of a fast and slow growing organism will result in very different detection times. The incubation temperature used in impedance measurements is usually considerably higher than the storage temperature of the food. For example, a temperature of 25°C may be used for the assessment of a product normally stored under refrigeration. If the product has a mixed microbiological flora, then the types of organism(s) causing impedance changes may not necessarily be those causing spoilage of the product under storage conditions.

Generally, there is an inverse relationship between $\log_{10}$ plate colony count per ml or per gram and the impedimetric detection time. The method has an inherent delay whilst growth occurs, generally taking >6 h to detect 10^5 bacteria ml^{-1}. Impedance measurements cannot be regarded as rapid in the strict sense of the word, but they are particularly useful for the screening of large numbers of samples as little sample preparation is required.

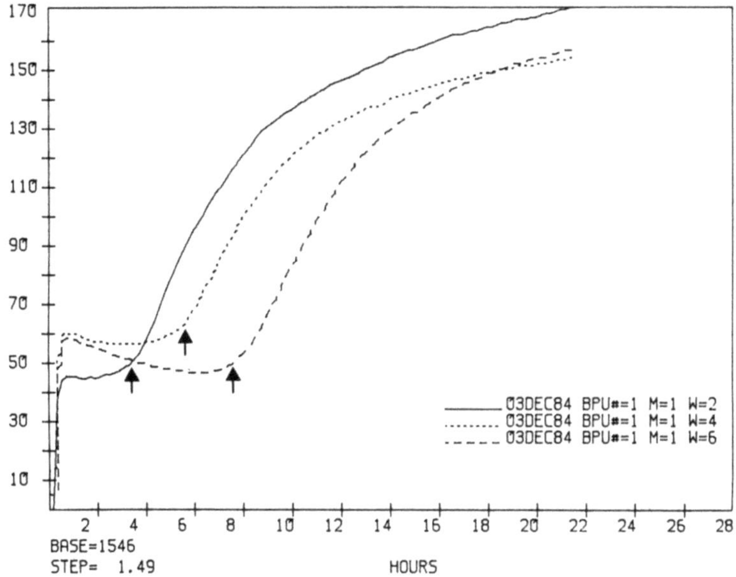

Fig. 11. Impedance curves for modules containing different numbers of *Escherichia coli*. Arrow indicates detection time. (———) 5×10^5; (····) 5×10^4; and (----) 5×10^3.

Automated impedance measurements have been used to assess the microbial content of raw and heat-treated milk (Cady *et al.*, 1978; O'Connor, 1979). Using an 8·5 h detection time, milks could be classified into two groups containing either more or less than 10^5 bacteria ml^{-1}. There was an 81% agreement between the classifications obtained by impedance and plate colony count methods, with the misclassifications equally distributed either side of the impedance detection limit (Fig. 12). Firstenberg-Eden and Tricarico (1983) have reported an improved correlation ($r = 0{\cdot}95$) between impedance detection time and place colony count. Samples containing $>10^5$ bacteria per ml (total plate colony count) or $>10^5$ bacteria ml^{-1} (mesophilic plate colony count) can be detected in *c.* 16 and 4 h, respectively.

Bioluminescence

The functional significance of adenosine-5'-triphosphate (ATP) in the metabolism of living cells suggests that its assay should be an excel--lent monitor of biological activity in the sample. The firefly luciferase reaction, where light is produced by an enzyme reaction, is frequently

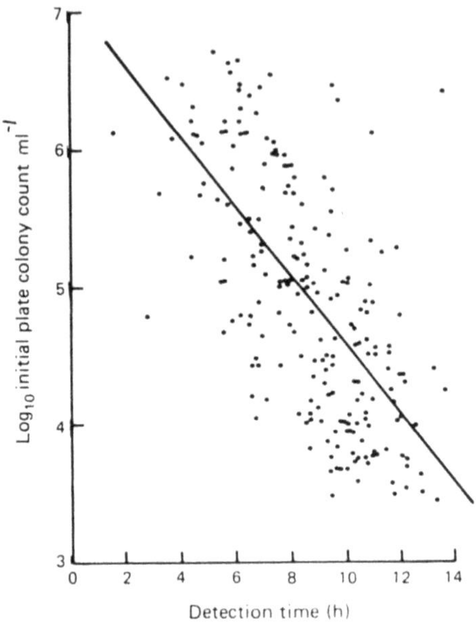

Fig. 12. Relationship between $\log_{10}$ initial plate colony count and impedance detection time for samples of refrigerated raw milk. (From O'Connor (1979), reproduced by permission of the Agricultural Research Institute, Republic of Ireland.)

used as an assay as it is specific for ATP. The reaction occurs as follows:

$$E + LH_2 + ATP \xrightarrow{Mg^{2+}} E.LH_2\text{-}AMP + PP$$

$$E.LH_2\text{-}AMP + O_2 \longrightarrow E + L + AMP + CO_2 + h\nu$$

where E is firefly luciferase; LH_2 is reduced luciferin; ATP is adenosine-5′-triphosphate; $E.LH_2$-AMP is luciferyl adenylate complex; PP is pyrophosphate; L is oxidised luciferin, and $h\nu$ is light. The quantity of light recorded by a photometer is proportional to the concentration of ATP.

There are two factors which adversely affect the estimation of bacterial numbers by measurement of ATP: first, non-bacterial ATP and, second, quenching of the emitted light. Many biological materials, such as milk, contain non-bacterial ATP which must be destroyed before an accurate estimate of bacterial ATP can be made. This necessitates lysing the somatic cells and then incubating with the enzyme apyrase to destroy the ATP released. Following this treatment, the bacteria are chemically disrupted and the released ATP measured. Failure to destroy the non-bacterial ATP will give an elevated estimate of bacterial ATP, and hence bacterial numbers.

There is a relationship between the total ATP content and bacterial numbers in milk, but the correlation is poor (Britz *et al.*, 1980). This is presumably due to the varying levels of somatic cell ATP. Selective measurement of bacterial ATP in milk can be made by first destroying the somatic cell ATP (Bossuyt, 1981, 1982). For raw milk, the relationship between bacterial ATP and the plate colony count in the range 10^4–10^7 ml^{-1} had a correlation coefficient of 0·93 (Fig. 13). There is little increase in apparent bacterial ATP over the rage 10^4–15^5 bacteria ml^{-1}, presumably because of residual somatic cell ATP influencing the result. Repeatability of this method is poor at levels of 10^5 bacteria ml^{-1} and below. Whilst the method may be useful for rapidly detecting raw milk of very poor bacteriological quality ($>10^6$ ml^{-1}), it is unlikely to be sufficiently accurate for estimating bacterial numbers in good quality milk.

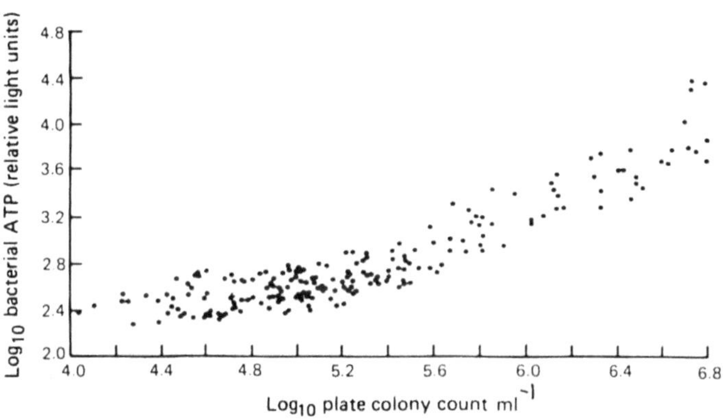

Fig. 13. Relationship between $\log_{10}$ plate colony count and $\log_{10}$ bacterial ATP content of samples or refrigerated raw milk. (From Bossuyt (1982), reproduced by permission of Verlag Th. Mann, Germany.)

Foss Electric have developed the BactoFoss, a fully automated instrument using ATP-bioluminescence for estimating the numbers of bacteria in raw milk. The use of membrane filtration to retain the bacteria while permitting the removal of ATP from somatic cells has considerably improved the sensitivity of the method. The manufacturers have reported a good correlation between the BactoFoss result and plate count ($r = 0{\cdot}92$) and a lower sensitivity of 10 000 bacteria ml^{-1}. The instrument can analyse *c.* 20 samples per hour and takes only 3 min to give a result.

Enzyme-linked immunosorbent assay (ELISA)

ELISA detects and amplifies antigen–antibody reactions by using covalently bound enzyme–antibody molecules. The presence of the enzyme (indicating presence of the antigen) is detected by addition of the appropriate substrate. Detection systems are usually designed to produce a colour change which can be quantified by a microtitre plate reader. ELISA tests can be used in two modes, qualitatively to determine presence or absence, or quantitatively to determine the amount of antigen present. ELISA kits often depend on the adsorption of either the antibody or antigen onto a solid phase, e.g. wells of a microtitre plate, the surface of plastic beads or plastic stick. The choice of antibody (or antibodies) used determines the specificity of the ELISA assay which can range from genus-specific to strain-specific.

The principle on which ELISA methods are based prevents them from being of use in determining total bacterial counts in milk and milk products. However, they can be used to detect pathogens in dairy products. In addition, ELISA methods can be used to detect viral diseases of cattle by analysing the milk, blood or serum from individual animals, and also the presence of aflatoxins in cattle feed.

DNA technology

There are a number of new rapid microbiological methods which use DNA technology, e.g. gene probes, polymerase chain reaction (PCR), *lux* genes and ice nucleation. By their nature, these tests are usually species or strain specific and have similar limitations and advantages as ELISA methods, i.e. being of little or no use for determining total counts of bacteria but being extremely useful for detecting specific pathogens. These are 'emerging' technologies that will become widely available over the next few years, probably in the form of commercial diagnostic kits.

The most widely accepted of these methods are gene probes, which utilise the principles of nucleic acid hybridisation (Wolcott, 1991). This enables probes to be tailored to detect and identify microorganisms at the family, species or sub-species level. DNA probes can be highly sensitive, specific and easy to use. Suitable target molecules for the probes can be genomic DNA, plasmid DNA, ribosomal RNA (rRNA) or messenger RNA (mRNA). The PCR has applications for the amplification of DNA within DNA probe-based assay systems. The PCR is dependent upon a series of processes which are repeated several times. Each reaction cycle doubles the amount of specific sequence DNA present at the beginning of that cycle. In a few hours a million-fold amplification of the original

DNA is possible. The PCR can be used to increase the sensitivity of DNA-based detection systems.

In some *Vibrio* spp., bacterial luciferase catalyses the flavin-mediated oxidation of tetradecanal with the concomitant emission of blue-green light. The genes responsible for reactions of the fatty acid reductase complex (*lux* genes) have been cloned and their functions identified (Hill and Stewart, 1991). The *lux* genes have now been cloned into a range of different bacterial species such as *Salmonella* spp., *Escherichia coli* and *Listeria monocytogenes.* Bioluminescent bacteria have been used to study the effect of heat, freezing and biocides, giving results in minutes which are comparable to those of conventional methods (Stewart, 1990). Ulitzur and Kuhn (1987) described a novel concept for the enumeration and identification of microorganisms. The principle requires the introduction of the *lux* genes into the genome of a bacteriophage. The recombinant phages lack the intracellular biochemistry necessary for light production and are therefore dark. Infection of the host bacteria by the phage leads to the expression of host phage genes and within 30–50 minutes the additional *lux* genes. The result of phage infection is bioluminescent bacteria which are then readily detectable.

Ice nucleating bacteria are found in few genera. Certain species of *Pseudomonas*, *Erwinia* and *Xanthomonas* possess ice nuclei which initiate the formation of ice in water at temperatures as high as -2°C, compared to most inorganic ice nucleators which do not show activity until -8°C. The phenotypic characteristic of ice nucleation is probably encoded in a single gene. Ice nucleation genes and proteins have been used as sensitive labels in immunoassays. The detection of freezing events is made simple by the use of dyes that fluoresce in supercooled water but are quenched and change colour when the water freezes.

Limulus lysate test

The Limulus test can be used to rapidly and specifically determine the cumulative content of Gram-negative bacteria in foods. Gram-negative bacteria produce a lipopolysaccharide (endotoxin) which is a high molecular weight complex; it is not produced by Gram-positive bacteria. The lipopolysaccharide is generally released from the bacteria into the surrounding medium after death and lysis of cells, although some may be released by viable cells.

Present in the blue blood of the horseshoe crab, *Limulus polyphemus*, is a nucleated cell, called an amoebocyte, the cytoplasm of which is densely packed with granules. Limulus blood clots in the presence of bacterial lipopolysaccharide. All the necessary clotting factors are contained in an extract of the amoebocyte granules, called Limulus lysate.

The Limulus test is specific for lipopolysaccharide and very sensitive. As little as 10^{-12} g lipopolysaccharide ml^{-1} can be detected, occasionally even 10^{-15} g ml^{-1}. A single Gram-negative bacterium contains approximately 10^{-14} g lipopolysaccharide; because of the extreme sensitivity of the test, all utensils must be absolutely free from lipopolysaccharide.

For the Limulus Test, a 10-fold dilution series of the sample is prepared and equal volumes of Limulus lysate and diluted sample are mixed in a test tube. The tube is then incubated at 37°C for 4 h, before being inverted and read. If the mixture remains unchanged and runs down the wall of the tube then that dilution of the sample does not contain lipopolysaccharide. If a firm opaque gel is formed which sticks to the bottom of the tube, then that dilution of the sample contains lipopolysaccharide. Generally, visual reading of 10-fold dilutions will give sufficient information about the level of lipopolysaccharide present in the sample. The accuracy of the method can be increased by using a two-fold dilution series.

There have been only a few reports of the use of the Limulus test for assessing bacterial numbers in raw milk (Terplan *et al.*, 1981; Hansen 1982; Hansen *et al.*, 1982). There is an indication that milks with higher numbers of psychrotrophic bacteria contain a higher level of lipopolysaccharide. However, for a given level of lipopolysaccharide the count of psychrotrophic organisms may vary by as much as 2 log cycles (Fig. 14).

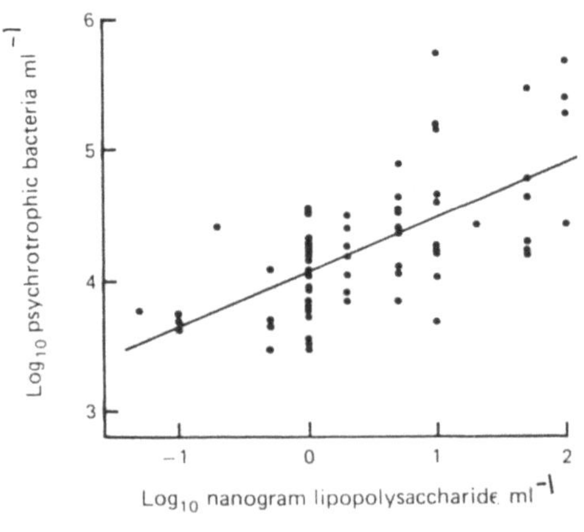

Fig. 14. Relationship between $\log_{10}$ psychrotrophic bacteria and $\log_{10}$ lipopolysaccharide for samples of refrigerated milk. (From Hansen *et al.* (1982), reproduced by permission of Cambridge University Press, UK.)

Pyruvate

The determination of pyruvate, which is an intermediary metabolite in bacterial metabolism, has been suggested as a method of assessing the bacteriological quality of milk (Tolle *et al.*, 1972). The estimation of pyruvate is rapid, inexpensive, accurate, and can be carried out automatically (Suhren, 1982). Immediately after production milk contains 0·5–1·5 mg pyruvate ml^{-1}, but this value does not reflect the initial viable count. Both the initial pyruvate level and the increase in pyruvate after storage correlates with the Wisconsin mastitis score, which suggests that somatic cells contribute to the pyruvate content of milk. For individual farm and silo milks stored at refrigeration temperatures, there is not a close relationship between pyruvate values and viable counts determined at intervals during storage (Cousins *et al.*, 1981). The relationship between the pyruvate content of milk and the bacterial count varies with both the temperature of storage and the age of the milk (Fig. 15). The pyruvate content of milk may be used to indicate milks with $>10^6$ bacteria per ml, but it does not give an accurate estimation of bacterial numbers.

Radiometry

Most radiometric methods are based on the principle that microbial growth in media can be monitored by measuring $^{14}CO_2$ released during

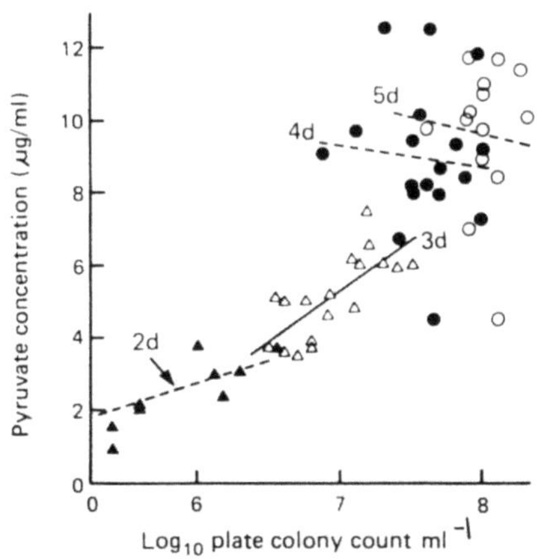

Fig. 15. Relationship between pyruvate concentration and $\log_{10}$ plate colony count for silo milks of (▲) 2 days, (△) 3 days, (●) 4 days, and (○) 5 days of age stored at 5°C. Line repesents fitted regression line. (———) Significant relationship ($p<0{\cdot}05$), and (------) non-significant relationship. (From Cousins *et al.* (1981), reproduced by permission of Cambridge University Press, UK.)

the metabolism of radio-labelled nutrients. These methods take from 2 to 24 h to complete depending upon the numbers of microorganisms present in the sample. As with impedance measurements, the detection time (the time taken for the measurement of a specific concentration of $^{14}CO_2$) is generally inversely related to bacterial numbers. Radiometric methods are perhaps best suited to determining the sterility of products as they are particularly sensitive, given a sufficient period of incubation. They can also be used to detect samples containing large numbers of organisms within a few hours. The concentration or rate of production of $^{14}CO_2$ does not, however, always reflect the initial number of viable micro-organisms.

The usefulness of $^{14}CO_2$ production from several radiometric substrates has been assessed as an indicator of the bacteriological quality of raw milk (Cogan and O'Connor, 1977). The method took about 10 h to detect $1{\cdot}9 \times 10^5$ bacteria ml^{-1} with very wide confidence limits of 2×10^4–$1{\cdot}6 \times 10^6$ bacteria ml^{-1} (Fig. 16). The method as described for milk seems to be of little practical use. The poor prediction of the initial viable count from the detection time may be due to a number of different factors; different species of bacteria in the milk have different growth rates, respiration of glucose yields $^{14}CO_2$ whereas fermentation may not, somatic cells may metabolise the radio-labelled substrates, and meta-

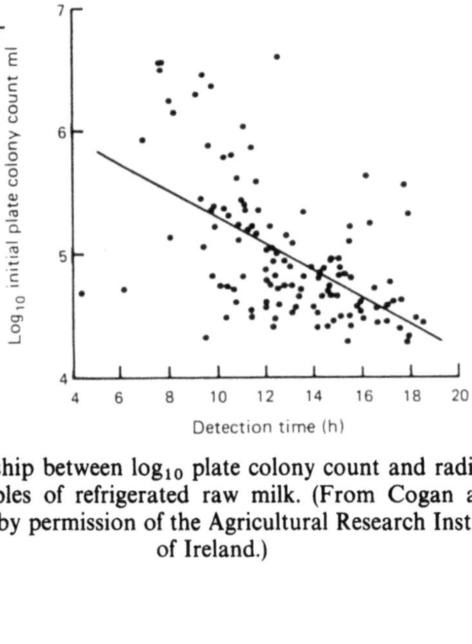

Fig. 16. Relationship between $\log_{10}$ plate colony count and radiometric detection time for samples of refrigerated raw milk. (From Cogan and O'Connor (1977), reproduced by permission of the Agricultural Research Institute, Republic of Ireland.)

bolism is related to the number of individual bacteria, whereas the plate colony count records the number of viable clumps of bacteria.

SPECIFIC MICROBIOLOGICAL METHODS FOR MILK AND MILK PRODUCTS

Raw Milk

The microbiological flora of raw milk varies considerably with source, and the time and conditions of storage. High numbers of specific groups of organisms often indicate the likely source of contamination and, therefore, selective enumeration may aid trouble-shooting.

Streptococci are frequently the causative organisms responsible for bovine mastitis. These bacteria can be selectively enumerated by plating on Selective Streptococci Medium (Cousins, 1972). If the total count is high and these organisms form >75% of the bacterial flora of a bulk farm supply, then it is likely that there is mastitis in the herd. Generally, once the infected animal is detected and its milk excluded from the bulk milk, then the counts return to normal levels.

Coliform bacteria are also responsible for mastitic infections in cows. Coliforms can be selectively enumerated on Violet Red Bile Agar (American Public Health Association, 1972). This group of bacteria is less useful as an indicator of mastitis than streptococci, as coliforms are also introduced into the milk from faeces via dirty teats. Increased numbers of coliforms in the milk may, therefore, indicate mastitis or poor udder preparation prior to milking.

The numbers of spore-forming bacteria present in raw milk can be determined by heating a sample of the milk at 80°C for 10 min, and then plating on Milk Agar. Spore-forming bacteria generally originate in the faeces and gain entry to the milk via dirty teats. The numbers of spore-forming bacteria in milk are a possible index of faecal contamination. Generally, the number of spores in the milk does not increase during storage, as these bacteria do not outgrow and sporulate under these conditions. However, the spore count may significantly increase as a result of contamination with milk and water residues, e.g. in tankers or equipment, where conditions may permit outgrowth and sporulation.

Psychrotrophic bacteria are mainly responsible for the deterioration in the microbiological quality of raw milk stored at refrigeration temperatures. Selective enumeration of this group indicates the time that the raw milk can be stored before processing. Various methods exist for enumerating psychrotrophs. The standard method is to plate on Milk Agar and enumerate colonies after incubation for 10 days at 7°C. As this method is time consuming, more rapid methods have been developed, e.g. enumeration of colonies after 24 h incubation at 25°C and the addition of crystal violet to inhibit the growth of Gram-positive bacteria.

Impedance methods can be used to determine the level of psychrotrophic bacteria in raw milk. The detection times for $10^5\,ml^{-1}$, $10^4\,ml^{-1}$ and $10^3\,ml^{-1}$ are *c.* 16, 19 and 23 h, respectively (Firstenberg-Eden and Tricarico, 1983).

Pasteurised Products

The major cause of spoilage of correctly refrigerated, pasteurised milk products are psychrotrophic bacteria which are post-pasteurisation contaminants. Assuming that the product is efficiently pasteurised and recontamination with the raw product is avoided, then these bacteria gain entry to the product from the equipment surfaces (notably the filler heads), the package or bottle, or from the air during filling. Very small numbers of psychrotrophic Gram-negative rods, mainly *Pseudomonas* species, can markedly reduce the shelf-life of these products. Commercial silo milks which were in-bottle pasteurised, thereby avoiding post-pasteurisation contamination, had a shelf-life of 14–32 days at 5°C whereas the commercially bottled pasteurised milk had a shelf-life of 8–17 days (Schröder *et al.*, 1982).

The total viable count of pasteurised milk products is of little use in detecting post-pasteurisation contamination due to the higher numbers of heat-resistant bacteria (mainly spore-formers). Plate counts made using Violet Red Bile Agar can be used to detect gross contamination; 1 ml of product should be free from coliform bacteria. The technique can be made more sensitive by plating 5 ml of product on a large Petri dish. As the initial numbers of post-pasteurisation contaminants responsible for reducing the shelf-life of refrigerated milk products are below the detection limit of most microbiological methods, it is generally accepted that some form of pre-incubation is required to increase the numbers prior to enumeration.

One method which is widely used, particularly in the USA, is the Moseley Test in which the plate count is used to measure an increase in bacterial numbers in the product stored at 7°C (American Public Health Association, 1972). This method is impractical in the context of a modern

Dairy Industry as the pasteurised product in the distribution chain is spoiling at the same rate as that in the laboratory, and then it takes a further 2–3 days for the plate colony count to give a result. It is, therefore, likely that the consumer will detect any problems arising from poor production methods before the quality control laboratory. The modern Dairy Industry needs a rapid method for detecting post-pasteurisation contamination which, ideally, predicts the eventual shelf-life of the products.

Pasteurised dairy products contain relatively large numbers of Gram-positive bacteria, *c.* 1000 ml^{-1}, and small numbers of post-pasteurisation contaminants, often less than 1 ml^{-1}. Pre-incubation of samples of the products with chemicals inhibitory to Gram-positive bacteria, e.g. benzylkonium chloride and crystal violet, permits the selective enrichment of Gram-negative bacteria (Langaveld *et al.*, 1976). Following selective enrichment, a number of microbiological methods have been used to detect an increase in bacterial numbers, electrical changes, or to predict the eventual shelf-life of the product. Bioluminescence methods have been used to determine the numbers of post-pasteurisation contaminants in pasteurised milk using a Most Probable Number approach (Waes and Bossuyt, 1982), i.e. presence or absence in 1 litre, 100 ml, 10 ml and 1 ml portions.

Impedance methods have been used to determine the presence or absence of post-pasteurisation contamination in pasteurised milk (Bossuyt and Waes, 1983). Using crystal violet, nisin and penicillin as the selective inhibitors, Griffiths *et al.* (1984) showed that pre-incubation for 25 h at 21°C followed by use of either the DEFT or bioluminescence method could indicate the extent of post-pasteurisation contamination of cream within 26 h of production. These methods could predict whether the product would keep for more or less than 7 days at 6°C. A pre-incubated DEFT count has been used successfully to predict the eventual keeping quality of pasteurised milk within 24 h of production (Rodrigues and Pettipher, 1984). Samples could be divided into keeping quality groups of $<4{\cdot}5$, >6, $>7{\cdot}6$, $>9{\cdot}1$ or $>10{\cdot}6$ days with 80–95% correctly classified.

The use of selective enrichment techniques followed by rapid microbiological methods offers the dairy industry the opportunity to predict the eventual quality of pasteurised milk products. The introduction of these methods should speed the detection of poor production techniques and reduce economic losses.

Products Treated at Ultra-high Temperature (UHT)

UHT products, such as milk and cream, should be free of viable microorganisms if correctly processed and aseptically packaged. Occasionally, poor processing or packaging techniques or contamination of the packaging material may lead to the presence of viable microorganisms which can considerably reduce the shelf-life of these non-refrigerated products. Microbiological methods are insufficiently sensitive to detect very low levels of contamination without prior pre-incubation. In the Standard Method (Statutory Instrument, 1977) the package is pre-incubated at 30–37°C for 24 h and then 0·01 ml is plated, and if more than 10 colonies appear, the sample is considered to be contaminated and must be re-tested for confirmation.

The result of sterility testing of UHT products could be obtained faster if a rapid microbiological method were used in place of a viable count. The DEFT count has already proved useful for UHT milk (Pettipher and Rodrigues, 1981), and it is probable that flow cytometry bioluminescence and electrical methods could also be used. The use of a rapid method would permit the release of products 2 days earlier than if the standard method was used.

The Limulus lysate method for detecting the lipopolysaccharide of Gram-negative bacteria can be used on UHT products to give an indication as to the quality of the raw milk prior to processing. The lipopolysaccharide survives the heat treatment, and the concentration in the product is proportional to that in the raw milk. The lipopolysaccharide concentration is, therefore, a cumulative index of contamination with Gram-negative bacteria. The Limulus lysate test can be used to detect UHT products made from raw milk containing high numbers of Gram-negative bacteria (Südi, 1982).

Starter Cultures

Starter cultures of single or mixed strains of streptococci an lactobacilli species are widely used in the Dairy Industry, particularly for cheese and yoghurt manufacture. Although pre-prepared, freeze-dried cultures are commercially available, many manufacturers prefer to produce their own starter cultures on grounds of cost. It is advisable to check on the viability or activity of the starter culture before use, as a poor fermentation can result in considerable economic loss.

Generally, mixed cultures are used on a rotational basis as inocula to reduce possible detrimental effects of phage build-up. Plate counts can be used to monitor the numbers of bacteria in starter cultures, but this is impractical as results are retrospective. Impedance detection times have been shown to correlate well with lactic acid production by starter cultures and, therefore, electrical methods may provide useful alternatives. The DEFT has been used to determine the numbers of bacteria in starter cultures, and it is probable that the bioluminescence method could also be used successfully.

The transfer of the *lux* genes into *Lactobacillus casei* and *Streptococcus lactis* provides a potential reagent for the detection of antibiotics and bacteriophage in milk prior to fermentation (Stewart, 1990). The ChemFlow system has been used successfully to assess the activity of brewing yeast prior to pitching. Flow cytometry could be used to measure the activity of dairy starter cultures prior to inoculation of milk.

Yeasts as Spoilage Organisms

Yeast are the primary spoilage organisms of yoghurt, especially fruit yoghurt, due to contamination of, or growth in, the fruit mix before blending. It is, therefore, necessary to determine the numbers of yeasts in the fruit mix and final products. Conventional methods, such as plating on Rose Bengal Agar, take 3–5 days to give a result, often after the batch of fruit mix has been used. Electrical methods, based on either the impedance or capacitance signals, can be used to detect yeast cells in spiked fruit mixes. It takes *c.* 14 h to detect 100 cells ml^{-1} (Fleischer *et al.*, 1984). Results suggest that pre-incubation to increase the initial numbers, combined with one of a number of detection methods, can be used to monitor fruit mixes and finished products, such as yoghurt, for yeast contamination. The pre-incubated DEFT method can detect as few as 1 yeast g^{-1} in some products (Pettipher, 1987). Factory trials showed that there was good agreement ($r = 0{\cdot}94$–$0{\cdot}98$) between the ChemFlow counts and plate counts for Fromage Frais containing fruit, and that the rapid method produced a result 2–3 days earlier than the conventional method (Dumain *et al.*, 1990).

Rapid Methods for the Detection of Pathogens

Using a microcolony-DEFT method, it is possible to get reliable estimates of numbers of specific bacteria, such as coliforms, *Pseudomonas* spp. and *Staphylococcus* spp., in products within a working day (Rodrigues and Kroll, 1988). Electrical methods can be used to detect the presence of *Salmonella* spp. in foods, taking *c.* 48 h to give a negative or suspect positive result (Pugh *et al.*, 1988). For flow cytometry, immunofluorescence has been used in conjunction with measurement of DNA content (by propidium iodine labelling) and size (by light scattering) for the specific identification of *Listeria monocytogenes* in milk (Donnelly and Baigent, 1986). ELISA methods can be used to detect the presence or absence of *Salmonella* and *Listeria* spp. in dairy products in 48 h, a saving of 1–3 days over conventional methods. Other antibody based methods, e.g. latex agglutination, can be used to rapidly detect the presence of *Staphylococcus aureus* toxins in dairy products.

Gene Trak have produced commercially available kits containing gene probes for *Salmonella* spp., *Listeria* spp. and *Escherichia coli.* Ice nucleation research has lead to the development of the Bind assay for the rapid detection of *Salmonella* spp. (Wobler and Green, 1990). The Bind assay uses attachment of a phage to its host and expression of an ice nucleation gene after the injection of phage DNA to achieve highly specific and sensitive detection of the target organism. A range of *Salmonella* spp. were detected at levels of less than $10\ ml^{-1}$ and no ice nuclei were formed when only non-target bacteria were present. The Bind assay principle could be applied to detecting a range of bacterial pathogens and it typically gives results in less than 1 h. Similar advances are being made in *lux* gene technology. Using a recombinant lux^+ enteric page, indicator bacteria could be detected in less than 1 h and without enrichment provided that they were present in food at levels of 1000 per g^{-1} or more (Stewart, 1990).

POSSIBLE FUTURE DEVELOPMENTS

In the first edition of this Chapter published in 1986, some predictions were made about developments in rapid microbiological methods. The following predictions have become a reality.

(a) The DEFT has been automated and miniaturised in the Cobra instrument with the advantages of more sample throughput and reduced disposable costs.
(b) Improvements have made the current Bactoscan more sensitive than earlier models.

(c) In the BactoFoss, an improved method is used for separating bacterial and somatic cell ATP, increasing the sensitivity of the technique over previous instruments.

In 1986 it looked as if there would be great advances made in the area of biosensors but this did not lead to products useful to the dairy microbiologist. Currently, biosensors are still mainly used for medical applications to detect low levels of enzymes, substrates or products (Lowe *et al.*, 1983) and for biomass measurement in fermenters (Matsunaga *et al.*, 1982).

Predicting the future from the starting point of 1992, I see two likely developments, the increased use of automated instruments for total counts, and a rapid advancement and increased use of the newer technologies for detecting specific bacterial food pathogens and indicator organisms. If laboratory testing is rationalised into a few large laboratories testing hundreds of samples a day, then the use of fully automated instruments with inherent lower labour costs, such as BactoFoss, Bactoscan and Cobra, will make commercial sense.

ELISA methods are (in 1992) probably the most rapidly growing sector of microbiological testing with a variety of specific uses, e.g. the detection of *Salmonella* spp., *Listeria* spp. and viral diseases. During the next few years an upsurge in the area of DNA technology should revolutionise microbiological testing as advancements already made are developed into commercial diagnostic kits. DNA probes, *lux* genes and ice nucleation are all technologies which have the potential to allow the dairy microbiologist to test products for specific bacteria in hours rather than days.

CONCLUSION

The Dairy Industry has been using conventional microbiological methods based on cultural techniques for years. There has been a considerable improvement in the microbiological quality of farm supplies as a direct result of regularly applying quantitative microbiological analyses and, importantly, changing the payment scheme to reward hygienic producers.

Under pressure from their major buyers, the dairies and creameries are changing from retrospective cultural techniques to more rapid methods. There are a variety of methods available based on different technologies and often with varying degrees of automation. Rapid microbiological methods have considerable advantages over their conventional counterparts, many providing results in less than an hour. This permits corrective action to be taken almost immediately, should this be required. These methods, perhaps with modifications or sample pre-treatment, can also be applied to milk products. They can, therefore, be used by the processor to monitor incoming raw supplies, processing and the microbiological quality of finished products (pathogens and shelf-life). The increased use of rapid microbiological methods by the Dairy Industry should lead to improved control of raw materials, processes and products, thereby reducing economic losses.

REFERENCES

American Public Health Association (1972). Standard Methods for the Examination of Dairy Products, Washington, DC, USA.
Bossuyt, R. G. (1981). *Milchwissenschaft*, **36**, 257.
Bossuyt, R. G. (1982). *Keiler Milchwirtschaftliche Forschungsberichte*, **34**, 129.
Bossuyt, R. G. and Waes, G. M. (1983). *J. Food Protect.*, **46**, 622.
Bradshaw, J. G., Francis, D. W., Peeler, J. T., Leslie, J. E., Twedt, R. M. and Read, R. B. (1973). *J. Dairy Sci.*, **56**, 1011.
Breed, R. S. and Brew, J. D. (1916). Technical Bulletin No. 49. New York Agricultural Experimental Station, Albany, NY, USA.
Britz, T. J., Bezuidenhout, J. J., Dreyer, J. M. and Steyn, P. L. (1980). *S. Afr. J. Dairy Technol.*, **12**, 89.
Cady, P., Hardy, D., Martins, S., Dufour, S. W. and Kraeger, S. J. (1978). *J. Food Protect.*, **41**, 277.
Claydon, T. J. (1975). *J. Milk Food Technol.*, **38**, 87.
Cogan, T. M. and O'Connor, F. (1977). *Irish J. Food Sci. Technol.*, **1**, 49.
Cousins, C. M. (1972). *J. Soc. Dairy Technol.*, **25**, 200.
Cousins, C. M. and Bramley, A. J. (1981). *Dairy Microbiol.*, **1**, 119.
Cousins, C. M., Sharpe, M. E. and Law, B. A. (1977). *Dairy Indust. International*, **42**, 12.
Cousins, C. M., Rodrigues, U. M. and Fulford, R. J. (1981). *J. Dairy Res.*, **48**, 45.
Donnelly, C. W. and Baignet, G. J. (1986). *Appl. Environ. Microbiol.*, **52**, 689.
Dumain, P.-P., Desnouveaux, R., Bloc'h, L., Leconte, C., Fuhrmann, B., De Colombel, E., Plessis, M.-C. and Valery, S. (1990). *Biotech. Forum. Europe*, **3/90**, 224.
Fifield, C. W., Hoff, J. E. and Proctor, B. E. (1957). *J. Dairy Sci.*, **40**, 588.
Firstenberg-Eden, R. and Tricarico, M. K. (1983). *J. Food Sci.*, **48**, 1750.
Fleischer, M., Shapton, N. and Cooper, P. J. (1984). *J. Soc. Dairy Technol.*, **37**, 63.
Fleming, M. and O'Connor, F. (1975). *Irish J. Agric. Res.*, **14**, 27.
Fondén, R. and Strömberg, A. (1978). *XX International Dairy Congress*, Publishers Congrilait, Paris, France, E 330.
Galbraith, N. S., Forbes, P. and Clifford, C. (1982). *Br. Med. J.*, **284**, 1761.

Gilchrist, J. E., Campbell, J. E., Donnelley, C. B., Peeler, J. T. and Delaney, J. M. (1973). *Appl. Microbiol.*, **25**, 244.
Griffiths, M. W., Phillips, J. D. and Muir, D. D. (1984). *J. Soc. Dairy Technol.*, **37**, 22.
Grogan, W. M. and Collins, J. M. (1990). *Guide to Flow Cytometry Methods.* Marcel Dekker Inc., New York, USA.
Hadley, W. K. and Senyk, G. (1975). *Microbiology*, **1975**, 12.
Hansen, K. (1982). *Kieler Milchwirtschaftliche Forschungsberichte*, **34**, 138.
Hansen, K., Mikkelsen, T. and Moller-Madsen, A. (1982). *J. Dairy Res.*, **49**, 323.
Hill, P. J. and Stewart, G. S. A. B. (1991). In: *Bioluminesence and Chemiluminescence Current Status.* John Wiley and Sons, Chichester, UK. p. 19.
Kielwein, G. (1982). *Kieler Milchwirtschaftliche Forschungsberichte*, **34**, 74.
King, J. S. and Mabbitt, L. A. (1984). *J. Dairy Res.*, **51**, 317.
Law, B. (1979). *J. Dairy Res.*, **46**, 573.
Langaveld, L. P. M., Cuperus, F., Van Breemen, P. and Dijkers, J. (1976). *Netherlands Milk Dairy J.*, **30**, 157.
Lowe, C. R., Goldfinch, M. J. and Lias, R. J. (1983). In: *Biotech. 83*, Online Publications, Northwood, UK, p. 633.
Lück, H. (1982). *Kieler Milchwirtschaftliche Forschungsberichte*, **34**, 108.Matsunaga, T., Karube, I., Teraoka, N. and Suzuki, S. (1982). *Euro. J. Appl. Microbiol. Biotechnol.*, **16**, 157.
McKinnon, C. H. and Pettipher, G. L. (1983). *J. Dairy Res.*, **50**, 163.
O'Connor, F. (1979). *Irish J. Food Sci. Technol.*, **3**, 93.
O'Connor, F. and Fleming, M. G. (1979). *Irish J. Food Sci. Technol.* **3**, 11.
Peeler, J. T., Gilchrist, J. E., Donnelley, C. B. and Campbell, J. E. (1977). *J. Food Protect.*, **40**, 462.
Peterkin, P. I. and Sharpe, A. N. (1980). *Appl. Environ. Microbiol.*, **39**, 1138.
Pettipher, G. L. (1991). *Lett. Appl. Microbiol.*, **12**, 109.
Pettipher, G. L. (1987). *Lett. Appl. Microbiol.*, **4**, 95.
Pettipher, G. L. (1983). *The Direct Epifluorescent Filter Technique.* Research Studies Press, Letchworth, UK.
Pettipher, G. L., Watts, Y. B., Langford, S. A. and Kroll, R. G. (1992). *Lett. Appl. Microbiol.*, **14**, 206–9.
Pettipher, G. L. and Rodrigues, U. M. (1981). *J. Appl. Bacteriol.*, **50**, 157.
Pettipher, G. L. and Rodrigues, U. M. (1982). *J. Appl. Bacteriol.*, **53**, 323.
Pettipher, G. L., Mansell, R., McKinnon, C. H. and Cousins, C. M. (1980). *Appl. Environ. Microbiol.*, **39**, 423.
Pugh, S. J., Griffiths, J. L. Arnott, M. L. and Gutteridge, C. S. (1988). *Lett. Appl. Microbiol.*, **7**, 23.
Rodrigues, U. M. and Kroll, R. G. (1988). *J. Appl. Bacteriol.*, **64**, 65.
Rodrigues, U. M. and Pettipher, G. L. (1984). *J. Appl. Bacteriol.*, **57**, 125.
Saarinen, K. (1984). *Meijeritietellinen Aikauskirja.* **XLII** (n:01) 33.
Schröder, M. J. A., Cousins, C. M. and McKinnon, C. H. (1982). *J. Dairy Res.*, **49**, 619.
Sharpe, A. N. and Kilsby, D. C. (1971). *J. Appl. Bacteriol.*, **34**, 435.
Sharpe, A. N. and Michaud, G. L. (1974). *Appl. Microbiol.*, **28**, 223.
Sharpe, A. N. and Michaud, G. L. (1975). *Appl. Microbiol.*, **30**, 519.
Sharpe, A. N., Biggs, D. R. and Oliver, R. J. (1972). *Appl. Microbiol.*, **24**, 70.

Sharpe, A. N., Peterkin, P. I. and Malik, N. (1979). *Appl. Environ. Microbiol.*, **38**, 431.
Statutory Instrument (1977). No. 1033, HMSO, London, UK.
Stewart, G. S. A. B. (1990). *Lett. Appl. Microbiol.*, **10**, 1.
Südi, J. (1982). *Kieler Milchwirtschaftliche Forschungsberichte*, **34**, 141.
Suhren, G. (1982). *Kieler Milchwirtschaftliche Forschungsberichte*, **34**, 117.
Suhren, G., Reichmuth, J. and Heeschen, W. (1991). *Bull. Inter. Dairy Fed.*, **256**, 24.
Terplan, G., Bierl, J., Von Grove, H. H. and Zaadhof, K. J. (1981). *Archiv fur Lebensmittelhygiene*, **32**, 15.
Tolle, A., Heeschen, W., Wernery, H., Reichmuth, J. and Suhren, G. (1972). *Milchwissenschaft*, **27**, 343.
Ulitzur, S. and Kuhn, J. (1987). In: *Bioluminescence and Chemiluminescence New Perspectives.* John Wiley and Sons, Chichester, UK, p. 463.
Waes, G. M. and Bossuyt, R. G. (1982). *J. Food Protect.*, **45**, 928.
Wobler, P. K. and Green, R. L. (1990). *Trends Food Sci. Technol.* **October**, 80.
Wolcott, M. J. (1991). *J. Food Protect.*, **5**, 387.

Chapter 8

Technology for the Developing Countries

J. T. Homewood

Nestec Ltd, Vevey, Switzerland

and

H. J. Hess

Nestec Ltd, Vevey, Switzerland

The reader, especially if without experience in developing countries, may well ask why a separate chapter, under such a title should be considered necessary in a textbook on the dairy industry. Fresh milk has been a household article, often obtained from one or more animals kept by each family, in much of Asia, the Middle East and Europe since earliest historical times. European immigrants brought cattle to North and South America, and to Australasia, from the 16th century onwards. With their herds came the early dairy technologies of curd and cheesemaking, butter or yoghurt production, long established in Asia and Europe.

What has happened to justify an existing technology differential, as implied in the title, for those countries having centuries of dairying tradition, yet still categorised as developing? To answer this question, it has been necessary to recall some major steps in the evolution of the dairy industry, and related technology, in the developed countries. For those confronted with problems of a specific stage in their own country's development, this may serve to provide parallels with earlier experience. Errors and waste of resource may thus be avoided. The significance in this earlier experience of rapid growth of urban population and its needs, stimulating collection, transport, new processing methods and better distribution systems, is important for those looking to the future in the developing world. The benefits derived by the farming community from these systems should also be kept in mind.

The example of developments in the last 30 or 40 years in India, which by itself constitutes an important part of the developing world, is useful to illustrate how timely measures can at least mitigate problems arising from rapid urban expansion, to the benefit of urban health and rural productivity.

HISTORICAL EVOLUTION OF THE DAIRY INDUSTRY

A Common Starting Point

Dairy cattle are believed to have been first domesticated some 6000–10 000 years ago, and are recorded in early documents and artefacts in India, Babylonia, Egypt and Old Testament Palestine.

Milk was known as a food of high value—but was also recognised as being highly perishable. Therefore its transformation into products of somewhat greater keeping quality, such as cheese, butter and curds has been known and practised for many centuries. In India, many products such as Ghee (Samna), Khoa or Rabri (Anon., 1941) are still widely produced. These are suitable for domestic or small-scale, artisanal production. Marco Polo (Latham, 1957) records the use of sun-dried skim-milk by Tartar cavalry in the mid-13th century. Domestic or artisanal producers can dispose of whey or other residues arising in small quantities, for human or animal feeding, so that wastage is reduced. However, the products can be of uncertain keeping quality. Butter and cheesemaking techniques were certainly transmitted to North and South America with the arrival of colonists in the 16th century, and later to Australasia and New Zealand. It is of interest, especially when considering the recent questioning of the value of milk products, to recall that the 'Mayflower' settlers in North America in 1625 lost half their numbers in their terrible first winter — and all their children under 2 years of age. They brought no cows with them; but subsequent parties were better advised, and were indeed obliged to bring a certain proportion of cattle, sheep and goats for each family; they fared better. Although many of the earliest importations into North and South America were mainly for meat and as draft animals, some were milked for domestic needs. These multiplied fast; in the Valley of Toluca, Mexico, their numbers are recorded as doubling every 15 months — first introduced in 1538, they numbered 150 000 20 years later. Impovement of dairy characteristics in the herds in Latin America followed in the late 18th and mid-19th centuries; artisanal cheese manufacture started in 1875 in Brazil (region Serra da Mantiqueira Minas Gerais Province).

What, therfore, took place in Europe and North America to differentiate these regions from the developing world? Improved dairy cattle had been bred in Holland, in the UK and elsewhere. There was, however, no significant difference in the technology used in dairying in the first surge of expansion into North and South America and Australasia.

The Differentiation of Technology

The differentiation of technology in the developed countries followed the industrial revolution from which their new 'developed' status stemmed. Demand was stimulated by population growth and concentration in urban manufacturing centres. Rapid improvements in road and rail systems made it possible first to bring fodder in increased quantities to urban cowsheds, and to remove manure. In England, these cowsheds were judged, anyway until to 1860s, to provide the best milk in towns, as their cows were better fed and often better housed than those on farms surrounding the towns. In the 1840s and 1850s, milk was retailed round the streets of London by men and women carrying, on a shoulder yoke, a pair of wooden or metal tubs holding 8 or 10 gallons in all (Whetham, 1964).

The reduction in transport costs brought about by railways had a complex effect on the market. It was cheaper to move fodder and dung to and from urban cowsheds; but quantities of milk from distant frams (often of poor quality due to unsatisfactory milking conditions, transport and lack of cooling) competed wherever people required milk for cooking rather than drinking. At that time, a premium would have been paid for milk 'warm from the cow', and many consumers would otherwise boil any milk before consumption (as is still done in many developing countries). The increasing volume of sales in towns in he 1860s led to the use of hand or horse-drawn carts, carrying bulk containers from which milk was usually ladled by a measured dipper into the consumer's jug. Whetham (1964) reported that 'First consumers from a churn got all the cream and the last only skimmed milk. Further, all dairymen were plagued with the dishonesty of their roundsmen, who sold more milk than they were allotted, with the help of water'.

An impulse towards the elimination of town cowsheds came from a combination of the recognition of the often poor hygienic state of these sheds, and from diseases which inflicted heavy losses on the urban dairies' herds in 1865–1866. Both factors led to a rapid increase in rail deliveries of 'fresh country milk', now held in higher esteem. Supplies to London by rail increased from 7 million gallons in 1866 to 20 million gallons in 1990 (Whetham, 1964).

A complex structure of contract purchase from farmers near to railway stations, and the purchase or sale of surplus quantities between retailers at the London railway terminal, led to the establishment of milk wholesalers, who provided churns, supervised transport and dealt with deficiencies or surpluses in supply. These wholesalers began (Whetham, 1964) in the late 1870s to set up depots at country railway stations, where milk delivered by farmers could be properly cooled. This cooling was carried out in the first place by well water, which in England probably permitted cooling milk to not more than 15°C. Private milk retailing firms, growing in size, and wholesalers, provided a basis for the formation of limited companies and the provision of capital necessary for three technological advances which occurred at about this time.

First, in 1879, the centrifugal cream separator was exhibited at the Royal Agricultural Society's show, and this made butter manufacture a factory rather than a farm trade. Instead of standing milk for 24 h for cream to rise, milk could be run through the separator to give a continuous supply of fresh cream. Steam-powered churns and mechanical butter-workers could finish the task. The second new technology available in the 1870s to 1880s was mechanical refrigeration. At first, this facilitated imports of dairy products from America or Australasia, which depressed prices and caused farmers to turn from butter and cheesemaking to liquid milk sales; butter and cheesemaking co-operatives in the Midlands switched to cooling bulk liquid milk for the urban trade, and mechanical refrigeration, cooling milk to 4°C, was the best guarantee of success wherever the capital and technical skills necessary could be assured. The third technological innovation was the milk condensing plant. In the 1860s, processes of vacuum concentration had been applied to milk in the USA, resulting in the development of so-called sweetened condensed milk and unsweetened (evaporated) condensed milk. These were at first distributed in cities in bulk exactly as was fresh milk. The American Civil War created a demand for army requirements, and here sweetened condensed milk, which was protected (like fruit jams) from bacterial development by the osmotic pressure exerted by its sugar content and packed in sealed, tin-plate containers, was extensively used.

The storable product, rapidly adopted for ships' stores, was also used to supplement fresh milk supplies in towns. Condenseries, built in country districts, could supply remote towns without insurmountable risks of spoilage, and many were constructed from the 1860s and 1870s onwards, in the USA, in Switzerland, England and France. These factories stimulated development in tin-making on a factory scale, and the use of steam (as in the wholesalers' depots) for cleaning and sterilisation of farmers' cans. Trade in condensed milk soon crossed national frontiers, and a new range of dairy products, later expanded by milk-drying techniques, first on steam-heated rollers and later by spray-drying techniques, was added to the existing international trade in butter and cheese.

These three technologies, separation, cooling and condensing/drying, enabled further use to be made of improved transport systems, and later of techniques such as pasteurisation (originating in the brewery industry). But their most important outcome was, perhaps, the increased requirement of capital involved by participation in the dairy business. Individual producers could (and still do) distribute their milk as producer-retailers. But the larger enterprises involved in wholesaling, transport and processing were obliged to develop tighter controls and suppress fraudulent practices, such as those described above by Whetham: increase in scale and capital involvement enhanced the importance of the assurance of a good reputation in order to retain the customer. The concentration of milk handling and retailing also facilitated the work of health, weights and measures, and other public authorities.

Whetham (1970) stated: 'By 1914, the London trade was dominated by four or five wholesalers, equipped with country depots, a huge supply of churns, regular contracts with the railways, pasteurising plant and cold stores in their town dairies'. This should be contrasted with the situation described by Whetham as existing 15 years earlier: 'Bottled or specially treated milk provided a small fraction of the trade before 1900; only the medical profession and a few educated families were aware of the gross bacterial infection which contributed to such widespread diseases as infant diarrhoea and tuberculosis. Yet this growing interest in hygiene encouraged consumers to be suspicious of cheap milk, and to favour the large firms with cooling depots in country districts with steam-powered plant for washing churns and delivery cans'.

The importance of the improvements achieved by 1915 and later can be assessed. Beaver (1973), commenting on the infant mortality rates for England and Wales (1840–1970) illustrated in Fig. 1 refers to the striking reduction recorded (from 150 to less than 20 per 1000) after 1900 and claims that this reduction can be attributed in large part to improved

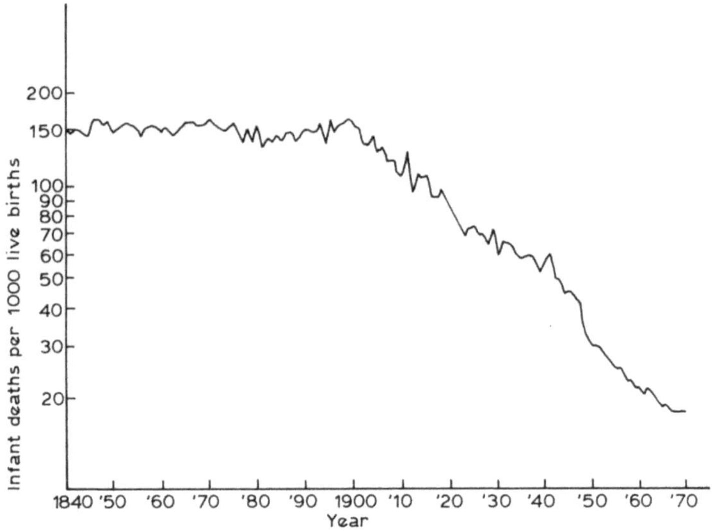

Fig. 1. Logarithmic scale chart of infant mortality rates, England and Wales, 1840–1970. (From Beaver (1973).)

availability of safe milk (sometimes as fresh milk, but also in the form of suitable infant foods) and the diminishing presence of poor-quality milk.

All in all, the result of concentration in the wholesaling, transport, retailing and processing of milk in England and Wales helped to bring about a marked — and justified — increase in public confidence in the end-product. This result, and the change in the image of milk vendor as at best a doubtful, and often a definitely untrustworthy, commercial partner, points the way for these wishing to build up dairy projects in countries where the conditions for such public confidence do not yet exist. Such confidence is as vital as the latest equipment, and can be harder to obtain.

The preceding paragraphs sketch only the broad outline of the main technological changes affecting the dairy industry in England and Wales. Something of the same sort occurred on a much larger scale in the US and also, with local differences due to dietary habits and other causes, in most of Europe. A host of smaller changes have added to the total result, including detailed refinement of machinery and equipment, and in recent years, changes in marketing structures which are still not complete. The simplified version of events given in this section is intended to help readers wishing to define the state of development elsewhere. It may also, perhaps, encourage those facing apparently adverse conditions for a new venture, to concentrate on the essentials, and not to forget the importance of building, and sustaining, public confidence in milk and milk products. As has been found in the developed countries, this calls not only for an increase in public control functions, but also and above all for the building of a sense of responsibility amongst those working in the industry, at every stage from the farm to the consumer's doorstep.

The Return to a Common Technology

In the developing countries, the technology of the 18th century European and North American producers and processors has sometimes persisted. This was, until 1947, broadly the state of affairs in India, which may be taken as an example of the rapid return to a common technology possible when circumstances are favourable, and Government policies also favour development.

Earlier conditions in India (Anon., 1941) can be judged from the following extracts from that report:

> 'Most of the village cattle are semi-starved and badly managed—about half of the milk required in urban areas is produced on the spot under most uneconomical and insanitary conditions ... duty-free import (of skimmed milk powder) is being used for the adulteration of whole milk—the measures at present used in the milk trade are both unsatisfactory and unsuitable—most milk is distributed under filthy conditions—vessels used have no lids and dirty straw, green grass and plugs made of old newspapers and rags are used instead ... pasteurised milk is at present available only at a few places but, even at these, the hospitals, etc., do not buy it always because of its high cost'.

It is certain that in India, with about one-sixth of the world's population, many of whom have a very low standard of living, conditions are still far from ideal. It has also happened that projects using modern technologies were badly sited or badly managed. Nevertheless, the centrally controlled use of available external aid in the form of low-cost supplies of skim-milk powder, and of modern technology in both the public and the private sectors, backed by adequate training at the national level, has made it possible to promote public health (Beaver, 1973), and to create the basis for transformation of an age-old industry. This illustrates the principle of self-help, where improvement in agriculture can help to solve problems of urban nutrition which are likely to become of increasing concern (Fox, 1984).

Other examples of modernisation have been based not only on a supplementation of local fresh milk supplies by toning quantities of skim-milk powder, but by large-scale use of imported constituents wherever little or no local fresh milk can be obtained. This has allowed the setting up of factories planned to replace imports of condensed or powdered full-cream milks, using recombining techniques first fully developed for supplies to American forces in the Pacific area in 1941–1945. Here, ice cream and fluid milk were the main requirements, but by the 1960s, trials had shown that sweetened condensed or evaporated milk, and even later milk powders and infant foods, could be manufactured by recombining butter oil (anhydrous milk fat) and suitable skim-milk powders. There were powerful economic arguments in favour of this, whenever dairy surpluses were being made available, as under PL 480 from the US, or later under the ECs surplus export scheme.

Such factories used existing modern technology with the few additional features necessary to ensure rapid and complete reconstitution of the dry powders or butter oil required. Plants of this type were set up in the 1960s in South-East Asia. They were also to be found from the 1970s in several West African countries. In the Far East, some increase in local milk production can be attributed to these plants, where governments have fixed progressively higher fresh milk prices. These can, perhaps, be borne for a time by local recombining industries if local fresh milk only constitutes a small part of their total raw material needs—in effect at the expense of the consumer of preserved milk products. Longer term, an equalisation scheme, allowing the burden of high local milk costs, if necessary to encourage an increase in the production, to be shared by all users of skim-milk powder (which goes to ice cream and yoghurt manufacturers) will be needed for further progress (Rampini, 1978).

As stated earlier, the area where specific technology for developing countries is called for is, above all, in fresh milk production and collection. The recombining plants themselves mostly use equipment and methods derived from their counterparts in the developed world.

THE ESTABLISHMENT OF A DAIRY INDUSTRY—PROBLEMS AND SOLUTIONS

In countries with some existing fresh milk production, the manufacture of local cheeses is generally the simplest way adopted for absorbing surplus milk at any distance from urban centres. This is usually a white soft cheese, farm produced, and with limited keeping quality, but requiring little investment for equipment or buildings. Such cheeses are consumed locally, and the whey can be fed to stock. The next development has often been that of a milk powder or condensed milk production, favouring the total milk production and its quality, but pushing the artisanal cheese production further away.

Urban development and the resulting demand for liquid milk and milk products can subsequently demand priority for pasteurised and ultra-high temperature (UHT) milk, soft cheese and yoghurts. The canned or dried milk operations then have to seek further afield for their milk supplies. Depending upon the scale of development in each particular instance, a check list which may be useful to those called upon to assess or plan a dairy project in a developing country, could be as follows. Only factors having a strong influence on ultimate success or failure of such a venture are considered.

Political and Economic Context

(a) Actual and estimated potential demand for end products of the venture; present and projected per capita consumption of such products and of alternative foods; present local production of potential raw or packing materials; possibility of procurement of individual ingredients.

(b) Existing trade in fresh milk; existing or potential price structures for fresh milk, and for milk constituents such as locally produced or imported butter, butter oil, and full-cream or skim-milk powders; availability of ancillary constituents, such as sugar or lactose, and probable incidence of other cost components such as labour, energy, packing materials and taxes.

(c) Which competing products, locally produced or imported, already exist? What are their volume, value and price structures?

(d) Existing or potential price structures for finished products forseen; do these permit economically self-sustaining operations or are subsidies or tariff protection necessary, and obtainable? In estimating the cost of local production, the effect of variances of, for example, $\pm 10\%$ or even 20% should be calculated for sales volumes, and for principal raw material and end product prices.

(e) Determine existing Government policies and practices affecting prices, particularly price controls or import constraints (quota systems) applied to finished products; do such constraints permit price or other adjustments in face of variations in raw material costs, fluctuations in consumer demand, inflation in local or imported cost elements? What other support will be available to the project if constraints mentioned above threaten its economic survival?

(f) What are Government, trades union, or other bodies' policies and practices affecting, in the shorter or longer term, the effective management of the project?

Fresh Milk

If the outcome of a preliminary investigation, along the lines of the check list above, is broadly favourable, the next step may be to investigate the availability of fresh milk. This may in some developing countries present no problem other than that of assessing competitive demand (from other established industries) for an existing resource. However, some other such countries may not yet possess an organised fresh milk production; and for these instances, a further and more extensive check list may be useful; this is outlined below.

(a) What are the characteristics of existing herds of milking animal (e.g. cow, buffalo, yak, chowrie, sheep, etc.)? What, if any, schemes exist for improving these characteristics? What are the customary levels of animal nutrition?

(b) What are the characteristics of cattle owners—are human nutritional levels adequate to allow collection of milk surplus to domestic requirements? What is their level of literacy, and their openness to motivation and innovation?

(c) What are the preconceptions of potential suppliers concerning the sale of milk? Concerning the slaughter of surplus stock and sale of meat? What is the price ratio beef (kg live weight)$^{-1}$ to milk kg^{-1} (should be around 4 or less)?

(d) Is there information concerning economics of milk production/stock raising in the form of comparisons of yield per hectare with alternative crops? Take into account suitability of terrain/climate for such alternatives and possible underemployment of family labour resources; what would be the social impact of regular payments for milk versus annual payments for other crops?

(e) What, if any, system of credit (Co-operative, Agricultural Credit Bank, etc.) is available to producers? What are the interest rates charged? What chance to secure loans do small or even landless owners of cows?

(f) What surfaces could be made available for fodder growing, or what possibilities exist for fodder collection by those lacking such surfaces? Resources such as crop residues or straw, cane tops, silage, molasses, trees and grasses from dividing strips in other cultivations, such as rice paddies, should be assessed; what new varieties of grasses or legumes or other fodder crops could be introduced?

(g) Is there potential for increased production through improved animal health brought about by education of producers and by increased availability of veterinary care and supplies? What is their present fertility and calving interval?

(h) Is there a possibility of integration of increased animal production into existing mixed or other farming patterns? Is the value of animal manure for fertilisation of other crops understood, or is this used for fuel, or neglected?

(i) Is there some form of custom or planning defining arable and grazing areas? Are producers nomadic or semi-nomadic? If so, consider problems of overstocking likely to arise if revenues from fresh milk sales are directed to herd increase—any new venture for meat processing may be a desirable complement rather than a competitor in such instances.

(j) Are main and farm or village access roads adequate for all weather use? If not, what plans or possibilities exist at State or local level to remedy this?

(k) Can the milk collection area (milk district) be extended in future times? Can such extensions reach into areas of varying pluviometry or irrigation, or with varying soil types and traditional crop patterns, to facilitate protection against dominance by the fortunes of one or two major crops which may contend with animal fodders? Do soil analyses reveal major or trace element deficiencies? Will there be an improvement of soil structures through increased animal manure?

(l) What quantities of milk are likely to be available for collection initially in the district (litres km^{-2} $year^{-1}$)? What competition is to be expected for supplied of fresh milk available in the district? What are the distances to major points of consumption of fresh milk from the district? Do the authorities concerned define a milk district for any project? Are there possibilities of agreements to limit uneconomic duplication of collection systems? What potential exists for long-term expansion of intake per square kilometre of a district, to meet increased sales requirements, or possible future reductions in the district for the benefit of other milk users?

MEASURES TO STIMULATE MILK PRODUCTION

It will be seen from these check lists that to establish a dairy industry in a developing country calls for more than a good knowledge of modern dairy technology. A grasp of politics will be needed—since this is an industry which for a given capital input can involve relatively large numbers of producers and very large numbers of consumers, linked by a product which is more perishable than most and, in many countries, is a basic foodstuff. Both producers and consumers are an element in politics. A sound commercial sense, some feeling for sociological questions, a good knowledge of animal husbandry and related animal health problems, and an appreciation of the evolution of infrastructural elements, will add to the chances of correct conclusions by a technologist assessing a potential milk district. Of course it will, in the end, be technology and economics which mainly determine the outcome of a project; but the technical man will do well to draw conclusions on the other aspects, from whatever resources are available. It is a great advantage if these resources are local ones, since people on the spot usually know best about their own region and its problems.

Analysis of the existing state of affairs may show adequate willingness by producers to sell surpluses of milk, promising potential for increased fodder production, and possibilities of increasing herd numbers by improved fertility (from better feeding) and by reducing calf mortality. There may be good future prospects for increased milk production per animal by improvements in veterinary care and through herd improvement (provision of bulls or artificial insemination with superior genetic characteristics). There may also be plans, or better, work in progress on infrastructures such as roads and irrigation systems.

Those preparing a project should not, however, assume that these necessary components of increased potential milk production will automatically be evoked by the creation of a new market for fresh milk, in the shape of a collection, processing and product marketing venture. A well conducted campaign to inform potential suppliers, a sound collection system, fair dealing over quality and prompt payment for supplies are essentials for success; but it will still be necessary to make a cool assessment of factors which will help or hinder such a development. The comprehensive survey by McDowell (1981) may suggest other positive or negative factors in a particular instance. Where the State or Local Government bodies provide services, their effectiveness should be checked, and the cost of providing any necessary additional back-up included in the venture costing. It may, for instance, become evident that supplied of high-quality seeds for fodder growing are not available (sometimes due to administrative, phytosanitary or import licensing obstacles). The possibility of overcoming such obstacles should then be assessed and costed.

After considering all likely obstacles to increased production, the additional cost of all necessary supports must be taken into account when estimating the total cost of fresh milk as a raw material in a dairy project. Typically, Field Services and Inspection may add between 2 and 5% to raw milk cost—and much more in the first years of a project.

Raw Milk Price Structures

In many developing countries, fat testing and payment per kilogram of milk fat is common. This helps to eliminate fraud (Rampini, 1978). However, where a rigid year-round price is paid, in a country where production drops seasonally in dry weather, suppliers are tempted by higher prices offered by itinerant milk vendors. This can best be met by a premium (which must be large enough to be perceptible—up to 15% on basic price) to those suppliers who maintain deliveries at such times (see Fig. 2). Payment on milk solids-non-fat is rarely encountered.

Social Factors

It has been sometimes observed that in close-knit, village farming communities there are frequently affinities and antipathies which can influence milk collection. Experience has shown that by setting up more than one collecting centre, at suitable points in a village, quantities collected can be substantially greater. Not only is the collection system physically more accessible to suppliers; they also have a choice as to whom they sell their milk.

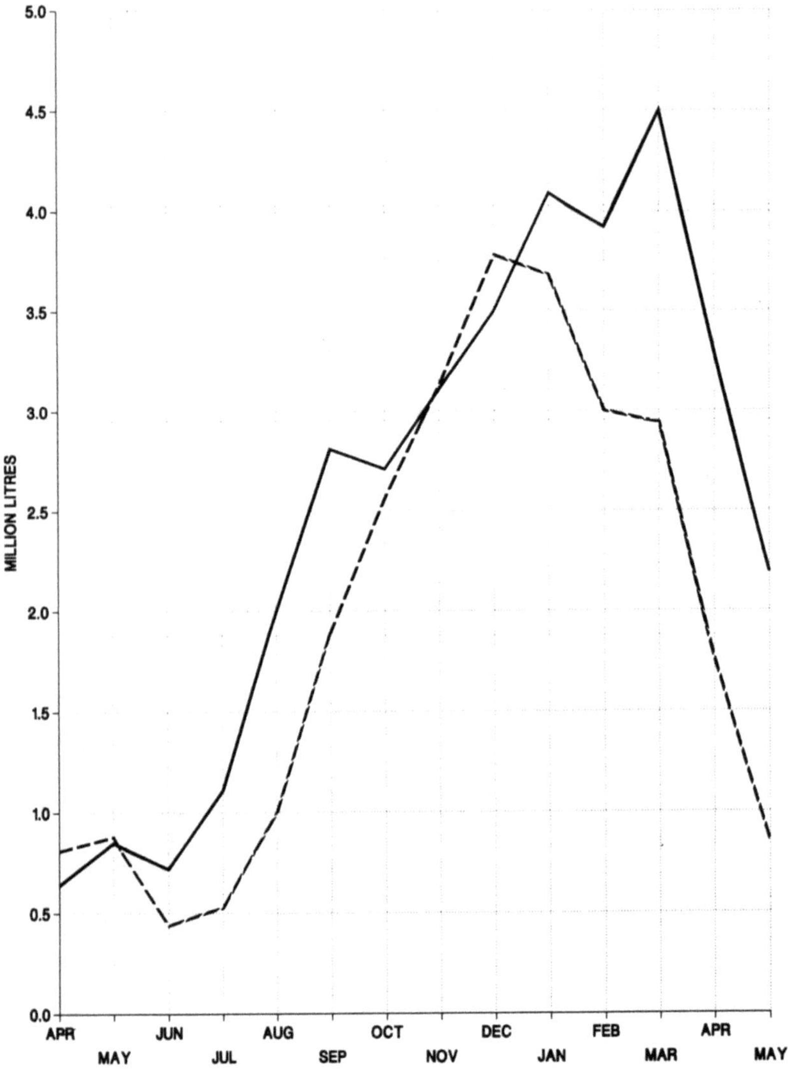

Fig. 2. Seasonal shift of peak and increase of intake over 1 year, caused by a 15% bonus paid for the first time in a low season, compared with pattern of intake collected in a larger area 3 years previously without premium. (---) Original intake from 19 000 suppliers in 9000 km^2; (———) intake developed from 22 500 suppliers in 6000 km^2 with premium payments.

COLLECTION OF RAW MILK

The movement of available supplies of raw milk from the producer's shed, yard or the point of the field where milking takes place, to the processor's plant, often takes place under very different conditions in developed and developing countries. In the latter, road systems are liable to be less developed (see Figs 3 and 4). The climate is often more arduous. Individual producers may supply only very small quantities, varying according to domestic needs. For all these reasons, methods adopted must be appropriate, as must the evaluation of possible technological aids. This evaluation may differ from that common in developed countries.

For those preparing a milk collection system, it must first be emphasised that this activity can cost more than 10% of the price paid for raw milk at the processing plant. This figure can be substantially higher if intermediate receiving and cooling stations, where quality is assessed and suppliers' deliveries are weighed, are to be provided and maintained. Even in those products with the highest added value, milk powders and concentrated canned milk, raw milk delivered to the processing plant often accounts for two-thirds of the total product cost ex factory. For pasteurised and sterilised whole milk in carton or bottle, the proportion

Fig. 3. Village road conditions in the wet season, Madagascar, encountered when prospecting for a milk district.

Fig. 4. Deterioration of road surfaces may prevent mechanisation of collection.

is even higher. The situation is therefore clear; over-investment, or inefficient organisation of raw milk collection will constitute a burden ultimately to be borne by the individual producer or consumer and may be too heavy to bear.

The milk collection system must correspond to the scale of the processing plant. This will be dictated in the first place by the market for its products; but also by the need to use sophisticated equipment, anyway for sterilised milks in carton or bottle, or for condensed or dried products. Here a certain minimum throughput must be maintained or else overheads and capital charges will be too heavy. Thus, the size of milk district will depend upon a calculation of available of potential litres $day^{-1}\ km^{-1}$, of collecting route, the road system, and the economic and marketable throughput necessary for the plant to prosper.

Very often, such simple calculations will be complicated by factors such as existing and potential competition for raw milk supplies, delimitation of collection areas by the responsible authorities, and legal prescriptions concerning milk collection and cooling, often based on those established in developed countries, but sometimes inappropriate to local conditions. All factors having been taken into account, a detailed survey of each route or sub-district, recording the locations of each potential supplier and the quantities anticipated, must be prepared. Where suppliers are far from an all-weather road, some means of intermediate transport, either to a village collecting station or roadside collecting point (see Figs 5 and 6)

Fig. 5. All-weather transport in Latin America—without all-weather roads.

Fig. 6. Village transport, transfer to truck from village roads, Chiapas, Mexico.

is necessary. This may involve hand-carried or head loads, bicycle or (in hilly districts) mule or horse transport suitable for the tracks available.

In many developing countries, it will very rarely be possible to collect large quantities of cooled, fresh milk from the farm gate, as in Europe, where power, water supplies and supplier's size and means combine to make this possible.

In a few exceptional instances, such as corral farms installed in certain Gulf states, the complete processing and carton or bottle packing of pasteurised or sterilised milk is done at the farm, using technologies identical with those common in the developed countries. For the other developing countries, an intricate system, often starting from village collecting stations, will be needed. But this may be more a matter of organsation than of direct investment. Indeed, before considering the glittering array of technologically advanced equipment available to the would-be investor at any dairy equipment exhibition, the need for technology to remain appropriate must be firmly kept in mind.

For a new venture in a developing country, where even the market for the end-product, and often the potential for raw milk production, is not yet certain, modesty in capital investment both in the milk collection activity and in the processing plant itself, is often imperative. The first indulgence should, perhaps, be in securing a site for processing large enough for extensions in the future.

At this point, it should be mentioned that practical experience seems to show that milk from relatively low-yielding animals in warmer climates, where there is little feeding of concentrates and extensive grazing, can be surprisingly resistant to souring when compared with that produced by typically high-yielding animals in developed dairying areas. Those dealing with projects where large numbers of improved animals are to be brought in to a hitherto little developed region should, therefore, not only plan for the necessary (and too often inadequate) feeding, housing and veterinary care, but also watch for an adverse change in milk quality at the plant, if timely steps are not taken to speed-up collection. Conversely, where no major genetic uplift is anticipated, measures based upon European or North American experience may prove to have been unnecessary in planning collection systems.

Large tanker collections are not always the best solution as the vehicles often cannot stand up to the prevailing road conditions; a conventional churn collection may prove more adequate, particularly where resistant buffalo milk is collected.

The village collecting station or roadside pick-up point (see Fig. 7) should also be the focus of quality control and quantity measurement, since it is vitally important that the supplier sees evidence that the quality of his milk is fairly judged. Therefore, after weighing, often in the presence of the supplier, the first quality checks should take place here. For these

Fig. 7. Roadside collection of village supplies, with testing, Madagascar, 1976.

purposes, the simplest equipment in the form of bottles with preservative for fat content samples (on the assumption that there are no great fluctuations in supply or quality, aliquot samples are not necessary) is sufficient. For keeping quality in cow's milk, after simple organoleptic checking, the alcohol test (Bodex tester) can be applied to churn quantities or suspect deliveries at this point. The village centre, operated by a reliable agent, who is himself usually also a producer, can be the source of advice to suppliers, of payment and as a link to the veterinary and field services (Anon., 1975).

Transport by contractors' vehicles (bicycles, carts, tractor/trailer and for hard roads, motor vehicles) can improve flexibility for seasonal fluctuations, as well as for extension into promising areas and curtailment of those which do not develop. This may also help to reduce initial investment in transport by the project. It has to be remembered that a once or twice daily employment of, say, 2–5 h is hardly sufficient to amortise the cost of a vehicle. Own investment in special vehicles (e.g. tankers) should be passed on proven possibilities of employment and economics compared with contractor's offers.

It is assumed that, on arrival at the processing plant, all supplies, either in churn or bulk, will be tested first for keeping quality, for example by alcohol, clot-on-boiling or reductase tests. Supplies intended for products having legally defined compositions will also be checked for compositonal quality. But it is the keeping quality tests which will give the first warning as to the need for measures to improve the handling of raw milk. These warning signs may come at times when the milk is liable to be unstable, that is, when changing from dry to green fodder (grass milk) and at periods of calving. Faced with these warnings, those responsible will have a range of options from which to choose. First should come reorganisation or improved discipline in the existing collection structure, more rigorous checks on suppliers milking practices and utensils, and on any intermediate handling. Any shortening of the time taken to bring milk to the plant will help. A review of road systems, collection routes, utensils and vehicles may suggest how to achieve this.

When no more, or not much more, can be expected from these measures, some means of checking bacterial development during inevitable transport delays must be sought.

For some years, United Nations agencies (principally the Food and Agricultural Organization and the World Health Organization) have considered the use of hydrogen peroxide as an additive which slows the development of acidity in cow's milk (Anon., 1980). Despite strong reservations in the report as to quantity, quality and method of use, the observation of which it might be difficult to ensure in practice, hydrogen peroxide is used by suppliers and collection systems in some developing countries, to allow collection in difficult terrain from small suppliers without capital investment in receiving and cooling stations.

Amongst milk processors there is, however, a residual suspicion that the action of hydrogen peroxide in suppressing development of acidity does not arrest other changes, enzymatic or chemical, which can take place progressively over time in raw milk. It is felt that the adverse effects of such changes may develop in products with a long storage life. *Therefore, it is felt better to avoid the use of hydrogen peroxide.*

A better-founded resistance to the use of this product is that resting on the feeling that the use of sterilising agents, before raw milk keeping quality is assessed, masks the effects of careless handling which should, reflected in acidity development, provoke follow-up and correction at farm or collecting centre, by the inspection and field service. In effect, hydrogen peroxide attacks the symptoms, not the disease.

When everything other than investment in cooling facilities, intermediate between the supplier and the processor, has been tried, the type of such investment likely to be most cost-effective must be considered. If electricity and water supply systems, refrigerant supplies and maintenance, are available and reliable at the village collecting centres, then the possibilities will be determined by the scale of collection. If, as was often the case in the past, such systems are absent or unreliable, the simplest cooling arrangements can be those using ice blocks to produce chilled water for circulation through a surface or a plate cooler, or even, in the most primitive circumstances where no electricity is available, for cooling of milk in an ice water tank, in churns with periodic agitation by hand. Ice blocks are often available in small towns in the developing countries, especially where cold stores are operated for perishable agricultural products such as poultry, dairy products and seed potatoes, and where itinerant soft drink vendors use ice for their delivery tricycles. Thus, additional investment for ice-making can often be avoided. Ice blocks can be distributed to village collecting centres by the collection vehicles.

Such simple cooling arrangements are the least costly. When, however, efficient cooling to 4°C is sought, and power is available or can be provided by diesel power, the farm cooling tank provides the first item of modern dairy technology in the collection system. These are available, with either integral or free-standing compressor units. Generally, up to 5 hp($\sim$ 3730W), these units are air-cooled, reducing complexity for maintenance. This complexity is further reduced if the direct expansion type is used.

The alternative ice-builder type, with better cooling characteristics and requiring less horse-power (which is important if a diesel-powered generator has to be provided) have the disadvantage of the additional water circulating system which also needs maintenance.

Cooling characteristics typical of direct expansion and ice-builder tanks are shown, under different conditions of usage, in Figs 8–11. It will be seen that these tanks usually require more than 2 h to bring the milk down to temperatures below which bacterial multiplication is substantially reduced. This is a drawback of this type of lightweight cooling set-up, when compared with the more expensive separate ice-building refrigeration systems using a plate cooler for heat exchange.

Approximate cost of wholly owned receiving and cooling stations, equipped with direct expansion farm tanks, are shown in Fig. 12, expressed in US$. It will be seen that site and buildings can form a substantial part of the investment. Where the cost of such stations has seriously limited expansion of a milk district, it has sometimes been found possible to interest groups of would-be suppliers in providing a suitable room or building, in which the processor can install equipment.

In some instances, rural co-operative societies have preferred to set up such facilities on their own account, perhaps with credits and technical assistance from the processing plant. In this way, a group of suppliers will

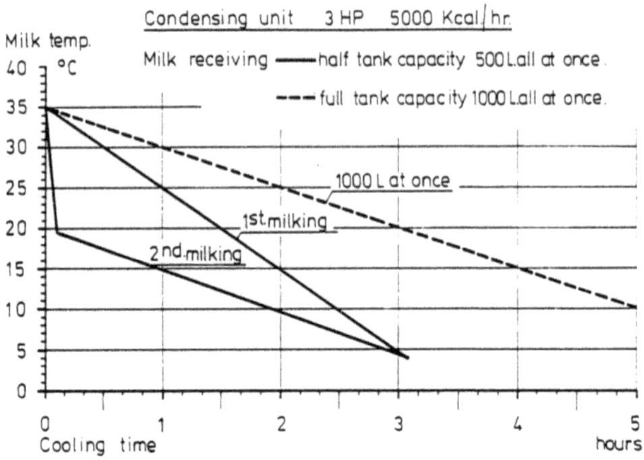

Fig. 8. Typical performance characteristics farm cooling tank, 1000 litres with direct expansion unit, under different loading conditions. (European and US standards specify two milkings in 1 day, 35–34°C in 3 h; (pick-up every 2 days); four milkings in 2 days, 35–34°C in 2 h. Direct expansion tanks cannot therefore be used for a single loading of warm milk to tank capacity.)

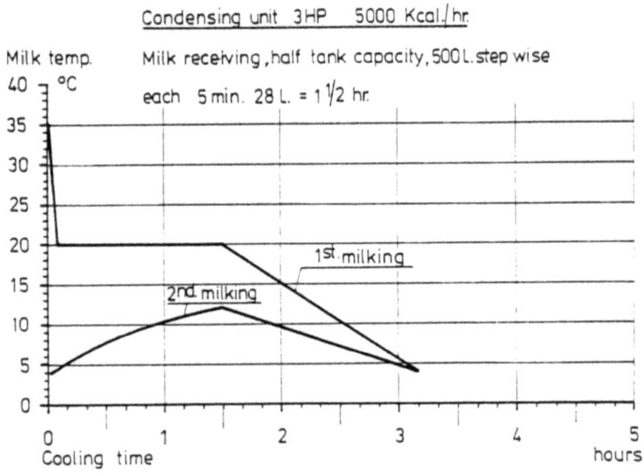

Fig. 9. Typical performance characteristics of a farm cooling tank, 1000 litres, with direct expansion unit with stepwise loading.

feel freer to negotiate sale of their milk to their best advantage. At the same time the processor who provides credit or technical assistance has every incentive to provide adequate continuing field service and veterinary support, as a means of retaining his milk supply.

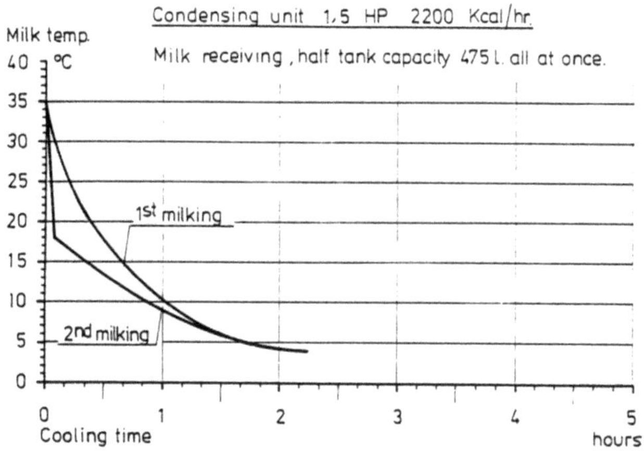

Fig. 10. Typical performance characteristics of a farm cooling tank, 950 litres, ice bank type, loaded in two stages, each half tank volume.

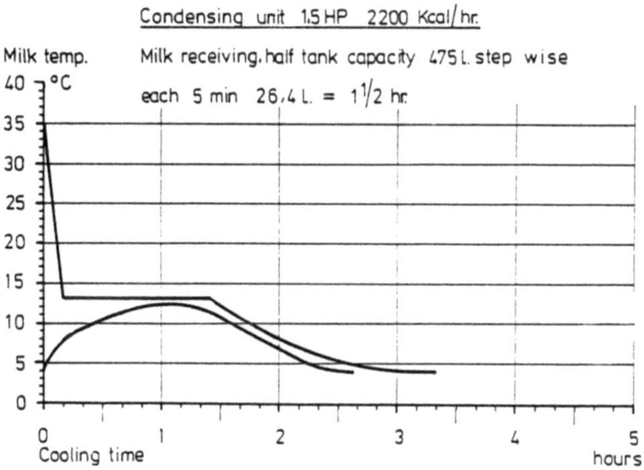

Fig. 11. Typical performance characteristics farm cooling tank, 950 litres, ice bank type, loaded stepwise.

More substantial receiving stations of the 'instant cooling' type are equipped with mechanical refrigeration (from 12 up to about 20 000 litres day^{-1}, with NH_3 systems for larger stations where the cost of refrigerant gas becomes more important) in air-cooled units up to 5 hp (~ 3730W) (water-cooling above this power), serving ice-builders with a water circuit to plate cooler all allowing immediate reduction of milk temperature from 35°C to 4°C. As a rough guide, and assuming that such stations are

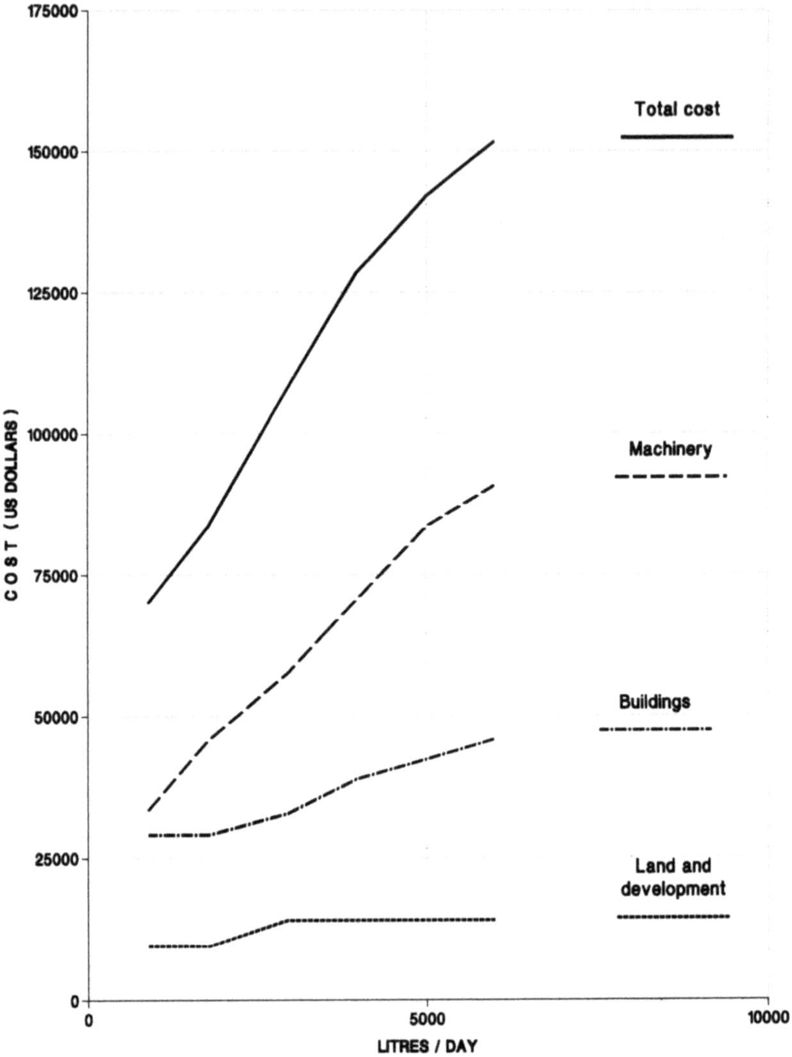

Fig. 12. Typical 1992 land, building and machinery costs for milk cooling stations using farm tanks, expressed in US$.

equipped with stand-by diesel generators, and in the larger sizes above 20 000 litres day^{-1} with reject sour milk segregation and separation facilities, as well as rotary, steam churn washers and related services (which accounts for their relatively higher cost per unit of throughput), costs of such instant cooling stations were estimated by one processor in early 1992 as given in Table 1.

Generally, advantages of scale are equally great in terms of operating cost, since the expense of reliable operating supervision will not greatly differ within the range of quantities handled, indicated above. These operating costs can vary in developing countries from less than 1 % to as much as 7% of the raw milk price paid to the producer.

TABLE I
Capital cost of milk cooling stations designed to handle the volume indicated

Litres day^{-1}	*Litres h^{-1}*	*Total capital cost in US$*	*Capital cost per litre per day US$*
5000	2000	175000	35·00
10000	4000	225000	22·50
20000	8000	455000	22·75
40000	14000	640000	16·00

The figures quoted in Table I, for different types and dimensions of receiving and cooling facilities, may, it is hoped, help those who have to confront cost calculations involving economies of scale in processing facilities versus increase in collection costs for increasing quantities of raw milk. For certain very large-scale operations, for instance, for large milk powder specialities plants, it may be found necessary to resort to pre-condensing (where feasible and permitted for the end-product in question) to reduce milk haulage costs. For rough calculations of the cost of providing pre-condensing facilities, a capital cost, excluding site, of

approximately 38US$ litre^{-1} for a daily volume of 100000 litres, and
approximately 25US$ litre^{-1} for a daily volume of 200000 litres
(1992 values)

This includes normal provision for steam, water and power supplies. No provision is included in these figures for effluent treatment, which should conform to specific local requirements.

In concluding this section, it should again be emphasised that once economic problems are solved, the most important factor in building up a raw milk supply in a developing country is not a matter of technology but of psychology. The confidence of the supplier, that he will receive fair treatment, prompt payment, and support from field and veterinary services, that the collection system will be reliable and that his supplies will be accepted within whatever quota or other limitations he has agreed to observe, takes time to build. Once achieved, this confidence must be maintained by the efficient organisation and adequate equipment of the field, verterinary, collection and accounting services; without it, such investments will count for little until confidence is established. Suppliers have long memories; every possible step should be taken to avoid betrayal of their confidence.

ROLE OF RECONSTITUTED MILK CONSTITUENTS

Economic and Political Background

In this chapter, much emphasis has been placed on means of developing a local raw material supply for a dairy industry. This has become increasingly a preoccupation of governments, after a period in which import substitution was the main aim, even at the continuing expense in foreign currency of milk constituent imports. In past years, the fact that some dairy products have constituents such as sugar, cereals or fruit fillings which can be locally purchased, and are packed in glass, tin-plate, paper complexes or cartons which can also be of local origin, has given a sufficient local content even when the main dairy constituents had still to be imported. If local energy, labour and services were taken into account, such products could show a substantial saving in foreign exchange if compared with the cost of imports. If such imports were of fluid products, such as sterilised or condensed milks, the high cost of transporting the water content would also favour local manufacture.

Historical Development

The basic dairy constituents used in recombining are skim-milk powder and butter or, more recently, butter oil or anhydrous milk fat. This 'oil' is a more than 99·8% pure product obtained from cream or butter by a process of emulsion-breaking, centrifugal separation and vacuum drying. Packed in airtight and usually nitrogen gassed drums, it will keep for months at ambient temperatures. As stated before, recombining resulted mainly from the needs of the US armed forces in the Pacific Area in 1940–1945.

In addition to pre-war processing plants, a number of large-capacity spray drying plants were erected in the US during the war years, originally for full-cream milk powder. The report on Milk Utilisation published by the British Productivity Council in 1953 after a team visit to the US in 1952, records the rapid extension of skim-milk powder production using these facilities. Already at that time, strict quality grading served to enhance the usefulness and acceptance of skim-milk powder for many purposes in different sectors of the food industry. The liquid skim-milk, previously returned to suppliers for stock feeding, was available from the butter manufacturers who handled a quarter of the more than 50×10^6 of fresh milk then produced in the US.

Skim-milk powder was exported first as food aid to devastated areas of Europe and Asia after 1945, and later under Public Law 480 in subsequent food aid programmes.

The competitive situation resulting first from the US exports, and subsequently from increased production in the European Community, in Australia and New Zealand, led to lower prices for the consitutents, easily stored and exportable in bulk, than for the conventional finished dairy products. This, coupled with the stimulus given by governments in importing countries for the establishment of local industries, often by tariff barriers against conventional imports of dairy products, quickly led to the setting up of a series of recombining plants first for evaporated and sweetened condensed milk. These were built in Malaysia, Philippines, Indonesia, Thailand, the Caribbean and Sri Lanka in the 1960s and 1970s, and were later complemented by recombining operations for sterilised milk in bottles or cartons, milk powders and infant foods. Yoghurts and ice cream also developed in these countries on the basis of recombining.

Present and Future Economic Situation of Recombining

Figure 13 shows the fluctuations experienced in the world market prices of butter oil (anhydrous milk fat) and skim-milk powders in the period 1975–1992. It will be obvious that this fluctuation must cause problems for industries dependent on these raw materials.

Suppliers from Europe are influenced by the levels of restitutions on exports. The EC substantially dominates somehow the exports and supplies of milk raw materials and by the export refunds influences to a great extent world market price levels (see Fig. 14).

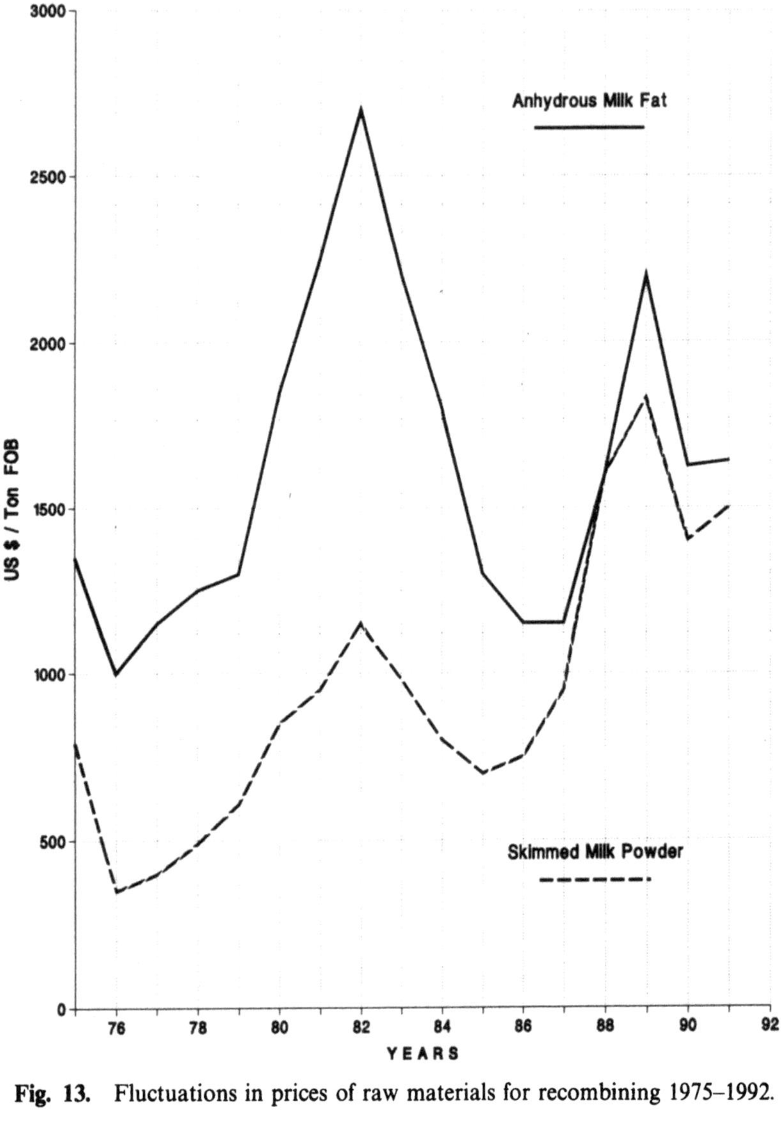

Fig. 13. Fluctuations in prices of raw materials for recombining 1975–1992.

Another economic influence, frequently exercised by governments on recombining industries in developing countries, involves an obligation to accept any quantities of fresh milk offered by suppliers, often at much higher cost than the equivalent imported constituents. This can be a considerable stimulus to fresh milk production. A problem often

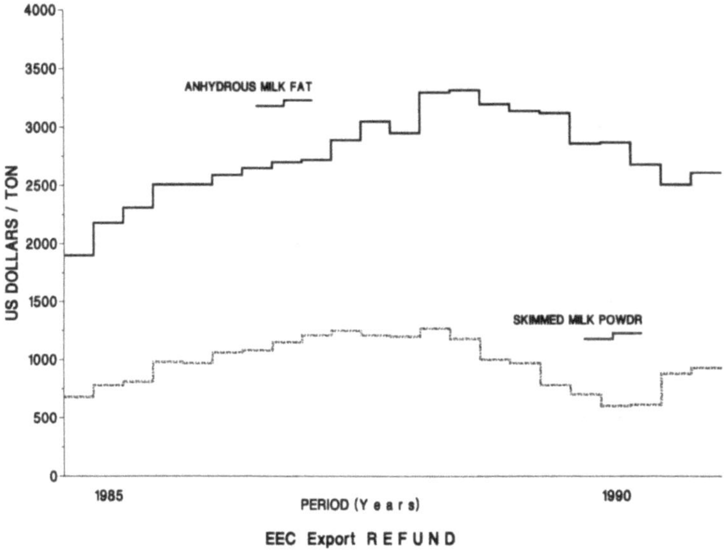

Fig. 14. World market price levels of raw materials are substantially influenced by EC export refunds.

experienced in such instances is the uneven response of individual processors to such obligations. Bakeries and ice cream producers, using substantial quantities of imported milk constituents, find it difficult to absorb fresh milk. Ultimately those industries properly complying with such obligations may find the financial burden intolerable. Therefore an equalisation scheme whereby *all* imports of milk constituents are taxed for the benefit of local fresh milk production for *all* purposes, is highly desirable. However, some guarantees that such taxes will indeed benefit fresh milk production, and not disappear into the revenue, are imperative.

Application of Recombining

Within the limits imposed by such factors, recombining can be invaluable in developing a modern dairy industry in developing countries.

There are two principal ways in which this can come about. The operation at economic capacity of a new plant can be ensured, up to the limit set by marketing opportnities, throughout the long period needed to build-up a sufficient fresh milk supply. If large seasonal fluctuations in fresh milk availability persist, as is often the case in developing countries, recombining can maintain supplies of products with relatively limited keeping quality. As a second possibility, those plants which cannot be based on local milk constituents, but which find sufficient justification from secondary raw and packing materials, energy, labour, or other locally available constituents, can produce a range of dairy products to reduce import costs whilst improving availability of milk products.

Specific Technology for Recombining

The first technological problem involved in recombining was that of dissolving large quantities of skim- or full-cream milk powder in water or milk. Dumping funnels and fluid recirculating systems were already recorded in use in the US in 1952 (Anon., 1953). The International Dairy Federation has in recent years published a series of monographs detailing appropriate technology for specific products (Anon., 1979).

Air entrainment was on difficulty encountered in recombining operations. This not only affects efficiency of subsequent homogenisation, but can also provoke coagulation under heat treatment of high-concentration pre-mixes, and adversely affect keeping quality of long-life products. To eliminate air, some form of deaeration is normally required if not already available in a concentration step under vacuum. However, most modern recombining plants for sweetened condensed milk, evaporated milk or milk powders no longer incorporate a concentration step, unless dilution to allow filtration, to remove impurities in local sugar supplies, for instance, dictates this.

Although full-cream milk powders are often used in recombining, on account of price advantage or simplicity of use, the shorter storage of life of these powders, due to the relatively high exposure to oxidation of their fat content, leads many processors to prefer the separate butter oil or anhydrous milk fat and skim-milk powder as constituents. For butter oil or other fats, simple heating devices—usually hot rooms—ensure a fluid state for pumping; this allows accurate dosage. The elimination of any dwelltime of warm fats with fresh milk, or milk constituents containing lipase is essential for good keeping quality of the end-product. Typical specifications for butter oil are available from the International Dairy Federation (Anon., 1983). For skim-milk powders, product requirements will dictate a suitable level of heat treatment, as defined in the American Dairy Products Institute (ADPI) standards. These are based on differentiation in the whey protein nitrogen index (WPNI).

In recent years, the merits of sweet buttermilk powder, which contains 8–10% of butter fat and about 1·5% of lecithin, have been appreciated, since this powder has often been available at about the same price as skim-milk powder with only about 1% butterfat. It is not only a price advantage which recommends buttermilk powder, as from a technological point of view, buttermilk is an emulsifying aid for the fats in a recombined milk product. Buttermilk can also, in appropriate proportions, favourably influence organoleptic quality. It also restores an element otherwise missing when recombining, in order to achieve a composition corresponding to the original whole milk used for butter and skim-milk manufacture, and, in certain instances, buttermilk powder can assist in stabilising premixes under heat treatment.

One other product range, in addition to those already mentioned, which can now be based on recombining, is that of cream and natural cheeses. Cheese base powders, produced by the removal of milk serum by ultrafiltration, can eliminate costly whey disposal problems in natural cheese production. This may also make it possible to avoid the need for imported natural cheese in the maufacture of processed cheese, usually based on a percentage of ripened natural cheese, with milk powders, caseinate and butter oil.

As mentioned earlier, economics are vital, and it should be repeated that, within the EC, considerable pressure has been exercised by dairy products manufacturers to prevent undue advantages in export restitutions for milk constituents as compared with finished milk products.

CONSIDERATIONS OF SCALE

Pasteurised, Chilled and Frozen Products

Artisanal methods permit home production, for instance, of yoghurts and soft cheese, and this is frequently found in developing countries, either using local fresh milk or milk powder. Cultures are carried over from day to day.

Small industrial operations can start up with a few hundred litres day^{-1}. However, the increasing difficulties of widening distrbution system, and the crippling cost of returned, spoiled goods, limit such ventures, which require skilled commercial and operational management. For larger-scale but still limited ventures, success may depend upon the possibility of adding to an existing similar activity in distribution, say of butter and pasteurised milk, where the knowledge and management ability required is already available.

For larger-scale yoghurt and dessert production, full industrial-scale can be reached at the approximate investment figures quoted in Table II (costs of distribution vehicles and chilled shop cabinets are excluded). These figures illustrate the effect of scale on capital investment. The largest units may also offer operating economies through the use of 'form and fill' machines, which give lower packaging material costs. There may be other scale advantages, especially in distribution and publicity, as the size of the operation increases.

TABLE II
Capital costs for yoghurts and fresh dessert production

	Capacity ($t\ year^{-1}$)		
	3000	*10 000*	*20 000*
Investment US$ t^{-1} year^{-1} (1992 values)	1400	750	600

For fresh cheese production on an industrial scale, the rough estimates given in Table III apply (under the same reservations), and show that here there is rather less advantage in capital cost differentials.

Ice cream production is also possible on an artisanal scale, and a wide range of small-scale industrial equipment is available, mainly from continental suppliers. The product is, however, demanding in terms of cold chain stability. This costs money and requires strict organisation down to the retail outlet or vendor (for street articles). Small ventures rely on street impulse sales—industrial ventures also, but to an increasing extent on so-called 'take-home' sales of larger packs, so far as domestic refrigeration permits. For relatively large-scale industrial

TABLE III
Capital costs for fresh cheese production

	Capacity ($t\ year^{-1}$)	
	2000	*4000*
Investment US$ t^{-1} year^{-1} (1992 vaules)	2800	2000

operations, capital investments for production, but excluding long-term storage space, distribution vehicles and freezer cabinets in retail outlets (allowing for 50% impulse items) will be of the order of those given in Table IV. It will be seen that the advantages of scale are somewhat greater in yoghurts than in ice cream, at least as far as capital investment for production is concerned.

TABLE IV
Capital costs for ice cream production

	Capacity (litres year^{-1})		
	12	*21*	*50*
Capital investment US$ litre^{-1} year^{-1} (1992 values)	3·50	2·70	1·70

Products with Longer Keeping Quality

Fluid sterilised milk

Processing lines for the UHT treatment of whole fresh or recombined milk do not differ greatly in cost for larger or smaller hourly throughputs. If, however, varying sizes of packs have to be filled, involving varying hourly throughputs on a given line, this may increase costs (use of variable speed drives, split heat treatment apparatus or sterile intermediate storage). There will also be increased charges for additional filling and packing equipment, which is generally not adjustable for different sizes of packs. Filling units are commonly only available on a rental basis plus per unit costs on the packing material, supplied or licensed by the machinery leasor. In order to make a rough estimate of scale effect in such plants, the base rental (payable on installation) has to be included as part of the capital cost.

Recombined sweetened condensed milk

The capital cost of such plants is often strongly influenced by the need to provide infrastructures not available in developed countries. Since some of these, such as container manufacture, are available in certain developing countries, a rather low cost per tonne (for a product containing a high percentage of sugar and at least 25% water) can be reached if container manufacture is excluded (Table V).

TABLE V
Capital costs for recombined SCM production

	Capacity (t year^{-1}) finished product	
	10 000	*30 000*
Capital cost US\$ t^{-1} year^{-1} (1992 values)	1600	800

Full-cream milk powder (see Table VI)

There is relatively less benefit from an increase in scale of such milk powder plants, due to the need for increased concentration, drying and container manufacturing equipment.

TABLE VI
Capital costs for recombined full-cream milk powder production (with own container manufacture)

	Capacity (t year^{-1}) finished product	
	7500	*15 000*
Capital cost US\$ t^{-1} year^{-1} (1992 values)	2200	1800

CHOICE OF MANUAL AND AUTOMATED SYSTEMS

Today, almost any fluid circuit, water, steam or refrigeration, in common use in dairy plants, will incorporate some simple, automatic controls. Thus, a degree of automation is already inherent to the service equipment available to those planning a dairy industry. In the milk circuits, circulation cleaning (cleaning in place—CIP) is now assumed in the design of equipment. The tendency is to construct systems with a minimum of joints, always a difficulty in cleaning, and to limit these to inlets and outlets of valves, pumps and other pieces of equipment, such as tanks, homogenisers or heat exchangers. For circulation cleaning, and for process control, simple sequential controllers are commonly employed. The maintenance of such limited automation has to be ensured by adequate stocks of spare parts, where no manufacturer's representation can be ensured.

Thus any new project will therefore almost certainly comprise some automation. The choice, for all but artisanal production, is between limited or advanced automation, where more elaborate sensing devices and feed-back systems are used to correct divergences from standard operating conditions. In general, a project should be designed to take advantage of the degree of sophistication of the infrastructure in the country concerned, especially those providing maintenance, for electrical and electronic systems. Often this will mean keeping to the simplest systems available.

The second factor encountered is manpower available. There may be an extreme shortage of skilled or unskilled industrial workers. Where this is the state of affairs, manual systems (particularly for handling of raw materials) would be too costly, and investment in bulk handling systems may be justified, as well as a higher degree of automation.

Elsewhere, governments are often inclined to favour labour-intensive investments. There are, however, strict limits to the increase in staffing possible inside a food factory, where every extra worker is a potential source of contamination and of errors in processing. Such operations as milk collection and transport are not so much under this handicap.

Within a factory, individual steps, such as vacuum drying, may be carried out by hand-loaded batch ovens in place of continuous band driers, in order to increase employment. But such choices in process are relatively limited. Thus, the choice of manual or automated systems of greater or lesser complexity will be conditioned by availability and cost of unskilled and skilled labour, by the type of infrastructure for maintenance, and by the process involved and the capital cost of installations in relation to their effective output.

PACKAGING, STORAGE AND DISTRIBUTION

The distinction between such activities carried on in a developed or a developing country is likely to arise from differences in ambient temperature (often higher in developing countries), in infrastructures such as transport, mechanical and electrical maintenance (often weaker in developing countries), and in density of sales outlets (less kg per km of van route in developing than in developed countries). Thus, it is evident that sterile or dried or preserved products, relatively unaffected by ambient temperature and requiring less frequent delivery to retailers, will not call for special precautions, at least so far as hard packs (tin-plate or adequate plastic–paper–foil complexes) are concerned. Rodent attack and damage from boring insects will set limits to the use of paper or complexes in replacement of tinplate. Many products will suffer if exposed to direct sunlight or very high temperature, and this will set limits to the use of lighter packing materials and the type of storage used. Condensation caused by exposure to air of high humidity and high temperatures, such as occurs when cargoes are shipped in winter from Europe (temperatures from 4°C upwards) to West Africa (80% humidity and 30°C air temperature) will cause rust spots on tinplate unless ventilation and heating are arranged during transport to bring cargo temperatures up.

This problem will not affect local distribution, but even for this there are similar problems when tropical seaboard plants ship to high plateaux.

Increased altitude can also cause apparent swelling of the sealing membranes used for tins of powder products, often a source of complaints or rejection, although there may be no defect in the contents.

For milk preserves, suitable for distribution at ambient temperatures, a comparison of the cost of packaging materials per litre of fresh milk equivalent has been derived from experience in certain developing countries in the 1990s (Table VII). These figures are for materials only, and exclude capital charges or rentals, operating and maintenance costs.

TABLE VII
Packing materials cost per litre of fresh milk equivalent

Pack	*Approx. cost litre^{-1} (US\$)*
Sweetened condensed milk (400 g tin)	0·07
Full-cream milk powder (800 g–1 kg tin)	0·05
Sterilised whole milk in cartons (1 litre)	0·14
Sterilised whole milk in cartons (¼ litre)	0·23

As stated earlier, the relative fragility of packs should be considered in relation to the type of distribution system foreseen. Thus, a lighter pack with shorter life and higher risk of product loss through leakage or souring may be acceptable in a well organised distribution, where the final vendor can eliminate defective units and can account for such losses.

There is constant pressure to diminish packaging costs, despite risks, and this is understandable in the context of low consumer purchasing power in a developing country. Thus, for liquid pasteurised and ever sterilised milk, pouch packs in plastic have been employed, despite some difficulties in handling both in distribution and in the home. Plastic pouches have also been used extensively for milk powders, despite a relatively rapid taste deterioration and ultimate rancidity due to inadequate barrier properties against light and oxygen.

For shorter life products, such as yoghurts and other chilled products, packaging will probably depend upon availability of preformed cups. In the larger projects, substantial economies can be achieved by the use of 'form and fill' machines, thermoformed pots from continuously fed reels of appropriate complexes. The use of thermo-adhesive labels allows a reduction of film substance, and can lead to packaging material economies of up to 10% or more compared with the cost of preformed cups.

In ice cream distribution, the limited availability of home freezers in most developing countries may make it necessary to concentrate on street, impulse-buy articles, bulk sales being limited to hotels and institutional outlets. Capital investment in distribution will often, at least in the early years, be almost equal to those in production facilities, with about 60% going to cabinets and 40% to vehicles. The annual amortisation and maintenance charges for cabinets and vehicles may amount to nearly half their original cost. Generally, for storage and distribution, conditions required depend upon the product and packaging chosen.

FUTURE OF MILK PRODUCTS IN DEVELOPING COUNTRIES

According to the '*World Market for Dairy Products*' issued by the GATT, Geneva in November 1989, the liquid milk consumption has increased at an average annual rate of 1% over the last 10 years.

The per capita consumption, however, has remained rather stable at close to 46 kg throughout this period. For obvious reasons, the per capita consumption of milk is largely different between countries and regions. For developed countries the liquid milk consumption is up to 160 kg and above, whereas in certain developing areas the consumption reaches not even 2·5 kg per capita per year.

In Europe and North America consumers show an increasing preference for so-called 'light' products. A further diversion of the consumption habits towards fermented and flavoured milks has been steadily increasing. The trend to switch from whole to partially skimmed and fermented milks is expected to continue, but the milk consumption in most developed countries seems to be stagnant.

A major growth of milk consumption is reported for Asia; this is mainly attributed to subsidised campaigns to promote milk consumption and school milk distribution.

In other areas of the developing world, especially where local fresh milk production is inadequate or non-existent, condensed and dried milk consumption continued to increase, with some development of UHT sterile recombined fluid milk where transport and distribution possibilities favour this bulkier and less resistant form of packing. Cheese consumption is still very little developed in these areas.

Thus, the total picture is one where considerable new investment to meet increasing demands, either through development and processing of local fresh milk or through import replacement by recombining, may be expected to continue in the developing world. It may also be expected that rapid growth in urban populations in the developing world (Fox, 1984), and the need for better structures in agriculture, may stimulate further fresh milk production where this is possible, and the development of new marketing systems, such as those for chilled and frozen products, where these offer economic and social advantages.

Whatever the means chosen, the examples quoted in this chapter of increased benefits in agricultural development, improved planning of processing structures and distribution, may, it is hoped, assist those who will be engaged in new ventures using modern, and appropriate, dairy technology in the developing world.

REFERENCES

Anon. (1941). *Agricultural Marketing in India.* Government of India.

Anon. (1953). *Milk Utilisation.* British Productivity Council, London, UK.

Anon. (1975). *Nestlé in the Developing Countries.* Nestlé SA, Vevey, Switzerland.

Anon. (1978). Projections of Product Development to 1985; Milk and Milk products Supply, Demand and Trade, FAO Rome ESC., PROJ/78/3rd.

Anon. (1979). Monograph on recombination of milk and milk products, IDF Bulletin No. 116, International Dairy Federation.

Anon. (1980). Evaluation of certain food additives, 24th Report of the joint FAO/WHO Expert Committee on Food Additives, WHO, Geneva, Switzerland.

Anon. (1982). *The Etah Development Programme.* Unilever, London, UK.

Anon. (1983). Milk fat products, Standard A2, IDF Publication D-DOC-188.

Anon. (1989). *International Dairy Arrangement, Tenth Annual Report, The World Market for Dairy Products.* GATT, Geneva, Switzerland.

Beaver, M. W. (1973). Population, infant mortality and milk. *Population Studies,* **27** (2), 243–54.

Fox, R. W. (1984). The world's urban explosion. *Nat. Geogr. Mag., August,* 179–85.

Khurody, D. N. (1977). Increased production and rationalised consumption of milk and Asia. *Indian Dairyman,* **29** (3), 149.

Latham, R. E. (1958). *Travels of Marco Polo.* Penguin Books. Harmondsworth, UK.

McDowell, R. E. (1981). Limitations for dairy production in developing countries, *J. Dairy Sci.,* **64**, 2463.

Rampini, F. (1978). Nestlé in Indonesia. *Politica ed Economica,* Anno IX no. 3, Rome (translated by Nestlé Vevey, Switzerland).

Whetham, E. H. (1964). The London milk trade 1860–1890. *Economic History Review,* **17**, 369–80.

Whetham, E. H. (1970). *The London Milk Trade 1900–1930,* Reading.

Index

MIX
Papier aus verantwortungsvollen Quellen
Paper from responsible sources
FSC® C105338

If you have any concerns about our products,
you can contact us on
ProductSafety@springernature.com

In case Publisher is established outside the EU,
the EU authorized representative is:
Springer Nature Customer Service Center GmbH
Europaplatz 3, 69115 Heidelberg, Germany

Printed by Libri Plureos GmbH
in Hamburg, Germany

MODERN DAIRY TECHNOLOGY

Volume 2

Advances in Milk Products
Second Edition